Encyclopaedia of
Mathematical Sciences
Volume 28

Editor-in-Chief: R.V. Gamkrelidze

R.V. Gamkrelidze (Ed.)

Geometry I

Basic Ideas and Concepts of
Differential Geometry

With 62 Figures

Springer-Verlag

Berlin Heidelberg New York
London Paris Tokyo
Hong Kong Barcelona
Budapest

Consulting Editors of the Series:
A.A. Agrachev, A.A. Gonchar, E.F. Mishchenko,
N.M. Ostianu, V.P. Sakharova, A.B. Zhishchenko

Editor: Z.A. Izmailova

Scientific Editors of the Subseries:
N.M. Ostianu, L.S. Pontryagin

Title of the Russian edition:
Itogi nauki i tekhniki, Sovremennye problemy matematiki,
Fundamental'nye napravleniya, Vol. 28, Geometriya 1
Publisher VINITI, Moscow 1988

Mathematics Subject Classification (1980):
53-01, 53A04, 53A05, 53A07, 53B21

ISBN 3-540-51999-8 Springer-Verlag Berlin Heidelberg New York
ISBN 0-387-51999-8 Springer-Verlag New York Berlin Heidelberg

Library of Congress Cataloging-in-Publication Data
Alekseevskiĭ, Dmitriĭ Vladimirovich, 1940–
[Geometriĭa I. English]
Geometry I: basic ideas and concepts of differential geometry / R.V. Gamkrelidze (ed.);
[D.V. Alekseevskij, A.M. Vinogradov, and V.V. Lychagin].
p. cm.—(Encyclopaedia of mathematical sciences; v. 28)
Translation of: Geometriĭa I, issued as v. 28 of the serial: Itogi nauki i tekhniki. Sovremennye problemy
matematiki. Fundamental 'nye napravleniĭa.
Includes bibliographical references and index.
ISBN 0-387-51999-8 (U.S.)
1. Geometry, Differential. I. Gamkrelidze, R.V. II. Vinogradov, A.M. (Aleksandr Mikhaĭlovich)
III. Lychagin, V.V. (Valentin Vasil 'evich) IV. Title. V. Title: Geometry one.
VI. Title: Basic ideas and concepts of differential geometry. VII. Series.
QA641.A59313 1991 516.3'6—dc20 90-45086

List of Editors, Authors and Translators

Editor-in-Chief

R.V. Gamkrelidze, Academy of Sciences of the USSR, Steklov Mathematical Institute, ul. Vavilova 42, 117966 Moscow, Institute for Scientific Information (VINITI), ul. Usievicha 20a, 125219 Moscow, USSR

Authors

D.V. Alekseevskij, Moscow Pedagogical Institute, Verkhnyaya Radishchevskaya 18, 109004 Moscow, USSR

V.V. Lychagin, Department of Mathematics, Institute of Civil Engineering, Kalinikovskaya 30, 109807 Moscow, USSR

A.M. Vinogradov, Department of Mechanics and Mathematics, Moscow State University, 119899 Moscow, USSR

Translator

E. Primrose, 12 Ring Road, Leicester LE2 3RR, England

Basic Ideas and Concepts of Differential Geometry

D.V. Alekseevskij, A.M. Vinogradov, V.V. Lychagin

Translated from the Russian
by E. Primrose

Contents

Preface

Apparently, at the basis of geometrical thinking there lie mechanisms, so far unknown, that enable us to extract and use structurally formed elements of information flow. This point of view is developed in more detail in Chapter 1. Starting from this, we make an attempt to realize the variety of modern geometrical theories, imitating the physicists who, starting from the "big bang" hypothesis, explain the existing state of the universe.

In putting together this survey of what seem to us the fundamental concepts, ideas and mthods of modern differential geometry, we have not intended that it should be read systematically from beginning to end. Therefore within each chapter and the book as a whole the presentation gradually speeds up, so that the reader can start or stop wherever it is natural for him. Any new theme begins with "general conversations"; the process of turning these into precise formulae is traced as far as possible. We have drawn attention to this aspect, since the art of a geometer is determined to a large extent by the ability to organize this process.

Our understanding of geometry as a whole has changed significantly in the process of writing this survey, and we shall be satisfied if the benefit that we ourselves have gained turns out to be not only the property of the authors.

In conclusion we wish to thank sincerely our friends and colleagues in the Laboratory of Problems of High Dimension of the Institute of Program Systems of the USSR Academy of Sciences for the very substantial help they have given us in preparing the manuscript for the press, and the Chief Editor of this series, Corresponding Member of the USSR Academy of Sciences R.V. Gamkrelidze.

Chapter 1
Introduction: A Metamathematical View of Differential Geometry

§1. Algebra and Geometry – the Duality of the Intellect

It is known from physiology that in the process of thinking the hemispheres of the human brain fulfil different functions. The left one is the site of the "rational" mind. In other words, this part of the brain carries out formal deductions, reasons logically, and so on. On the other hand, imagination, intuition, emotions and other components of the "irrational" mind are the product of the right hemisphere. This division of labour can have the following explanation.

The process of solving some problem or other by a human being or an artificial mechanism involves the need to draw correct conclusions from correct premises. The logical computations that carry out these functions in various specific circumstances can easily be formulated algorithmically and thereby carried out on modern computers. There are good reasons for supposing that the human brain acts in a similar way and the left hemisphere is its "logical block". However, the ability to argue logically is only half the problem, and apparently the simpler half. In fact, to solve any complicated problem it is necessary to construct a rather long chain, consisting of logically correct elements of the type "premise-conclusion". However, from given premises it is possible to draw very many correct conclusions. Therefore it is practically impossible to find the solution of a complicated problem by randomly building up logically correct chains of the form mentioned, in view of the large number of variants that arise. Thus the problem we are posing is: in which direction should we reason? We can solve it only if there are various mechanisms of selection and motivation, that is, mechanisms that induce the thinking apparatus to consider only expedient versions. Man solves this problem by using intuition and imagination. Thus, we can think that the process of evolution of nature has led to the two most important aspects of any thought process – the formally logical and the motivational – being provided by the two functional blocks of the brain – its left and right hemispheres, respectively.

Mathematics is the science that deals with pure thought. Therefore the two main aspects of mental activity mentioned above must be revealed in its structure. In fact, this is familiar to everyone from the school division of mathematics into algebra and geometry. Apparently we can regard it as established experimentally that an algebraist or analyst is a mathematician with a pronounced dominant left hemisphere, while for a geometer it is the right hemisphere. Thus, if we consider the body of mathematicians linked by various lines of communication, having a common memory, in which there are mechanisms of stimulation and repression, etc., as a thinking system, then geometry is the product of its right hemisphere.

It is clear a priori that successful functioning of the intellect (natural or artificial) can be ensured by balanced interaction of its right and left hemispheres, acting on different levels. Hypertrophy of the function of the left hemisphere leads to a phenomenon that can appropriately be called thought bureaucracy (formalism, scholasticism, and so on). On the other hand, hypertrophy of the right hemisphere leads to unsound fantasies and wandering in the clouds. For example, F. Klein (Klein [1926]) writes: "We state here as a principle that we will always combine the analytical and geometrical treatment of problems, and we shall not, like many mathematicians, take a one-sided point of view. An analytical treatment does not give a visual idea of the results obtained, while a geometrical examination can only give an approximate basis for proof ... ".

We liken the development of thought in the solution of a problem to the process of propagation of an electromagnetic wave, as a consequence of the inductive connection of an electric field (E) and a magnetic field (H) (Fig. 1). We venture to compare the left hemisphere (algebra) with E, and the right hemisphere (geometry) with H, since a magnetic field does not have sources. Short-wave oscillations of algebra-geometry type inevitably arise when one or several investigators are working on a specific problem. Long waves of this kind, which often interfere with one another in an odd way, are the waves of mathematical history.

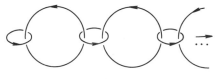

Fig. 1

If 20 years ago the formal algebraic spirit, personified by Nicolas Bourbaki, dominated, nowadays the geometrical spirit is on the crest of the wave, with the typical growth of interest shown by mathematicians in problems of physics and nature in general. The chain

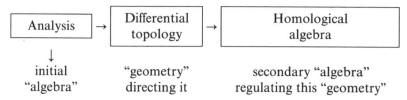

Analysis	→	Differential topology	→	Homological algebra
↓				
initial "algebra"		"geometry" directing it		secondary "algebra" regulating this "geometry"

is an example of a large-scale wave in modern mathematics.

§ 2. Two Examples: Algebraic Geometry, Propositional Logic and Set Theory

2.1. In ancient manuscripts of geometrical content, instead of a proof of a theorem there is often displayed a convenient diagram together with the instruction "look". Thereby it is implicitly assumed that the contemplation of the diagram is capable of arranging the thoughts of the observer into a conclusive chain of deductions. Modern algebraic geometry is an approach to the solution of problems of commutative algebra by way of an intelligent and systematic construction of the necessary visual images. The source of this approach is an idea that goes back to Descartes: it is possible to obtain a visual representation of the set of solutions of a system of polynomial equations, interpreting it by introducing coordinates as an "algebraic subvariety" of affine space. In modern algebraic geometry the main way of visualizing is to interpret an arbitrary commutative algebra as the algebra of functions on some set. Its description in general outline reduces to the following (the definitions of the simplest concepts of commuatative algebra used below can be found, for example, in Volume 11 of the present series).

Let A be a commuatative algebra with unity over a field k. Consider the set Spec A of all prime ideals of this algebra. An element $a \in A$ can be thought of as a "function" on this set, whose value at a point $\mathfrak{p} \in$ Spec A is equal to the image of a under the natural homomorphism $A \to A/\mathfrak{p}$. As a result we obtain the following picture: to each point $\mathfrak{p}$ of Spec A we "attach" an algebra $A/\mathfrak{p}$ without divisors of zero (since $\mathfrak{p}$ is a prime ideal) and the "function" mentioned above, associated with the element $a \in A$, is the map $\mathfrak{p} \mapsto [a] \in A/\mathfrak{p}$. The algebra $A/\mathfrak{p}$, generally speaking, depends on $\mathfrak{p}$. If this dependence did not exist, that is, all the algebras $A/\mathfrak{p}$ were the same, then the construction we have described would lead us to the usual $A/\mathfrak{p}$-valued functions on Spec A.

Example. Let $k = \mathbb{C}$ (the field of complex numbers) and let $A = \mathbb{C}[x]$ be the algebra of polynomials with complex coefficients in the variable x. Any non-zero prime ideal $\mathfrak{p} \subset A$ consists of polynomials divisible by $x - c$ (the number $c \in \mathbb{C}$ is fixed). Dividing a given polynomial $p(x) \in A$ by $x - c$ and taking the remainder, we obtain a unique representation in the form $p(x) = p_0 + (x - c)h(x)$, where $p_0 \in \mathbb{C}$. Then $A/\mathfrak{p} = \mathbb{C}$ if $\mathfrak{p} \neq \{0\}$, and we can represent the operation of factorization $A \to A/\mathfrak{p}$ as the operation $p(x) \mapsto p_0$, that is, the value of the polynomial $p(x)$ in the ideal $\mathfrak{p} \in \operatorname{Spec} A$ is equal to p_0. If we now identify the ideal $\mathfrak{p}$ with the number c that determines it uniquely, we see, in view of the equality $p_0 = p(c)$, that the usual meaning of the words "value of a function" and the one described above coincide.

Example. If $k = \mathbb{R}$ (the field of real numbers), then in the algebra $A = \mathbb{R}[x]$ of polynomials with real coefficients there are two types of nonzero prime ideals. An ideal of the first type consists of polynomials divisible by the binomial $x - c$ (the number $c \in \mathbb{R}$ is fixed). Ideals of the second type consist of polynomials divisible by a fixed trinomial $x^2 + px + q$, where $p, q \in \mathbb{R}$ and $4q > p^2$. If $\mathfrak{p}$ is an ideal of the first type, then $A/\mathfrak{p} = \mathbb{R}$, and if it is of the second type, then $A/\mathfrak{p} = \mathbb{C}$. In particular, the set of prime ideals of the first type can be identified with the "real" line $\mathbb{R}$. If in the usual way we consider a polynomial $p(x) \in A$ as a a function on $\mathbb{R}$, then this function coincides with the restriction of the "function" associated with it by means of the construction carried out above to the set $\mathbb{R} \subset \operatorname{Spec} A$ of ideals of the first type.

The latter example shows that we can diminish the dependence of the algebra $A/\mathfrak{p}$ on $\mathfrak{p}$ by considering not the whole set $\operatorname{Spec} A$, but only a suitable part of it. We illustrate this by an example that is fundamental for differential geometry.

Example. Let $\Omega \subset \mathbb{R}^n$ be a domain, let $k = \mathbb{R}$, and let $A = C^\infty(\Omega)$ be the algebra of all infinitely differentiable functions defined on Ω. With any point $a \in \Omega$ we can associate an ideal μ_a of A by putting

$$\mu_a = \{f(x) \in A \mid f(a) = 0\}.$$

It is not difficult to see that μ_a is a maximal ideal, and so $\mu_a \in \operatorname{Spec} A$ and $A/\mu_a = \mathbb{R}$. The value of f on μ_a, according to the basic construction, is equal to its usual value $f(a)$ at the point a. For this reason any function f as an element of A is uniquely determined by its values on the subset $\operatorname{Specm}^{\mathbb{R}} A \subset \operatorname{Spec} A$ consisting of all maximal ideals $\mu \subset A$ such that $A/\mu = \mathbb{R}$. In fact, we have the fundamental equality

$$\Omega = \operatorname{Specm}^{\mathbb{R}} C^\infty(\Omega).$$

This equality remains true if Ω is replaced by an arbitrary smooth manifold (see below).

The set $\operatorname{Spec} A$, and also the subsets of it that it is expedient to consider, are equipped in a natural way with additional geometrical structures, for example a topology, that visually reflect the properties of the original algebra A. Working on this basis, we can construct visual models for the basic concepts of commutative algebra. For example, an ideal in $A \Rightarrow$ "submanifolds" in $\operatorname{Spec} A$, a module over $A \Rightarrow a$ "bundle" (see p. 59 below) over $\operatorname{Spec} A$, and so on. Below we give some visual interpretations of the "differential" aspects of commutative algebra. A systematic use of such models enables us to construct visual interpretations of problems of commutative algebra, which play the same role in the proof as the diagram in classical geometry.

The spirit of the interaction of the algebraic and geometrical origins in algebraic geometry was expressively conveyed by A. Weil, who said that geometrical intuition is "invaluable so long as we recognize its limitations".

2.2. It is very instructive to apply the principle of visualization in algebraic geometry to the analysis of propositional logic.

The simplest method of organizing a certain set of propositions into a system is based on the following rule for combining them. Firstly, from two propositions P and Q we can form new ones: $P \vee Q$ (conjunction) and $P \wedge Q$ (disjunction). Secondly, with any proposition P we can associate its negation $\bar{P}$. If a certain set of propositions $\mathscr{A}$ is closed under the operations of conjunction, disjunction, and negation, then we can define on it the structure of a commuatative ring, introducing the operations of addition "$+$" and multiplication "$\cdot$" according to the rule

$$P + Q = (P \vee Q) \wedge (\overline{P \wedge Q}), \qquad P \cdot Q = P \wedge Q.$$

The empty proposition is the zero of this ring, and its negation is the unity. Obviously,

$$P^2 = P.$$

A ring K with unity in which $a^2 = a$ for all $a \in K$ is called a *Boolean algebra*. If a, $b \in K$, then $a + b = (a + b)^2 = a^2 + ab + ba + b^2 = a + b + ab + ba$. Hence $ab + ba = 0$. In particular, if $a = b$, then it follows that $2a = a + a = a^2 + a^2 = 0 \Leftrightarrow a = -a$.

These identities show that any Boolean algebra is commutative and is an algebra over the field $\mathbb{F}_2$ of residues mod 2.

Thus, any closed universe of propositions is a Boolean algebra. Since any Boolean algebra is commutative, we can visualize it by the method described above.

It is not difficult to verify that any prime ideal μ of a Boolean algebra $\mathscr{A}$ is maximal and $\mathscr{A}/\mu = \mathbb{F}_2$. For this reason any element $P \in \mathscr{A}$ can be understood as an $\mathbb{F}_2$-valued (that is, taking values 0 and 1) function on the set Spec A. The subset $M_P \subset \text{Spec } A$ consisting of those points where this function is equal to unity uniquely determines this function, and hence the element P itself. We have

$$M_{P+Q} = (M_P \cup M_Q)\backslash(M_P \cap M_Q),$$

$$M_{P \cdot Q} = M_P \cap M_Q,$$

$$M_{P \vee Q} = M_P \cup M_Q, \qquad M_{P \wedge Q} = M_P \cap M_Q.$$

These relations show that any Boolean algebra $\mathscr{A}$ is isomorphic to a certain algebra of subsets of Spec A with respect to the operations of symmetric union (or symmetric difference) and intersection (product). (The symmetric union of subsets M_1 and M_2 is the subset $(M_1 \cup M_2)\backslash(M_1 \cap M_2)$.) This isomorphism is carried out by the map $P \to M_P$. The operation $(P, Q) \to P + Q + PQ$ in a Boolean algebra corresponds to the more usual operation of union for subsets. If A is finite, then any subset of Spec A has the form M_P for some $P \in A$. If A is infinite, then only subsets of Spec A that are closed under some discrete topology, whose description we omit, have the form M_P. What we have said admits the following interpretation.

The set Spec A is the set of objects consisting of the subject of the statements that occur in A. The subset M_P consists of those objects for which the statement P is true. This, in turn, gives a basis for postulating that

set theory is geometry controlled by formal logic.

In other words, set theory is a visualization of formal logic. Now, bearing this in mind, it is not difficult to understand the origin of the problem of founding all mathematics on the basis of set theory. As we see, it is simply a visually geometrical reflection of the fact that formal logic is constantly used in mathematical reasoning.

But, is such a formulation of the foundation problem well grounded?

Attempts to answer it by the methods of mathematical logic have been very fruitful for the development of this science and have led to the establishment of results of principal importance. Of course, it is inappropriate to discuss these questions here, and we refer the reader to the relevant literature (Manin [1977]). Our aim is merely to emphasize that the very fact of the existence of the right hemisphere of the human brain forces us to have serious doubts about whether this formulation is well grounded. The existence of differential calculus is another such fact. We can be convinced of that by taking a look at the history of geometry from the corresponding angle.

§ 3. On the History of Geometry

The history of modern geometry is clearly divided into three periods: ancient geometry, systematized by Euclid, next the analytical geometry of Fermat and Descartes, and finally differential geometry. These three periods of its development are fixed by the existing system of teaching mathematics, just as the history of form is reflected in the process of embryogenesis. From the viewpoint of § 1, this fact can be explained naturally. Namely, each of these three periods corresponds to a definite level of interaction between the "right and left hemispheres". From this point of view we now take a glance at the history.

3.1. Classical, or elementary, geometry studies the simplest idealized visual forms, principally linear (points, straight lines, planes) and piecewise-linear (angles, polygons, and so on), by the methods of ordinary logic. In other words, truth in elementary geometry is established by logical argument. Assembling elementary deductions into a proof is carried out by the geometer, starting from a contemplation of a diagram.

The masterpiece of elementary geometry is the theorem of Pythagoras, as well as the proof of it given below.

Let Φ_1 and Φ_2 be similar plane figures. From the theory of similarity it follows that $S_1 : S_2 = d_1^2 : d_2^2$, where S_i are the areas of the figures Φ_i, and d_i are the lengths

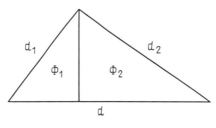

Fig. 2

of some characteristic elements of these figures that correspond to each other. Therefore $S_i = kd_i^2$, where the coefficient k does not depend on i. Now suppose that a right-angled triangle Φ is split by its altitude into two triangles Φ_1 and Φ_2 similar to it (Fig. 2). Let S_i (respectively, S) be the area of Φ_i (respectively, Φ), and d_i and d the lengths of its legs and hypotenuse, respectively. Then $S = S_1 + S_2$, $S_i = kd_i^2$, $S = kd^2$, so

$$kd^2 = S = S_1 + S_2 = kd_1^2 + kd_2^2 \Rightarrow d^2 = d_1^2 + d_2^2.$$

The possibilities of elementary geometry as a method of studying visual forms are very limited, at least in two respects. Firstly, these forms are not logical mosaics, composed of a bounded set of elements. Hence it is impossible to describe them adequately by combining the few indefinable concepts that are the starting point of any actual working axiomatic system. Secondly, everyday logic, which is a means of deduction in elementary geometry, is a very universal language, equally adapted to the needs to descriptive geometry and say cookery. In this connection it is appropriate to draw attention to the fact that the "everyday logic" of geometry, as we have seen (see 2.2), is set theory, which is too poor in itself to express the simplest facts of descriptive geometry without additional means. Thus, a new step in the development of geometry must be to construct it on the basis of a specialized language, better adapted to its needs. This is how analytical geometry arose.

Geometrical theories constructed "according to Euclid", that is, by an axiomatic method, are usually called synthetic, emphasizing the fact that within the bounds of this theory all the concepts and theorems are logically synthesized from the basic ones.

From what we have said it is clear that synthetic geometries can be useful only locally, that is, in order to consider a limited range of questions within the bounds of a wider non-synthetic theory. As an example we mention the so-called finite geometries, which have only recently found useful applications in coding theory (see Vlehduts and Manin [1984], Karteszi [1976]).

3.2. The analytical geometry of Fermat and Descartes differs essentially from any synthetic system of geometry in that truth in it is established by means of algebraic, not logical, calculations. The expressive possibilities of analytical geometry are much greater than those of elementary geometry. Apart from anything else, this is because the logical constructions of elementary geometry are interpreted in a natural way in the language of commutative algebra. For example, the logical connective $\wedge$ can be used in elementary geometry to describe the intersection of two figures. In analytic geometry figures are described by systems of equatons. As we know, the system of equations that specifies their intersection is obtained by joining together the systems that specify the individual figures. From the modern point of view the left-hand sides of the equations that specify the figure under consideration must be interpreted as generators of some ideal in the corresponding commutative algebra (for example, the algebra of polynomials). Therefore the connective $\wedge$ in this context is realized as the operation of adding ideals.

The principal achievement of the analytical method of Fermat and Descartes consists, on the one hand, in mechanizing the arguments used in elementary geometry, and on the other hand in the possibility of involving the investigation of a significantly wider class of objects (for example, algebraic curves of any order). But apparently most important here is the fact that only with the appearance of analytic geometry has the development of mathematics taken on the character described in § 1 of an "electromagnetic wave". In other words, from this moment the right and left hemispheres of the mathematical body joined in valuable interaction.

The masterpiece of analytical geometry, which illustrates its spirit and possibilities, is, for example, the following argument of Plücker, which proves that on a plane algebraic curve of the third order there are three points of inflexion (possibly complex) that lie on a straight line.

We observe, first of all, that an arbitrary equation of the third degree in two variables can be reduced to the form

$$\prod_{i=1}^{3} f_i - f^3 = 0, \tag{1}$$

where $f_i = a_i x + b_i y + c_i, f = ax + by + c$. In fact, the set of functions f_1, f_2, f_3, f is characterized by twelve constants. The left-hand side of (1) is not changed by the substitution $f_1 \to \alpha f_1, f_2 \to \beta f_2, f_3 \to (\alpha\beta)^{-1} f_3, \alpha, \beta \in \mathbb{C}$, and is multiplied by λ^3 by the substitution $f_i \to \lambda f_i, f \to \lambda f$. Hence it is obvious that a family of equations of the form (1) is characterized by nine parameters. On the other hand, any polynomial of the third degree in two variables is determined by its ten coefficients. Since the equation is not changed if we multiply by a constant, the family of equations of the third degree is also nine-parameter. On the basis of this we can conclude that equations of the form (1) exhaust all equations of the third degree.

Furthermore, representing the equation of the curve under consideration in the form (1), we see that the required points of inflexion are the points of intersection of the lines $f_i = 0$ with the line $f = 0$.

The analytical geometry of our day is called algebraic geometry. If, in the time of Descartes and Fermat and later, algebra served geometry, now in the framework of algebraic geometry the latter plays the role of a spider's web in the labyrinths of commutative algebra. For example, the outstanding achievement of our day is the proof by Faltings of Mordell's conjecture that the number of solutions of Diophantine equations of degree higher than two is finite – this was accomplished largely thanks to the application of the methods of algebraic geometry.

The limitation of analytic geometry as a method of investigating visual geometrical forms reveals itself in at least two respects. Firstly, it is very ineffective in the investigation of curves and surfaces specified by transcendental equations. Secondly, such important problems as, for example, the problem of constructing the tangent or finding the measure of the curvature of a curve, which are elementary in nature, cannot be expressed in the language of algebraic operations. Thus, the operational basis of geometry must be extended and this is why differential calculus appears on the scene.

3.3. Properly speaking, differential calculus arose under the pressure of a number of problems of descriptive geometry, in particular the problem of the tangent mentioned above. Its introduction into geometry made it possible to give a precise meaning to such intuitively obvious concepts as "tangent", "curvature", "area of a curved surface", and so on, and to make them the object of mathematical research[1]. As a result, the bounds of geometry were significantly extended.

Commutative (initially simply polynomial) algebra and differential calculus have existed for a long time and have been perceived as absolutely independent mathematical disciplines. A significant time was needed for the suspicion to arise, only in the middle of this century, that this was not so. Today we know that differential calculus is a natural extension of the language of commutative algebra. The explanation of this is given in the next section. Roughly speaking, commutative algebra is differential calculus of order zero. We emphasize that we use the words "differential calculus" here in a much wider sense than, say, in a first course at the university, taking as its elements tensor analysis, the theory of differential operators, and so on.

Thus there arises a new differential geometry in which truth is established by means of both algebraic and "differential" calculi. The famous "theorema egregium" of Gauss is a real masterpiece of the earlier differential geometry. It asserts that the product of the principal curvatures of a surface in three-dimensional Euclidean space is not changed by bendings of the surface, that is, it is uniquely determined by its intrinsic metric (see Ch. 2, 2.8). The proof of this theorem is very characteristic of differential geometry: successive differentiation of the original data (in the given case the first and second quadratic forms) and then algebraic working of the resulting material.

It is now appropriate to draw attention to how great are the possibilities of differential geometry in comparison with synthetic geometry by recalling the

---

[1] From the modern point of view "integral calculus" is part of "differential calculus", namely the theory of de Rham cohomology.

dramatic history of the discovery of the non-Euclidean geometry of Lobachevskij and Bolyai. The exceptionally fine synthetic constructions of these pioneers of the new geometry, which led to a complete verificaton of their authenticity, are completely replaced by simple constructions of differential geometry such as the Beltrami pseudosphere, the Klein model or the Poincaré model (for details see Ch. 3, § 4). In this history we observe an unlimited process of successive derivation of true consequences from true premises of which we spoke in § 1 (we recall that Lobachevskij and Bolyai at first derived all possible consequences of Euclid's system of axioms, in which the fifth postulate is replaced by its opposite), and also the regularizing role of the "right hemisphere" in the form of a differential-geometric model.

The very words "differential geometry" in the rise of this science are understood as the art of solving problems of descriptive geometry by means of analysis. However, in the course of time the meaning of these words underwent an evolution closely analogous to what happened with the words "analytical geom-etry". Namely, modern differential geometry is increasingly taken to be the con-struction and study of visual forms that enable us to find our bearings among the intricacies of differential calculus as such. We see how Riemann understood geometry as an illustration of analysis in his theory of Riemann surfaces, which as a result led to the formation of the concept of a smooth manifold, which is fundamental for modern differential geometry. Still further in this direction went Sophus Lie, who created contact geometry – historically the first complete geometrical theory of differential equations. Symplectic geometry, which is so fashionable nowadays, is a systematic visual interpretation of those constructions of differential calculus that are used in classical mechanics in studying systems with finitely many degrees of freedom.

§ 4. Differential Calculus and Commutative Algebra

Now is the time to justify the assertion made above that differential calculus is an extension of the language of commutative algebra. We do this by showing that the basic concept of differential calculus, that of a linear differential operator, has a purely algebraic nature.

To this end we consider the $\mathbb{R}$-algebra $A = C^\infty(\mathcal{U})$ of all infinitely differentiable functions in a domain $\mathcal{U}$ of the arithmetic space $\mathbb{R}^n$. Then, according to the standard definition, a differential operator $\Delta: A \to A$ (of order at most k) is an operator of the form

$$\Delta = \sum_{|\sigma| \leq k} a_\sigma \frac{\partial^{|\sigma|}}{\partial x_\sigma}, \qquad a_\sigma \in A, \tag{2}$$

where $\sigma = (i_1, \ldots, i_n)$ is a multi-index, $|\sigma| = i_1 + \cdots + i_n$, and

$$\frac{\partial^{|\sigma|}}{\partial x_\sigma} = \frac{\partial^{|\sigma|}}{\partial x_1^{i_1} \ldots \partial x_n^{i_n}}.$$

With any function $f \in A$ we associate the operator $f \colon A \to A$ of multiplication by $f \colon g \mapsto fg$ for all $g \in A$. Starting from the obvious commutator relation $\left[\dfrac{\partial}{\partial x_i}, f \right] = \dfrac{\partial f}{\partial x_i}$, by induction on $|\sigma|$ it is not difficult to verify that

$$\left[\frac{\partial^{|\sigma|}}{\partial x_\sigma}, f \right] = \sum_{\substack{\tau + \nu = \sigma \\ |\tau| > 0}} c_\tau \frac{\partial^{|\tau|} f}{\partial x_\tau} \frac{\partial^{|\nu|}}{\partial x_\nu},$$

where c_τ are certain constants. Hence it follows that the order of the operator

$$[\Delta, f] = \sum_{|\sigma| \leqslant k} a_\sigma \left[\frac{\partial^{|\sigma|}}{\partial x_\sigma}, f \right]$$

does not exceed $k - 1$. Thus, for any $f_0, \ldots, f_k \in A$

$$[f_0, [f_1, \ldots, [f_k, \Delta] \ldots]] = 0. \tag{3}$$

We can show that the converse is also true, that is, if an $\mathbb{R}$-linear map $\Delta \colon A \to A$ is such that (3) holds for any $f_0, \ldots, f_k \in A$, then Δ has the form (2). This shows that the concept of a linear differential operator can be defined in the following general algebraic situation.

Let K be a commutative ring with unit, A a commutative K-algebra, and P, Q certain A-modules. A K-linear map $\Delta \colon P \to Q$ is called a K-linear differential operator of order at most k if for any set of elements $a_0, a_1, \ldots, a_k \in A$ we have

$$[a_0, [a_1, \ldots [a_k, \Delta] \ldots]] = 0.$$

Here for any A-module R an element $a \in A$ is understood as an operator $a \colon R \to R$ that acts according to the rule $r \mapsto ar, r \in R$.

Starting from this definition, we can construct a whole set of concepts of both elementary analysis (tangent vector, the differential of a function, and so on) and "higher" analysis (vector fields, differential forms, integral, jets, and so on). Moreover, in this way we can discover essentially new important concepts of differential calculus (for example, the "higher" analogues of differential forms, generalized Spencer cohomology, and so on), which are apparently impossible to discover by the traditional approach.

We now illustrate by several examples how the geometrical interpretation of differential calculus arises. The latter, as we have just said, should be regarded as differential calculus in commutative algebras. Fundamental for this interpretation is the representation of elements of an algebra A (see above, § 2) as functions on the space Spec A (or Specm A). The latter symbol denotes the totality of all maximal ideals of A.

Interpretation 1

Differentiation of the algebra A with values in the A-module $A/\mathfrak{p}$, $\mathfrak{p} \in$ Spec A	$\Rightarrow$	Tangent vector to the "manifold" Spec A at the "point" $\mathfrak{p}$

A *differentiation of the algebra A with values in the A-module P* is a K-linear map $\Delta: A \to P$ such that $\Delta(ab) = a\Delta(b) + b\Delta(a)$, $a, b \in A$. From this it is obvious that any differentiation is a differential operator of order at most 1 satisfying the condition $\Delta(1) = 0$. The converse is also true.

This interpretation is motivated by the following. Let $A = C^\infty(\mathscr{U})$, where $\mathscr{U}$ is a domain of the arithmetic space $\mathbb{R}^n$, $a \in \mathscr{U}$, and v is a vector starting from the point a. With the vector v we can associate the operator $\xi_v: A \to \mathbb{R}$ of differentiation in the direction of the vector v. More precisely, if $v = (\alpha_1, \dots, \alpha_n)$, then

$$\xi_v = \Sigma\, \alpha_i \frac{\partial}{\partial x_i}\bigg|_a,$$

where $\dfrac{\partial}{\partial x_i}\bigg|_a$ is the operator of partial differentiation at the point a. If we consider the maximal ideal $\mu_a \subset A$ consisting of all functions $f \in A$ that vanish at the point a, then there is a natural isomorphism $A/\mu_a = \mathbb{R}$. Therefore, if by means of this isomorphism we identify the A-module A/μ_a with $\mathbb{R}$, then the image $[f]$ of the function $f \in A$ in A/μ_a is identified with the number $f(a)$. In other words, we can identify the operation of taking values of functions at the point a with the factorization operation $A \to A/\mu_a$. Now understanding the operator ξ_v as an operator that acts from A to A/μ_a, identifying v with ξ_v and putting $\mathfrak{p} = \mu_a$, we see that the vector v is a differentiation of the algebra A with values in the A-module $A/\mathfrak{p}$.

Interpretation 2

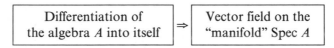

Differentiation of the algebra A into itself	$\Rightarrow$	Vector field on the "manifold" Spec A

A differentiation $\Delta: A \to A$ of the algebra A enables us to associate with any $\mathfrak{p} \in \operatorname{Spec} A$ the differentiation $\Delta\mathfrak{p}: A \to A/\mathfrak{p}$, which is a composition of the maps $A \overset{\Delta}{\to} A \to A/\mathfrak{p}$. Now understanding the operator $\Delta\mathfrak{p}$ as the "tangent vector to Spec A at the point $\mathfrak{p}$", we see that the differentiation Δ enables us to associate a tangent vector with each point $\mathfrak{p} \in \operatorname{Spec} A$, that is, to obtain a "field" of vectors on Spec A.

The fact that the coordinates of geometrical objects are transformed "in some direction" under transformations of the coordinates enables us to split them into "covariant" (in the same direction) and "contravariant" (in the opposite direction). Vectors and vector fields are examples of contravariant objects. In this connection it is appropriate to give an example of the geometrical interpretation of a covariant object. We do this by an example of differential forms of the first degree – the simplest covariant concept of differential calculus.

Covariant objects are objects that are in some sense dual to contravariant ones. Therefore, the geometrical interpretation of covariant quantities, if we are talking about their most general features, consists in interpreting them as functions of one kind or another on the corresponding contravariant object.

Moreover, in the framework of differential calculus covariant objects arise in a natural way as universal realizations of some operations or other. Therefore, the "functional" interpretation of covariant quantities is essentially a direct consequence of their "regular" definition. The next example serves to illustrate the exact meaning of what we have said.

Interpretation 3

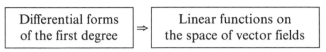

Differential forms of the first degree	$\Rightarrow$	Linear functions on the space of vector fields

First of all we must give a "regular" definition of differential forms of the first degree over Spec A, that is, in differential calculus over the algebra A. A differentiation $d: A \to \Lambda^1$ of A with values in the A-module Λ^1 is said to be universal if for any A-module P and any differentiation $\Delta: A \to P$ there is a unique homomorphism $h: \Lambda^1 \to P$ such that $\Delta = h \circ d$. In other words, any differentiation Δ can be obtained from the universal one by means of a suitable homomorphism $h = h_\Delta$.

The existence of a universal differentiation is established as follows. Consider a free A-module $\tilde{\Lambda}$ whose basis consists of elements denoted by ∂a for all $a \in A$. We generate a submodule $R \subset \tilde{\Lambda}$ by all possible elements

$$\partial(a + b) - \partial a - \partial b, \qquad \partial(ka) - k\partial a, \qquad \partial(ab) - a\partial b - b\partial a,$$

for all $a, b \in A$ and all $k \in K$. If we now put $\Lambda^1 = \tilde{\Lambda}/R$ and denote by da the image of the element $\partial a \in \tilde{\Lambda}$ under the projection $\tilde{\Lambda} \to \Lambda^1$, then it is easy to see that we obtain a universal differentiation $a \mapsto da$.

The universal differentiation is unique up to isomorphism. For let $\delta: A \to \mathscr{L}$ be another universal differentiation. Then $\delta = h_\delta \circ d$, since d is universal, and $d = h_d \circ \delta$, since δ is universal. Hence it follows that

$$d = (h_d \circ h_\delta) \circ d \qquad \text{and} \qquad \delta = (h_\delta \circ h_d) \circ \delta.$$

The uniqueness of the homomorphism h that represents a given differentiation Δ shows that, in the case $\Delta = d$, $h_d \circ h_\delta: \Lambda^1 \to \Lambda^1$ is the identity automorphism of the module Λ^1. We can do the same for the composition $h_\delta \circ h_d: \Lambda \to \Lambda$. This shows that h_δ and h_d are mutually inverse isomorphisms of the A-modules Λ^1 and Λ.

The elements of Λ^1 will be called *differential forms of the first degree over A* (or Spec A). If $\Delta: A \to A$ is a differentiation (= vector field on Spec A) and $\omega \in \Lambda^1$, then $h_\Delta(\omega) \in A$, that is, $\Delta \mapsto h_\Delta(\omega)$ is an A-valued function on the set of all vector fields on Spec A. Obviously

$$h_{\Delta_1 + \Delta_2} = h_{\Delta_1} + h_{\Delta_2} \qquad \text{and} \qquad h_{a\Delta} = ah_\Delta, \qquad a \in A,$$

so this function is A-linear. This gives us the required interpretation.

Thus, a systematic development of the arguments presented above enables us to discover the purely algebraic nature of differential calculus as such (see [5]).

In this connection we shall understand the words "differential calculus" in this algebraic spirit.

§ 5. What is Differential Geometry?

5.1. Now all we have said above leads us in a natural way to the following answer to the question we have posed:

> it is useful to treat differential geometry, both from the viewpoint of the essence of the matter and from the viewpoint of the tendencies of its modern development, as the science whose aim is the construction and investigation of "visual" forms, controlled by differential calculus in the sense stated above.

In taking this formulation and the motivating arguments preceding it, we have not sought to pay tribute to metaphysical tradition. On the contrary, our aim has been to draw attention to those new possibilities that arise in connection with the discovery of methods of systematic geometrization of commutative algebra and differential calculus. For example, analysing which aspects of differential calculus serve as classical "spontaneously" developed branches of differential geometry turns out to be very useful both for a deeper understanding of the aspects themselves and for more meaningful applications, especially to problems of modern fundamental physics. Moreover purposeful geometrization of new domains of differential calculus enables us to extend the bounds of traditional differential geometry in a natural and useful way.

This point of view on the nature of differential geometry at first glance does not agree with the widespread opinion that the object of differential geometry is the study of "geometrical structures" (see Ch. 6) on smooth manifolds. In fact, this is not so and it can be shown that each of the structures studied by geometers today is a "picture" of some aspect or other of differential calculus. For example, a Riemannian metric (see Ch. 3) is a symbol of a scalar elliptic differential operator of the second order.

Differential calculus, being the natural language for the whole of modern mathematical science, undergoes with it an interesting and dynamic development. Apparently the central problem here is that of constructing a mathematical apparatus on which quantum field theory can be constructed as strictly and consistently as the classical theory. In this connection it is natural to suppose that there is a later ("secondary") level of differential calculus that is the natural language for quantum field theory and corresponds to "ordinary" differential calculus in the same way as quantum physics corresponds to classical physics. In fact, the primitive elements of this secondary differential calculus are known today (Vinogradov [1986], Vinogradov, Krasilshchik and Lychagin [1986]), so the birth of a secondary differential geometry is on the agenda.

5.2. As we have seen in § 3, the main result of the development of geometry from the time it was formed into a science in Ancient Greece up to the present day is the transformation of its archaic synthetic form into the modern differential form. It would be imprudent to ignore this fact. We therefore ask, what is the underlying cause of this transformation, in particular, is the apparently indissoluble link between modern geometry and differential calculus accidental?

In this connection let us recall the words uttered by Riemann in his famous inaugural lecture "On the hypotheses that lie at the basis of geometry" (see Riemann [1868]); "Progress in the knowledge of the mechanism of the external world, acquired in the course of recent centuries, has been conditioned almost exclusively by the accuracy of the construction that has become possible as a result of the discovery of infinitesimal analysis ... ". In particular, we can establish the complete adequacy of the language of differential calculus for an exact description and investigation of what we see (that is, for the aims of "descriptive geometry"). Conversely, most of the problems of "descriptive geometry" (for example, the problem of the tangent) cannot be expressed by means of analytic (algebraic) geometry, and even less by synthetic geometry. Therefore it would hardly be reasonable to regard the link mentioned above as accidental, and we attempt to find a regular direction of thought to explain it, by returning to the considerations of § 1.

Let us imagine a thinking system (for example, the brain of an animal) observing the motion of a point. If it fixes the position of the point, say 10 times a second, then in order to have a description of its motion during one minute it is necessary to keep in the memory of this system 1800 real numbers. Bearing in mind that the animal must follow billions of points simultaneously, we can see that its brain is not in a position to preserve and rework such a huge mass of information. Since there is no way of fixing the motion of a chaotically moving point other than directly remembering its coordinates, we must conclude that such a motion is unobservable by the thinking system. However, we need a substantially smaller volume of memory to describe the motion of a point whose successive displacements display a certain tendency[2]. The existence of such a tendency is obviously equivalent to the fact that a point has a definite velocity at each instant of time, in other words, its trajectory is a differentiable curve[3].

These arguments, of course, remain in force every time the thinking system tries to make an observation in conditions when the volume of information coming forward about the object under observation substantially exceeds the possibilities of this system for preserving and reworking it. In this situation what is observable from the viewpoint of the thinking system turns out to be equivalent to what is differentiable (it would be dangerous to separate the deduction we have made from the context surrounding it). In our opinion this is the reason why differential calculus is the language of modern geometry or rather the whole of modern physics, and also the "accuracy" of which Riemann spoke.

---

[2] Rhythm, melody and rhyme, which make music and poetry that are so attractive to us, are ways of organizing information based on the appearance of tendencies that exist in its structure. Therefore it is easier for us to remember poetically formed thought or melody than the chaos of sounds.

[3] In particular, thanks to this it is possible to preserve information about the motion of a point in such an elegant and compact form as the equations of classical mechanics.

5.3. The pragmatic spirit of the present century has given rise to a sceptical attitude to any kind of philosophy, especially among those who have successfully solved difficult concrete problems. This is natural, because philosophy, being a long-wave approximation to the phenomena under consideration, does not have the necessary resolving ability that makes it possible to distinguish those small hardly perceptible details that form the essence of concrete problems. However, it enables us to reveal the contours of a large-scale target. Bearing this in mind, we should like to subject to experimental verification the high quality of the general arguments that form this chapter. To this end we mention the following three cycles of questions; the advisability of investigating them obviously follows from these arguments, and they belong to domains in which the authors are not experts.

1. We accept the proposition that sets are the visual images of Boolean algebras (see 2.2). As we have seen, the concepts of a finite set and a finite Boolean algebra are equivalent. Infinite sets are poorer objects than infinite Boolean algebras. This is connected with the fact that the spectrum of a Boolean algebra is naturally endowed with the Zariski topology. Closed subsets in this topology are identical to those that can be completely characterized by the propositions that constitute the original Boolean algebra. Therefore it is more satisfactory to consider, instead of sets with structures, categories of modules over Boolean algebras and possibly other formations related to them. It would be interesting to investigate the consequences of this step systematically, especially in connection with those logical difficulties that follow from the use of infinite sets. For now we mention that in this approach such things as the "Boolean algebra of all Boolean algebras" does not arise.

2. Differential calculus, understood in the spirit of § 4, is still not the instrument of commutative algebra and algebraic geometry that it could be as the natural supplier of localized constructions for these domains. It seems an attractive alternative to the apparatus of sheaves and it would be useful to reinterpret the foundations of algebraic geometry systematically on this basis. We do not see any reason why it is impossible to carry out its construction over an arbitrary field of characteristic zero in the same spirit as, for example, in the book of Griffiths and Harris [1978] for the field of complex numbers. As a hopeful example we mention the fact that Spencer cohomologies are very subtle invariants of singularities of algebraic varieties. In addition, we can draw attention to the fact that from the apparatus of Spencer cohomologies, which is a natural part of differential calculus over arbitrary commutative algebras, there follow the fundamental cohomological systems of complex algebraic geometry (Vinogradov, Krasilshchik and Lychagin [1986]).

3. With the intention of adding specifically human elements to the algorithmic "thinking" of modern computers, we need to understand the working principle of the mechanisms that work out differential calculus, understanding the latter as an instrument for investigating those components of informational chaos that reveal various tendencies. Here we can imagine interesting computer experiments.

Chapter 2
The Geometry of Surfaces

> "A surface is an exceptionally thin substance."
>
> Hugh Valery

§1. Curves in Euclidean Space

The fact that it is useful to differentiate and integrate in geometry was first realized in the study of curves. The theory of curves enables us to understand more easily the role of transformation groups in geometry and, in particular, the fundamental fact that geometrical concepts themselves are *differential invariants* of certain *groups*. Finally, in the theory of curves there arises a *moving frame method*, a very general approach to the solution of many geometrical problems.

1.1. Curves. A *parametrized curve* in Euclidean space of n dimensions E^n is a smooth (infinitely differentiable) regular map $x: I \to E^n$, where I is an interval of the number line $\mathbb{R}$. In coordinates such a curve is given by a vector-valued function $x(t) = (x_1(t), \ldots, x_n(t))$, where $t \in I$, and $(x_1, \ldots, x_n)$ are Cartesian coordinates in E^n. *Regularity* means that the *velocity vector* $\dot{x}(t) = (\dot{x}_1(t), \ldots, \dot{x}_n(t))$, $\dot{x}_i = dx_i/dt$, is nonzero. The parametrized curves $x(t)$ and $y(\tau)$ represent the same (*geometrical*) *curve* if they reduce to each other by a change of parameter, that is, $x(t) \equiv y(\tau(t))$, where $\tau(t)$ is a smooth function and $\dot{\tau}(t) \neq 0$. Two (geometrical) curves have *the same form* or are *metrically equivalent* if they reduce to each other by some motion of the space E^n.

The main problem in the metrical theory of curves consists in establishing when they have the same form.

1.2. The Natural Parametrization and the Intrinsic Geometry of Curves. To solve the main problem we need, if possible, to get rid of the arbitrariness in the choice of parametrization of the curve. This is achieved by introducing the natural parameter. Namely, the length of arc of the curve $x(t)$ between the points $A = x(t)$, and $B = x(t_2)$ is given by the well-known elementary formula

$$s(A, B) = \int_{t_1}^{t_2} (\dot{x}(t), \dot{x}(t))^{1/2} \, dt.$$

Fixing the point A on the curve enables us to introduce on it a new, geometrically natural parameter $s(t) = s(A, t)$ called the *natural* parameter. It is uniquely defined up to a transformation $s \mapsto \pm s + \text{const}$, which reflects the arbitrariness in the choice of the initial point A and the orientation of the curve.

We can give the form of any curve by means of an inextensible thread. This means that *intrinsically* curves are indistinguishable "in the small". By intrinsic (metrical) properties we mean those that can be discovered on the basis of measurements "inside" the curve. A map $s \mapsto x(s)$, $s \in I$, where s is the natural parameter, is an isometry of a segment of the Euclidean interval I and the curve under consideration, that is, for all $s_1 < s_2 \in I$

$$s_2 - s_1 = s(A_1, A_2), \qquad A_i = x(s_i).$$

If s is the natural parameter, then $|\dot{x}(s)| = 1$, $\dot{x}(s) = dx/ds$.

1.3. Curvature. The Frenet Frame. The simplest extrinsic characteristic of a curve is its *curvature*. If s is the natural parameter, then $x'(s)$ and $x'(s + \Delta s)$ are unit vectors. Therefore the angle $\Delta \varphi$ between them is equal to $\Delta \varphi = x''(s)\Delta s + o(\Delta s)$, and the curvature $k(s)$ of the curve at the point $x(s)$, defined as the *rate of rotation of the tangent*, is equal to

$$k(s) = |x''(s)|.$$

Obviously the curvature does not depend on the choice of natural parameter and at corresponding points it is the same for curves obtained from each other by a motion. Therefore the curvature is a *metrical invariant of the curve*. Further invariants of this type can be obtained by constructing the family of *Frenet frames* along the curve.

To this end we consider the family of subspaces $\Phi_k = \Phi_k(s)$ spanned by the vectors $x'(s)$, $x''(s)$, ..., $x^{(k)}(s)$ for some fixed s. Obviously $\Phi_1 \subset \Phi_2 \subset \cdots \subset \Phi_n$. We shall assume that the vectors $x', \ldots, x^{(n)}$ are linearly independent. Otherwise the curve lies in a subspace of E^n of lower dimension. In view of this, $\dim \Phi_k(s) = k$. $\Phi_k(s)$ is called the *k-th osculating space* of the curve at the point $x(s)$. We define vectors $e_i(s)$, $i = 1, \ldots, n$, assuming that

$$e_i(s) \in \Phi_i(s), \quad e_i(s) \perp \Phi_{i-1}(s), \quad |e_i(s)| = 1, \quad (e_i(s), x^{(i)}(s)) > 0.$$

By these conditions the vectors $e_1(s) = \dot{x}(s)$, ..., $e_n(s)$ are defined uniquely and form an orthonormal frame, called the *Frenet frame* of the curve at the point $x(s)$

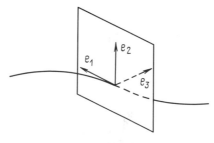

Fig. 3

(Fig. 3). The rate of change of the Frenet frame under a motion along the curve is described by the following *Frenet formulae*:

$$\frac{de_i}{ds} = -k_{i-1}(s)e_{i-1} + k_i(s)e_{i+1}, \qquad i = 1, \ldots, n,$$

where we assume that $k_0 = k_{n+1} = 0$, $e_0 = e_{n+1} = 0$. The functions $k_i(s)$ defined by these formulae are called the *higher curvatures* of the curve. Obviously $k_1(s) = k(s)$, $k_i(s) \neq 0$. When $n = 3$ the function $k_2(s)$ is called the *torsion of the curve*.

The main result of the theory of curves follows from the existence and uniqueness theorem for ordinary differential equations.

Theorem. *Under the assumptions we have made the functions* $k_1(s), \ldots, k_{n-1}(s)$ *uniquely determine a curve up to a motion. Moreover, an arbitrary system of functions* $k_i(s)$, $i = 1, \ldots, n - 1$, *satisfying the conditions* $k_1(s) > 0$, $k_i(s) \neq 0$ *for all s, determines in* E^n *a curve for which they are the higher curvatures.*

In other words, the higher curvatures form a *complete system of metric invariants* of a curve in the sense that all others can be expressed in terms of them. We can also say that the form of a curve is uniquely determined by the functions $k_i(s)$.

1.4. Affine and Unimodular Properties of Curves. An affine transformation $F: E^n \to E^n$, $(x_i) \mapsto (y_i)$ has the form $y_i = \sum_j a_{ij}x_j + c_i$, where $a_{ij}, c_i \in \mathbb{R}$. It is said to be *unimodular* if $\det \|a_{ij}\| = 1$. Properties of a curve that do not change under an affine (respectively, unimodular) transformation are said to be *affine* (respectively, *unimodular*). In this sense we talk about *affine (unimodular) equivalence* of curves or about their *affine (unimodular) invariants*. We describe them, assuming that for the curve $x(t)$ in question the vectors $\dot{x}(t), \ldots, x^{(n)}(t)$ are independent for all t.

In the unimodular case the "natural" parameter s is chosen from the condition $\langle x'(s), \ldots, x^{(n)}(s) \rangle = 1$, where $\langle a_1, \ldots, a_n \rangle$ denotes the *oriented volume* of the parallelepiped formed from the vectors $a_1, \ldots, a_n$. Differentiating this equality, we see that $\langle x'(s), \ldots, x^{(n-1)}(s), x^{(n+1)}(s) \rangle = 0$, or equivalently that

$$x^{(n+1)}(s) = \mu_1(s)x'(s) + \cdots + \mu_{n-1}(s)x^{(n-1)}(s). \qquad (1)$$

The unimodular natural parameter is defined uniquely up to a transformation $s \mapsto s + \text{const}$ if the number $\frac{1}{2}n(n + 1)$ is odd, and up to a transformation $s \mapsto \pm s + \text{const}$ if it is even. Therefore, up to signs that we could easily make precise, the functions $\mu_i(s)$, $i = 1, \ldots, n - 1$, are unimodular invariants of the curve. It is not difficult to show that they uniquely determine the curve up to a unimodular transformation and, moreover, the curve can be constructed with preassigned invariants $\mu_i(s)$.

We choose the *affine "natural" parameter s* starting from the condition that (1) is satisfied (on the right the term proportional to $x^{(n)}(s)$ is missing). It is defined uniquely up to a transformation $s \mapsto ps + q$, $p, q \in \mathbb{R}$, $p \neq 0$. Under such a transformation the system $(\mu_1, \mu_2, \ldots, \mu_{n-1})$ goes into the system $(p^n\mu, p^{n-1}\mu_2, \ldots, p^2\mu_{n-1})$. Bearing this in mind, from the quantities μ_i we can form others invariant with respect to a change of the affine natural parameter. One of the possible choices is as follows:

$$d/ds(\mu_{n-1})^{-1/2}, \quad \mu_i^2, \mu_{n-1}^{n+1-i}, \quad i = 1, \ldots, n-2,$$

$$\mathrm{sgn}\ \mu_{n-2k-1}, \qquad k = 0, 1, \ldots, \left[\frac{n-2}{2}\right].$$

Thus these quantities are affine invariants of the curve. We can show that they determine it uniquely up to an affine transformation and we can choose them arbitrarily.

Example 1. All spirals are affinely equivalent. Two spirals are unimodularly equivalent if they twist in the same direction and the quantity R^2h, where R is the radius of the spiral and h is its pitch, are the same for both of them.

Example 2. Suppose that $n = 2$. It is appropriate to call the invariant μ_2 the unimodular curvature of a plane curve. For the ellipse $x^2/a^2 + y^2/b^2 = 1$, the hyperbola $x^2/a^2 - y^2/b^2 = 1$ and the parabola $y^2 = 2px$, $\mu_2 = -a^{-2}b^{-2}, a^{-2}b^{-2}$ and 0, respectively.

§2. Surfaces in E^3

As in the case of curves, the main problem in the theory of surfaces in E^3 is the construction of a complete system of invariants that determine the surface uniquely up to a motion. The existence for surfaces of a nontrivial intrinsic geometry and the associated lack of an analogue of the natural parameter make this theory much more complicated and rich.

The study of surfaces in E^3 led to three fundamental discoveries of general geometrical significance. Firstly, the discovery of Gauss in his famous "Disquisitiones generales circa superficies curvas" of the intrinsic geometry of surfaces and, in particular, of the *intrinsic* or *Gaussian curvature* of surfaces. This discovery ranks with the discovery by Einstein of the bending of real physical space. It is the basis on which the geometry of Riemann was founded (see Ch. 3). Secondly, the Gauss-Bonnet formula, which establishes the connection between the local and global characteristics of a surface, served as the source from which the modern theory of *characteristic classes* arose (see Ch. 8). Finally, the theory of parallel transport of tangent vectors along a surface led to the rise of the *theory of connections*, a most important object of modern geometry, without which the modern theory of physical fields (the Yang-Mills field) would be unthinkable today.

From the tools of differential geometry in the theory of surfaces there crystallized the *moving frame method* (the work of Darboux), which was later significantly developed and generalized by Cartan.

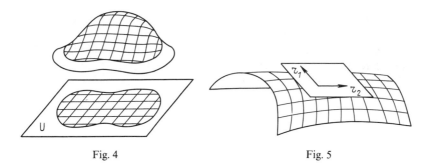

Fig. 4   Fig. 5

In this section we begin to use Einstein's rule, according to which the summation sign for repeated indices is omitted. For example, $h_{ij}\, du^i\, du^j$ means $\sum_{i,j=1}^{n} h_{ij}\, du^i\, du^j$, where n is the number of independent variables.

2.1. Surfaces. Charts. Let U be a domain in the plane of the variables $u = (u^1, u^2)$. A smooth map $r: U \to E^3$ is called a *parametrized surface* in E^3 if the partial derivative vectors $r_i = \partial r / \partial u^i$, $i = 1$, 2, are linearly independent for all $u \in U$. A subset $M \subset E^3$ is called a *smooth surface* if there is a family of parametrized surfaces $r_\alpha: U_\alpha \to E^3$ such that

$$r_\alpha(U_\alpha) \subset M, \qquad \bigcup_\alpha r_\alpha(U_\alpha) = M.$$

If $r(U) \subset M$, we say that the parametrized surface (r, U) is a *chart* of the surface M, and u^1, u^2 are *(local) curvilinear coordinates* on M (Fig. 4). In this case the vectors $r_1(u)$, $r_2(u)$ form a basis of the tangent plane to M at the point $x = r(u)$, which is denoted by $T_x M$ (Fig. 5).

2.2. The First Quadratic Form. The Intrinsic Geometry of a Surface. Any vector ξ that touches the surface M at the point $r(u)$ can be written in the form $\xi = \xi^i r_i(u)$. Therefore its scalar square is equal to $g_{ij}(u)\xi^i\xi^j$, where $g_{ij}(u) = (r_i(u), r_j(u))$. Thus on M there arises a "field" of quadratic forms $r(u) \to Q_u$, where $Q_u(\xi) = g_{ij}(u)\xi^i\xi^j$ is a quadratic form on $T_{r(u)}M$. This field of quadratic forms is called the *first quadratic differential form* (or *first fundamental form*) of the surface M and denoted by $g_{ij}(u)\, du^i\, du^j$. More precisely, this expression is the notation for the first quadratic form in the curvilinear coordinates u^1, u^2. The convenience of this notation is that it enables us to find automatically the form of the first quadratic form in any other curvilinear coordinates, say v^1, v^2. For this it is sufficient to substitute in $g_{ij}(u)\, du^i\, du^j$ the functions that express the u^i in terms of v^1 and v^2. In particular, from this there follows the transition rule

$$g'_{ij}(v) = g_{kl}(u(v)) \frac{\partial u^k}{\partial v^i} \frac{\partial u^l}{\partial v^j},$$

where $g'_{ij}(v)\, dv^i\, dv^j$ is the notation for the first quadratic form in the coordinates v^1, v^2.

The first quadratic form enables us to calculate in an "intrinsic" way lengths, angles and areas on the surface M.

If a curve on M is given by a rule describing how its curvilinear coordinates $u(t) = (u^1(t), u^2(t))$, $a \leqslant t \leqslant b$, depend on t, then its length L is expressed by the formula

$$L = \int_a^b \sqrt{g_{ij}(u(t))\dot{u}^i\dot{u}^j}\ dt.$$

The area of a domain $D \subset M$ is calculated from the formula

$$S = \iint_D \sqrt{|g|}\ du^1\ du^2, \qquad |g| = \det \|g_{ij}\|,$$

and the angle φ between the vectors $\xi, \eta \in T_x M$ from the formula

$$\cos \varphi = \frac{g_{ij}(u)\xi^i\eta^j}{(g_{ij}\xi^i\xi^j)^{1/2}(g_{ij}\eta^i\eta^j)^{1/2}}.$$

A two-dimensional being, living on the surface in question and having no suspicion of the existence of a third dimension, could, by measuring the lengths of various curves on it, find its first quadratic form experimentally. Therefore the properties of a surface that can be described by means of its first quadratic form are said to be *intrinsic* and refer to the *intrinsic geometry of surfaces*.

A map of one surface on another is called an *isometry* if curves that lie on them and are mapped into each other have the same lengths. Under an isometry the first quadratic forms of the surfaces are mapped into each other.

A motion of the space E^3 that makes one surface coincide with another is an isometry between them. However, isometries are not exhausted by motions, as is shown by the process of rolling a sheet of paper into a roll.

Above (in 1.2) we have seen that "in the small" curves are indistinguishable from the point of view of their intrinsic geometry. Is this true for surfaces? In particular, is it possible, to some extent, to discover the fact that a given surface is bent by intrinsic means? The answers to these basic questions are given in the course of the following exposition.

The scalar product of vectors $\xi, \eta \in T_x M$ is denoted below by $g_x(\xi, \eta)$. This is a symmetric bilinear form associated with the first quadratic form.

2.3. The Second Quadratic Form. The Extrinsic Geometry of a Surface. In curvilinear coordinates u^i any curve on a surface M can be described by a pair of functions $u^i(t)$, $i = 1, 2$; as a curve in E^3 it is given by the vector-valued function $\rho(t) = r(u^1(t), u^2(t))$. Expanding $\rho(t)$ in a Taylor series at a point t_0, we can easily verify that the value of the inclination of this curve to the tangent plane to M at the point $x = \rho(t_0)$ has the form

$$\frac{\Delta t^2}{2} \cdot h_{ij}(u_0)\dot{u}_0^i\dot{u}_0^j + o(\Delta t^2),$$

where $u^0 = u(t_0)$, $\dot{u}_0^k = \dot{u}^k(t_0)$, $h_{ij} = (m_x, r_{ij})$, m_x is the unit normal vector to M at the point x and $r_{ij} = \partial^2 r / \partial u^i \partial u^j$. Thus, the quadratic differential form $h_{ij}(u) \, du^i \, du^j$, being restricted to the curve $\rho(t)$, expresses the principal part of the inclination of this curve to the tangent plane. It is called the *second quadratic differential form* (or *second fundamental form*) of the surface. The symmetric bilinear form on $T_x M$ corresponding to it is denoted by h_x. The second quadratic form changes sign for the opposite choice of the normal vector, that is, it is uniquely determined by the orientation of the surface. From what we have said it is obvious that the quantity

$$k_n(\xi) = h_x(\xi, \xi)/g_x(\xi, \xi), \qquad \xi = \dot{\rho}(t_0) \in T_x M,$$

is equal to the *normal curvature* $\left(m_x, \dfrac{d^2\rho}{ds^2}(t_0) \right)$ of the curve $\rho(t)$ at the point $x = \rho(t_0)$. The normal curvatures at a point $x \in M$ of curves that touch each other at this point are equal and are called the *normal curvature of the surface in the direction of the tangent vector* ξ. This normal curvature coincides with the curvature at x of the *normal section* of M by a plane parallel to the vectors m_x and ξ (Fig. 6). The extreme values of the function $k_n(\xi)$, $\xi \in T_x M \backslash 0$, are called the *principal curvatures of the surface M at the point x*, and the values of the argument ξ corresponding to them are called the *principal directions*. They can be interpreted as the eigenvalues and eigenvectors, respectively, of the *Peterson operator* $A_x: T_x M \to T_x M$, defined by the formula

$$g_x(A_x \xi, \xi) = h_x(\xi, \xi).$$

Hence it is obvious that if the function $k_n(\xi)$ is not a constant (otherwise x is called an *umbilic*), then there are exactly two principal curvatures $k_+ > k_-$ and corresponding to them exactly two principal directions p_+, p_-, which are mutually perpendicular (Fig. 7).

In many questions it is more convenient to consider symmetric functions of the principal curvatures: $K(x) = k_+ k_- = \det A_x = \det \|h_{ij}\|/\det \|g_{ij}\|$, the *Gaussian curvature*, and $H(x) = \frac{1}{2}(k_+ + k_-) = \frac{1}{2} \operatorname{tr} A_x$, *the mean curvature*. The second quadratic form describes the extrinsic geometry of a surface in the same sense as the first quadratic form describes the intrinsic geometry. For a full

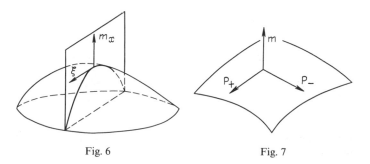

Fig. 6   Fig. 7

explanation of which properties of a surface are intrinsic and which are extrinsic, we must reveal the connections that exist between the first and second quadratic forms. We shall do this later, in 2.8.

2.4. Derivation Formulae. The First and Second Quadratic Forms. Let $x = r(u)$. The vectors $r_1(u)$, $r_2(u)$, $m_x = m(u)$ form a frame of E^3 at x. The variation of this frame for an infinitesimal change of the point x on the surface is described by the so-called *derivation formulae*:

$$\begin{cases} \partial_j r_i = \Gamma_{ij}^k r_k + h_{ij} m, \\ \partial_j m = -A_j^k r_k. \end{cases} \tag{2}$$

Here $\partial_j = \partial/\partial u_j$, and $\|A_j^k\|$ is the matrix of the Peterson operator in the basis r_1, r_2. The functions $\Gamma_{ij}^k = \Gamma_{ij}^k(u)$ defined by these formulae are called the *Christoffel symbols* and are expressed in the following way in terms of the coefficients of the first quadratic form:

$$\Gamma_{ij}^k = 1/2 g^{kl}(\partial_i g_{1j} + \partial_j g_{1i} - \partial_l g_{ij}). \tag{3}$$

The formulae (2) and (3) show that all the partial derivatives of the vector-valued function $r(u^1, u^2)$ can be expressed in terms of the coefficients of the first and second quadratic forms and their derivatives. Moreover, the following basic result is true, which follows immediately from the derivation formulae.

Theorem. *Surfaces M and M' are obtained from each other by a motion of E^3 if and only if there is a map $f: M \to M'$ that identifies their first and second quadratic forms, respectively.*

2.5. The Geodesic Curvature of Curves. Geodesics. Let $\rho(s) = r(u(s))$ be a naturally parametrized curve on a surface M. The projection $\ddot{\rho}_T$ of the acceleration vector $\ddot{\rho}$ on the tangent plane $T_{\rho(s)} M$ is called the *geodesic curvature vector*, and its length is called the *geodesic curvature* k_g of the curve $\rho(s)$. From (2) it follows that

$$\ddot{\rho}_T = (\ddot{u}^i + \Gamma_{jk}^i \dot{u}^j \dot{u}^k) r_i. \tag{4}$$

Since, by (3), the Christoffel symbols can be expressed in terms of the first quadratic form, the quantities $\ddot{\rho}_T$ and $k_g = |\ddot{\rho}_T|$ are concerned with the intrinsic geometry of the surface.

Curves on a surface M that are not bent from the point of view of its intrinsic geometry are defined by the condition $k_g = 0 \Leftrightarrow \ddot{\rho}_T = 0$, or in coordinates,

$$\ddot{u}^i(s) + \Gamma_{jk}^i(u(s))\dot{u}^j(s)\dot{u}^k(s) = 0, \qquad i = 1, 2.$$

They are called *geodesic curves* of the surface M. In ordinary Euclidean geometry unbent curves are shortest curves and conversely. This still holds in the intrinsic geometry of surfaces.

Theorem. *Geodesic curves on a surface M, and only they, are (locally) shortest curves on this surface.*

2.6. Parallel Transport of Tangent Vectors on a Surface. Covariant Differentiation. Connection. If we interpret the geodesic curvature of a curve on a surface M in the usual way as the "rate of turning of the tangent" (see 1.3), then we must assume that in the transition from a point $\rho(s)$ to a point $\rho(s + \Delta s)$ the tangent vector $\dot{\rho}(s)$, and together with it the whole tangent space $T_{\rho(s)}M$, rotates through an angle $\Delta \varphi = k_g(s)\Delta s + o(\Delta s)$. In other words, if at each point of the curve we introduce an orthonormal "Frenet frame" $\{\tau, n\}$, where $\tau = \dot{\rho}(s)$, $n = k_g^{-1}(s)\ddot{\rho}_T(s)$, we must postulate the 2-dimensional "Frenet formulae"

$$V_\xi \tau = k_g h, \qquad V_\xi h = -k_g \tau, \tag{5}$$

where V_ξ is the hypothetical infinitesimal operator of "parallel transport" of tangent vectors to M in the direction of the vector $\xi = \dot{\rho}(s)$.

At this point we must observe that there is no natural mechanism that enables us to identify tangent vectors to a surface M that have different points of contact. In plane Euclidean geometry a similar identification can be carried out by parallel transport of vectors from one point to another. Since this is not so in the situation we are considering, we have used inverted commas above.

Thus, our aim will be achieved if we take the formulae (5) as the definition of an *infinitesimal operator of parallel transport* V_ξ. In order to have the possibility of considering parallel transport with an arbitrary velocity, and not only with unit velocity, we define the operator V_ξ for any ξ, assuming that $V_{\lambda\xi} = \lambda V_\xi$, where $\lambda \in \mathbb{R}$. The rule of "addition of velocities" will be satisfied: $V_{\xi+\eta} = V_\xi + V_\eta$.

Thus, for any $\xi \in T_x M$ we have defined the operator

$$V_\xi: D(M) \to T_x M,$$

where $D(M)$ is the totality of all smooth tangent vector fields on M, called the *operator of covariant differentiation* in the direction of the vector ξ. In view of the fact that the geodesic curvature has an intrinsic meaning, from the defining formulae (5) it follows that the operators V_ξ are also determined by the intrinsic geometry of the surface M.

It is useful to go further and associate with any tangent vector field X on M the operator

$$V_X: D(M) \to D(M)$$

of *covariant differentiation along* X, assuming that the value of the vector field $V_X Y$ at a point $x \in M$ is equal to $V_\xi Y$, where ξ is the vector of the field X at the point x.

The operation V has the following properties, which follow immediately from its definition:

$$V_{fX+gY} = fV_X + gV_Y,$$

$$V_X(Y_1 + Y_2) = V_X Y_1 + V_X Y_2, \tag{6}$$

$$V_X(fY) = X(f)Y + fV_X Y,$$

where $f, g \in C^\infty(M)$, $X, Y \in D(M)$, and $X(f) = X^i(u)\partial f/\partial u^i$ if $X = X^i(u)r_i$.

From the derivation formulae (2) it follows that

$$V_i r_j = \Gamma_{ij}^k r_k,$$

where $V_i = V_{r_i}$. From this and (6), taking into account that $X = X^i r_i$ and $Y = Y^i r_i$ it follows that

$$V_X(Y) = X^i(\partial_i(Y^k) + \Gamma_{ij}^k Y^j) r_k. \qquad (7)$$

In particular, if $\xi(s) = \xi^i(s) r_i$ is a vector field along the curve $\rho(s)$, then

$$V_{\dot\rho(s)} \xi(s) = (\dot\xi^k + \Gamma_{ij}^k \dot u^i \xi^j) r_k. \qquad (8)$$

Taking account of the origin of the operator $V_{\dot\rho}$, we say that $\xi(s)$ is a *vector field of parallel vectors along the curve* $\rho(s)$ if $V_{\dot\rho(s)} \xi(s) = 0$. In this case we also say that a vector $\xi(s_1) \in T_{\rho(s_1)} M$ is *parallel* to a vector $\xi(s_2) \in T_{\rho(s_2)} M$ along the curve $\rho(s)$. Thus, on any surface M we have defined in an intrinsic way the *operation of parallel transport of vectors along curves*. The "intrinsic bending" of a surface M is shown by the fact that parallel transport of a vector along two different paths at the same point gives different results, generally speaking.

Theorem. *If parallel transport of tangent vectors on a surface does not depend on the choice of path, then this surface is locally isometric to a Euclidean plane.*

Fig. 8 shows a field of parallel vectors on a sphere along one of its parallels. If this parallel is at an angle α to the plane of the equator, then under a parallel transport round it the tangent vectors rotate through the angle $\varphi = 2\pi \sin \alpha$.

From (8) it follows that the vector field $\xi(t)$ is parallel along the curve $u(t)$ if and only if

$$\dot\xi_k + \Gamma_{ij}^k(u(t)) \dot u^i \xi^j = 0.$$

We imagine the tangent plane $T_{\rho(t_1)} M$ made of a sheet of plywood and "roll" it over the surface M so that its point of contact describes the curve $\rho(t)$ (Fig. 9). Marking successive points of contact on this sheet, we obtain a plane curve $\tilde\rho(t)$, called the *development* of the curve $\rho(t)$. Any tangent field $\xi(t)$ along $\rho(t)$ is "unrolled" into some field $\tilde\xi(t)$ along $\tilde\rho(t)$. It turns out that the field $\xi(t)$ is parallel along $\rho(t)$ if and only if the unrolled field $\tilde\xi(t)$ consists of vectors in the plane of

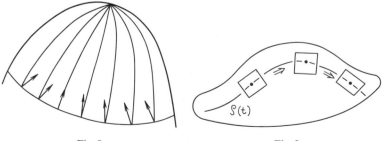

Fig. 8   Fig. 9

the development that are equal in the ordinary sense. Moreover, the geodesic curvature of the original curve and the "ordinary" curvature of its development at corresponding points are equal.

If we fix on the surface a way of transporting tangent vectors along curves "in a parallel way", we say that a *connection* is specified on this surface. The need to compare various geometrical quantities at different points of a "curved" space makes the concept of connection one of the most important in geometry and physics. Here is a physical interpretation of the operation of parallel transport described above. We shall transport a gyroscope along a curve $\rho(t)$, imposing on it an ideal ("frictionless") constraint that forces its axis to touch the surface all the time. Then the positions of this axis specify a parallel vector field along $\rho(t)$ in the sense described above.

2.7. Deficiencies of Loops, the "Theorema Egregium" of Gauss and the Gauss-Bonnet Formula. The operation of parallel transport of vectors from a point $x = \rho(t_1)$ to a point $y = \rho(t_2)$ along a curve $\rho(t)$ generates an orthogonal transformation

$$\|_{x,y,\rho} \colon T_x M \to T_y M.$$

Let $\rho(t)$ be a loop on M with origin at x, that is, $x = \rho(t_1) = \rho(t_2)$. If the surface M is orientable, then the action

$$\|_{x,x,\rho} \colon T_x M \to T_x M$$

of going round the loop ρ is an orthogonal transformation that preserves orientation. Hence it is a rotation through some angle $\varphi = \varphi(\rho)$, reckoned by taking account of the given orientation of the surface. It is called the *deficiency of the loop ρ*.

On an oriented surface we can define the *oriented geodesic curvature $\tilde{k}_g$* of a curve, which differs in sign from k_g, showing in which direction its tangent is rotating. For a piecewise smooth curve (Fig. 10), from direct visual observations we have the following formula:

$$\varphi(\rho) = 2\pi - \Sigma\alpha_i - \int_{t_1}^{t_2} \tilde{k}_g \, dt. \tag{9}$$

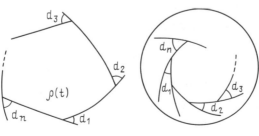

Fig. 10   Fig. 11

Fig. 12

As an example let us consider a geodesic polygon D on the unit sphere. Its sides are arcs of great circles. Let us extend them, as shown in Fig. 11. Then the sphere is represented as the union of the polygon D, n lunes with angles $\alpha_1, \ldots,$ α_n, and the polygon D' antipodal to D. Therefore 4π = area of the sphere = 2 (area of D) + $2\Sigma\alpha_i$. Since $\varphi(\partial D) = 2\pi - \Sigma\alpha_i$, we see that $\varphi(\partial D)$ = area of D = $\int\int_D K\, d\sigma$, since for the unit sphere the Gaussian curvature $K = 1$.

It is remarkable that the last formula is true for an arbitrary surface, that is,

$$\varphi(\partial D) = \int\int_D K\, d\sigma, \qquad (10)$$

where the closed curve ∂D is the boundary of a domain D on the surface, and K is its Gaussian curvature. It can be reduced to the formula already proved for the sphere by means of a *spherical* or *Gaussian map* of the surface M in question onto the sphere, under which to each point $x \in M$ there corresponds the unit normal vector m_x, laid off from the origin (Fig. 12). We need to use the fact (which follows from the derivation formulae (2)) that the Peterson operator A_x is the principal linear part of the spherical map at the point $x \in M$. Therefore since $K = \det A_x$, we have $K\, d\sigma = d\Sigma$, where $d\sigma$ (respectively, $d\Sigma$) is the element of area of the surface M (respectively, the sphere).

The infinitesimal version of (10) is obviously the following: the deficiency of an infinitesimal loop at a point x is equal to the area of the piece of the surface bounded by it, multiplied by $K(x)$. This proves the following remarkable fact, discovered by Gauss, which he called the "theorema egregium".

Theorem. *The Gaussian curvature of a surface is completely determined by its intrinsic geometry.*

In fact, the geodesic curvature of curves, as we have seen, is expressed in terms of the first quadratic form, and hence in terms of the deficiencies of the loops on the left-hand side of (10).

From (9) and (10) there follows the famous Gauss-Bonnet formula:

$$\int\int_D K\, d\sigma + \int_{\partial D} \tilde{k}_g\, dl + \Sigma\alpha_i = 2\pi \qquad (11)$$

for a simply-connected domain D on the surface in question with boundary ∂D.

Let M be a closed orientable surface of genus p (a sphere with p handles). Its Euler characteristic χ is equal to $2(1 - p)$. Then we have the following variant of (11):

$$\iint_M K \, d\sigma = 2\pi\chi \qquad (12)$$

(the *global Gauss-Bonnet formula*). It is proved by means of elementary combinatorial arguments if the surface M is cut into triangles and (11) is used for each of them.

Formula (12) is the ancestor of a whole series of other relations that connect the global topological structure of surfaces with various properties of their curvature (see Ch. 8, § 15).

2.8. The Link Between the First and Second Quadratic Forms. The Gauss Equation and the Peterson-Mainardi-Codazzi Equations. In 2.3 we defined the Gaussian curvature in terms of the first and second quadratic forms: $K = |h| \cdot |g|^{-1}$, where $|h| = \det \|h_{ij}\|$, $|g| = \det \|g_{ij}\|$. On the other hand, the theorema egregium shows that it can be expressed merely in terms of the first quadratic form. This implies that there are nontrivial interconnections between these two forms. The most important of them can be obtained immediatley as the compatibility conditions for the derivation equations (2). For example, since $\partial_2(\partial_1 r_1) = r_{112} = \partial_1(\partial_1 r_2)$, we have

$$\partial_2(\Gamma_{11}^k r_k + h_{11} m) = \partial_1(\Gamma_{12}^k r_k + h_{12} m).$$

Carrying out the necessary differentiations in this equality and expressing the resulting derivatives $\partial_i r_j$ and $\partial_i m$ by means of (2) again in terms of r_k and m, we obtain some differential equations that connect the first and second quadratic forms. Considering other combinations of differentiations, by similar arguments we can obtain the following independent equations:

$$\frac{|h|}{|g|} = -\frac{1}{|g|^2} \begin{vmatrix} E & E_1 & E_2 \\ F & F_1 & F_2 \\ G & G_1 & G_2 \end{vmatrix} - \frac{1}{2|g|^{1/2}} \left\{ \frac{\partial}{\partial u^2}\left(\frac{E_2 - F_1}{|g|^{1/2}}\right) - \frac{\partial}{\partial u^1}\left(\frac{F_2 - G_1}{|g|^{1/2}}\right) \right\},$$

$$2|g|(L_2 - M_1) - Q(E_2 - F_1) + \begin{vmatrix} E & E_1 & L \\ F & F_1 & M \\ G & G_1 & N \end{vmatrix} = 0,$$

$$2|g|(M_2 - N_1) - Q(F_2 - G_1) + \begin{vmatrix} E & E_2 & L \\ F & F_2 & M \\ G & G_2 & N \end{vmatrix} = 0,$$

$$(13)$$

$$g_{11} = E, \quad g_{12} = F, \quad g_{22} = G; \quad h_{11} = L, \quad h_{12} = M, \quad h_{22} = N;$$

$$Q = EN - 2FM + GL; \quad E_i = \partial E/\partial u^i, \quad \text{etc.}$$

The first of them was obtained by Gauss (1826) and expresses the Gaussian curvature explicitly in terms of the first quadratic form. The other two were obtained independently by Peterson (1852), Mainardi (1857) and Codazzi (1868).

From the following fundamental theorem of Peterson (1852) and Bonnet (1867) it follows that all the relations between the first and second quadratic forms are exhausted by the Gauss-Peterson-Mainardi-Codazzi equations (13) and consequences of them.

Theorem. *Suppose that in a domain U of the variables u^1, u^2 we are given the quadratic forms $g_{ij}\,du^i\,du^j$ and $h_{ij}\,du^i\,du^j$, the first of which is positive definite. Then if they satisfy the system (13), there is a parametrized surface $r\colon U \to E^3$ that realizes them as its first and second quadratic forms, respectively.*

In view of the theorem in 2.4, such a surface is defined uniquely up to a motion.

2.9. The Moving Frame Method in the Theory of Surfaces. In the study of some geometrical object by the "moving frame" method, with each point in some way or other we associate a system of frames. In the framework of this approach the basic objects are the functions that describe the change in the chosen frame when the point to which it is attached is changed. After this we establish equations that these functions must satisfy. For example, in the theory of curves it is convenient to consider the system of Frenet frames (See 1.3). In this case the curvatures $k_i = k_i(s)$ describe the rate of change of the Frenet frame under a motion along a curve (the Frenet formulae), and there are no relations between them.

The construction of the theory of surfaces described above is a version of the moving frame method, when with each point of the surface we associate a frame r_1, r_2, m. The functions Γ_{ij}^k, h_{ij}, A_j^k that occur in the derivation formulae (2) describe the rate of change of it, and the Gauss-Peterson-Mainardi-Codazzi equations give all the relations between them (note that g_{ij} can be expressed in terms of the h_{kl} and the A_q^p).

We can associate frames with a surface in another way. For example, Darboux considered all possible orthonormal frames τ_1, τ_2, m, where the vectors τ_i have the principal direction (2.3), and $m = [\tau_1, \tau_2]$. The advantage of this choice is the fact that the moving frame is invariant, that is, it is independent of the choice of curvilinear coordinates on the surface.

Let δ_i be the operator of differentiation in the direction of the vector τ_i. This operator is invariantly connected with the surface in question. Therefore $\delta_i I$ is a (metric) invariant of it if I is.

Since τ_i are the principal directions, we have

$$\delta_i m = -k_i \tau_i, \qquad (\delta_i \tau_i, m) = -(\tau_i, \delta_i m) = k_i,$$

where k_i are the principal curvatures (see 2.3). From these relations we easily obtain the invariant "derivation formulae":

$$\delta_1 \tau_1 = \kappa_1 \tau_2 + k_1 m, \qquad \delta_2 \tau_1 = -\kappa_2 \tau_2,$$

$$\delta_1 \tau_2 = -\kappa_1 \tau_1, \qquad \delta_2 \tau_2 = \kappa_2 \tau_1 + k_2 m, \qquad (14)$$

$$\delta_1 m = -k_1 \tau_1, \qquad \delta_2 m = -k_2 \tau_2,$$

where $\kappa_1 = (\delta_1 \tau_1, \tau_2)$, $\kappa_2 = (\tau_1, \delta_2 \tau_2)$.

The operators δ_i do not commute with each other. From (14) it is not difficult to deduce that

$$[\delta_1, \delta_2] = -\kappa_1 \delta_1 + \kappa_2 \delta_2.$$

Applying this operator equality to the vectors τ_i and m and using (14), we can obtain the equations

$$k_1 k_2 = \delta_1(\kappa_2) + \delta_2(\kappa_1) - (\kappa_1^2 + \kappa_2^2),$$

$$\delta_1(k_2) = \kappa_2(k_2 - k_1), \qquad \delta_2(k_1) = \kappa_1(k_1 - k_2),$$

which are the analogue of the Gauss-Peterson-Mainardi-Codazzi equations in this version of the theory.

2.10. A Complete System of Invariants of a Surface. We now describe how to construct the set of all metric invariants of a surface. To this end, on the surface in question we choose as coordinates the functions k_1 and k_2 (if this can be done) and consider the invariants $k_{ij} = \delta_j(k_i)$. The operators $\partial/\partial k_i$ of partial differentiation in this coordinate system are invariantly connected with the surface. Therefore $\partial I/\partial k_i$ is an invariant if I is. In particular, there is defined an infinite system of invariants

$$k_{i,j}^{st} = \frac{\partial^{s+t} k_{i,j}}{\partial k_1^s \partial k_2^t}.$$

Every invariant of the surface is a function of the variables k_1, k_2 and finitely many invariants $k_{i,j}^{st}$. This follows from the following result.

Theorem. *The invariants* $k_{i,j}$, $i, j = 1, 2$, *as functions of the variables* k_1, k_2 *determine the surface uniquely up to a motion.*

In other words, specifying the functions

$$k_{i,j} = f_{i,j}(k_1, k_2), \qquad i, j = 1, 2,$$

uniquely determines the form of the surface. However, the functions $f_{i,j}$ cannot be chosen independently of one another. In fact they satisfy a system of three first-order differential equations, which is a reformulation in invariants of the system of Gauss-Peterson-Mainardi-Codazzi equations. This system and consequences of it exhaust the relations between the metric invariants of surfaces.

§ 3. Multidimensional Surfaces

The construction of multidimensional geometrical theories is always beset with the characteristic difficulty that results from the extremely unwieldy nature of the coordinate descriptions of the basic geometrical quantities. This difficulty, which at first glance seems to be purely technical, is actually fundamental, since it substantially limits the possibilities of our mind to substitute conceptual thinking by calculations. Improving the notation in this situation is a palliative, although it can be useful to a limited extent. Einstein's summation rule, used in the previous section, is an example of this kind. Here the main solution is undoubtedly the development of a conceptual, and therefore coordinate-free, language of differential geometry (in this connection see the "definition" of differential geometry in Ch. 1, § 5).

In this section, extending to "many dimensions" the ideas and methods of the previous sections, apart from the presentation of factual material, we wish to demonstrate:

1) the elegance and effectiveness of a coordinate-free language (for example, the derivation of the multidimensional derivation equations and Gauss-Peterson-Mainardi-Codazzi equations);

2) the necessity of introducing new concepts that are not obvious (for example, the introduction of the curvature tensor, bivectors, and so on, in connection with the investigation of the structure of the multidimensional Gauss-Peterson-Mainardi-Codazzi equations);

3) the transition from quantity to quality (for example, the effect of rigidity of hypersurfaces).

3.1. n-Dimensional Surfaces in E^{n+p}. In this section we consider smooth n-dimensional surfaces in $(n + p)$-dimensional Euclidean space E^{n+p}. Since we are interested in local questions we assume that the surface M in question is covered by a chart $r: U \to E^{n+p}$, $U \subset \mathbb{R}^n = \{u = (u^1, \ldots, u^n)\}$, that is, $M = r(U)$. The linearly independent vectors $r_i(u) = \partial_i r(u)$, $\partial_i = \partial/\partial u^i$, $i = 1, \ldots, n$, form a basis of the tangent space $T_x M$ of the surface M at a point $x = r(u)$.

The restriction of the Euclidean metric $dx_1^2 + \cdots + dx_{n+p}^2$ to a tangent space $T_x M$ is called a *first quadratic form*. In "curvilinear coordinates" u^i it is written in the form

$$g(dr, dr) = g_{ij}(u)\, du^i\, du^j, \qquad g_{ij}(u) = (r_i(u), r_j(u)).$$

If $\xi, \eta \in T_{r(u)} M$, $\xi = \xi^i r_i(u)$, $\eta = \eta^i r_i(u)$, then

$$|\xi|^2 = g(\xi, \xi) = g_{ij}(u)\xi^i \xi^j,$$

$$\cos \alpha = \frac{g_{ij}(u)\xi^i \eta^j}{|\xi| \cdot |\eta|},$$

where α is the angle between ξ and η. An l-dimensional "subsurface" P of M can be specified by the functions $u^i = u^i(t^1, \ldots, t^l)$, $t = (t^1, \ldots, t^l) \in V \subset \mathbb{R}^l$. If vol P

is its (l-dimensional) volume, then

$$\text{vol } P = \int_V \det \left\| \frac{\partial u^k}{\partial t^i} \frac{\partial u^s}{\partial t^j} g_{ks}(u(t)) \right\|^{1/2} dt^1 \dots dt^l.$$

In particular, when $l = 1$, P is a curve and

$$(\text{length of an arc of } P) = \int_\alpha^\beta \left[\frac{du^k}{dt} \frac{du^l}{dt} g_{kl}(u(t)) \right]^{1/2} dt, \qquad (V = [\alpha, \beta]),$$

where $l = n$, $t^i = u^i$, P is a domain in M, and

$$(\text{volume of the domain } P) = \int_K \det \|g_{ij}(u)\|^{1/2} du^1 \dots du^n.$$

3.2. Covariant Differentiation and the Second Quadratic Form. The operation of parallel transport of tangent vectors on multidimensional surfaces can be defined and motivated by repeating word for word the arguments in 2.6. However, it is more convenient to take the infinitesimal point of view, as we do below, and regard the operation of covariant differentiation as of paramount importance. We denote by $D(M)$ the totality of all tangent vector fields on M, and understand functions on M as functions of the variables u^i. If $X - X^i(u)r_i(u) \in D(M)$, we put $\partial_X f(u) = X^i(u)\partial_i f(u)$.

A (not necessarily tangent) vector field Y in E^{n+p} along M can be differentiated componentwise as a vector function by means of ∂_X. Notation: $\partial_X Y = X^i \partial_i Y$. If $Y \in D(M)$, then generally speaking $\partial_X Y \notin D(M)$. We therefore put

$$\partial_X Y = \nabla_X Y + h_X Y, \tag{15}$$

where $\nabla_X Y \in D(M)$, and the field $h_X(Y)$ is perpendicular to M. The field $\nabla_X Y$ is called the *covariant derivative* of Y along X. We can verify immediately that the operation ∇_X thus defined satisfies the relations (6) of 2.6. We put $\nabla_i = \nabla_{\partial i}$ and

$$\nabla_i r_j = \Gamma_{ij}^k r_k.$$

Then the functions Γ_{ij}^k defined by this equality and called the *Christoffel symbols* are expressed in terms of the g_{ij} as follows:

$$\Gamma_{ij}^k = \tfrac{1}{2} g^{kl}(\partial_i g_{lj} - \partial_j g_{li} - \partial_l g_{ij}). \tag{16}$$

The operator ∇_X can be restricted to a curve if the field X touches it. This gives the possibility of defining the operator $\nabla_{\dot\rho(t)}$, which acts on the vector fields $\xi(t) = \xi^k(t)r_k(u(t))$ along the curve $\rho(t) = r(u(t))$. Then

$$\nabla_{\dot\rho(t)} \xi(t) = (\dot\xi^k + \Gamma_{ij}^k(u(t))\dot u^i \xi^j) r_k. \tag{17}$$

The field $\xi(t)$ is said to be *parallel along* $\rho(t)$ if $\nabla_{\dot\rho(t)} \xi(t) = 0$ or

$$\dot\xi^k + \Gamma_{ij}^k(u(t))\dot u^i \xi^j = 0, \qquad k = 1, \dots, n. \tag{18}$$

Just as in 2.6, this enables us to define a parallel transport procedure for tangent vectors along curves on the surface M.

Next, we define the *geodesic curvature* of the curve $\rho(t)$ as $k_g(t) = |\nabla_{\dot{\rho}(t)}\dot{\rho}(t)|$. If $k_g = 0$, then the curve is called a *geodesic*, and in view of (18) it satisfies the equations

$$\ddot{u}^k + \Gamma_{ij}^k(u(t))\dot{u}^i\dot{u}^j = 0, \qquad k = 1, \ldots, n. \tag{19}$$

Locally geodesics are *shortest curves* on the surface M.

All these facts refer to the intrinsic geometry of the surface M and result from consideration of the tangential part of (15). Information about its extrinsic geometry is partially contained in the term $h_X Y$. Since $\partial_i r_j = \partial_j r_i$, from the defining formula (15) it follows that $\nabla_i r_j = \nabla_j r_i$ and $h_{ij} = h_{ji}$. Hence $h_X Y = X^i Y^j h_{ij} = h_Y X$, so, putting $h(X, Y) = h_X(Y)$, we see that $h(X, Y) = h(Y, X)$. We now define a symmetric bilinear form h_x on $T_x M$, which takes values in its orthogonal complement $T_x^\perp M$, assuming that

$$h_x(\xi, \eta) = h(X, Y)(x), \qquad \xi, \eta \in T_x M,$$

where ξ and η are the values of the fields X and Y at the point x, respectively. The field of forms $x \mapsto h_x$ on M is called the *second quadratic form*. Fixing the normal vector fields m_a, $a = 1, \ldots, p$, along M, which generate a basis of $T_x^\perp M$ for each $x \in M$, we can "scatter" the vector-valued form h into p "ordinary" (scalar) forms:

$$h = h^a m_a, \qquad h^a(X, Y) = (h(X, Y), m_a).$$

The second quadratic form describes the principal part of the deviation of the surface M close to the point x from the tangent plane $T_x M$.

3.3. Normal Connection on a Surface. The Derivation Formulae.

The first and second quadratic forms, generally speaking, do not determine a multidimensional surface uniquely up to a motion (cf. 2.4). In order that this should be so, we need to add to them one more ingredient – the *normal connection*.

Let $D^\perp(M)$ be the totality of all normal (to M) vector fields along M. Let $X \in D(M)$, $N \in D^\perp(M)$. We define the operators

$$\nabla_X^\perp : D^\perp(M) \to D^\perp(M), \qquad h_X^* : D^\perp(M) \to D(M),$$

by means of the equality

$$\partial_X N = \nabla_X^\perp N - h_X^* N, \tag{20}$$

where $\nabla_X^\perp N \in D^\perp(M)$, $h_X^* N \in D(M)$. The field $\nabla_X^\perp N$ is called the *covariant derivative of the normal field N along the tangent field X*.

The operators $\nabla_X^\perp$ satisfy the same relations (6) as the operators ∇_X. Following the scheme of 3.2, we can define the operator $\nabla_{\dot{\rho}(t)}^\perp$ of *covariant differentiation of normal fields to M* along the curve $\rho(t) = r(u(t))$ on M. If $v(t) = v^a(t)m_a$ (see 3.2), then

$$\nabla_{\dot{\rho}(t)}^\perp v(t) = (\dot{v}^a + \Gamma_{ib}^{\perp a}(u(t))\dot{u}^i v^b)m_a,$$

where the functions $\Gamma_{ib}^{\perp a}$ on M are called the *Christoffel symbols of the normal connection* on M. The normal field $v(t)$ is said to be *parallel* along the curve $\rho(t)$ if $\nabla_{\dot\rho}^{\perp} v = 0$. We have thus defined the operation of parallel transport of normal vectors along curves on M, and in this sense we say that a *normal connection* is specified on M.

The operator h_X^* that occurs on the right-hand side of (20) is conjugate to the operator h_X in the sense that

$$(h_X Y, N) = (Y, h_X^* N).$$

Finally, for each $v \in T_x^{\perp} M$ we define the Peterson operator $A_v: T_x M \to T_x M$ (cf. 2.3), putting

$$(h_x(\xi, \eta), v) = g_x(A_v \xi, \eta), \qquad \xi, \eta \in T_x M, \quad v \in T_x^{\perp} M.$$

The formulae (15) and (20) are essentially the *derivation formulae of the surface* M, written in coordinate-free form (cf. 2.4). Fixing the basis m_a, $a = 1, \ldots, p$, of normal fields to M, we obtain the following coordinate interpretation of them:

$$\begin{cases} \partial_i r_j = \Gamma_{ij}^k r_k + h_{ij}^a m_a, \\ \partial_i m_a = \Gamma_{ia}^{\perp b} m_b - A_{ai}^j r_j, \end{cases} \tag{21}$$

where h_{ij}^a are the matrix elements of the quadratic form h^a, and A_{aj}^i are the matrix elements of the Peterson operator A_{m_a}.

From the derivation formulae (21) we have the following important fact (cf. 2.4).

Theorem. *Two n-dimensional surfaces M and M' in E^{n+p} go into each other by a motion of the space E^{n+p} if and only if there are charts $r: D \to M$, $r': D \to M'$ with common domain of definition $D \subset \mathbb{R}^n$ in which the corresponding coefficients of their first and second quadratic forms and normal connections coincide.*

Suppose that $n = 1$, that is, M is a curve, and that $m_a = e_{a+1}$, $a = 1, \ldots, n$, where e_i are the vectors of its Frenet frame (see 1.3). Comparing the Frenet formulae with the formulae (21), we can verify that the coefficients $\Gamma_{ib}^{\perp a}$ (in the given case these are the curvatures k_i, $i > 1$, or zero) cannot be expressed in terms of its first and second quadratic forms. This demonstrates why the notion of normal connection has to be introduced.

3.4. The Multidimensional Version of the Gauss-Peterson-Mainardi-Codazzi Equations. Ricci's Theorem. In this subsection we shall give a scheme for the coordinate-free derivation of an analogue of the Gauss-Peterson-Mainardi-Codazzi equations for a multidimensional surface M. First of all we observe that the operator $[\partial_X, \partial_Y] = \partial_X \cdot \partial_Y - \partial_Y \cdot \partial_X$, $X, Y \in D(M)$, has the form ∂_Z for $Z = (X^j \partial_j Y^i - Y^j \partial_j X^i) r_i$. The field Z is called the *commutator of the vector fields X and Y* and denoted by $[X, Y]$. Thus, by definition

$$[\partial_X, \partial_Y] = \partial_{[X,Y]}. \tag{22}$$

Any vector field V along M can be uniquely represented in the form $V = X + N$, $X \in D(M)$, $N \in D^{\perp}(M)$. If such fields V are represented as columns $(X, N)^{\tau}$

(τ denotes transposition), then the operator ∂_X defined on them is represented by some operator 2×2-matrix that acts on these columns. From (15) and (20) it follows that

$$\partial_X = \begin{pmatrix} \nabla_X & -h_X^* \\ h_X & \nabla_X^\perp \end{pmatrix}.$$

Now substituting similar matrix expressions for ∂_X, ∂_Y and $\partial_{[X,Y]}$ in (22), we obtain the *basic equations* of the theory of multidimensional surfaces. Namely, comparison of the 1×1-terms of the matrix equality thus obtained leads to a multidimensional analogue of the Gauss equations (see 2.8):

$$R(X, Y) = h_X^* h_Y - h_Y^* h_X, \tag{23}$$

where

$$R(X, Y) = [\nabla_X, \nabla_Y] - \nabla_{[X,Y]}.$$

Next, the 2×1-terms lead to an analogue of the Peterson-Mainardi-Codazzi equations (see 2.8):

$$(\tilde{\nabla}_Y h)_X = (\tilde{\nabla}_X h)_Y, \tag{24}$$

where

$$(\tilde{\nabla}_X h)_Y = \nabla_X^\perp \circ h_Y - h_Y \circ \nabla_X - h_{\nabla_X Y}.$$

The 1×2-terms also lead to the equations (24), in view of the skew-symmetry of the antidiagonal terms in the matrices for ∂_Z. Finally, comparison of the 2×2-terms gives the equations

$$R^\perp(X, Y) = h_X h_Y^* - h_Y h_X^*, \tag{25}$$

where

$$R^\perp(X, Y) = [\nabla_X^\perp, \nabla_Y^\perp] - \nabla_{[X,Y]}^\perp.$$

These equations do not have an analogue in the classical theory of surfaces.

The value of the vector field $R(X, Y)Z$ at a point $x \in M$ is uniquely determined by the values ξ, η, ζ of the vector fields X, Y, Z respectively at this point. We have thus defined the operator

$$R_x(\xi, \eta): T_x M \to T_x M, \qquad \xi, \eta \in T_x M,$$

which acts according to the rule $R_x(\xi, \eta)\zeta = (R(X, Y)Z)(x)$. This operator is linear and depends linearly on ξ and η. It is usual to give the name *tensors* to quantities that depend linearly on the tangent vectors (for a precise definition see below, Ch. 3, § 3). Therefore R_x is called the *curvature tensor of the surface M at the point x*. The field $R: x \mapsto R_x$ is called the *curvature tensor of the surface M*; the l-th component of the vector $R_x(r_i, r_j)r_k$ in the basis $r_1, \ldots, r_n$ is denoted by R^l_{ijk} and called the $(^l_{i\,jk})$-*component* of the tensor R_x.

In exactly the same way we define the operator

$$R_x^\perp(\xi, \eta): T_x^\perp M \to T_x^\perp M.$$

The object $R_x^\perp$ is called the *curvature tensor of the normal connection* at the point x.

The equations (23)–(25), written in coordinates, represent a system of differential equations for the functions $g_{ij}(u)$, $h_{ij}^a(u)$ and $\Gamma_{ib}^{\perp a}(u)$, which is the analogue of the classical Gauss-Peterson-Mainardi-Codazzi equations (13).

The multidimensional analogue of the classical Peterson-Bonnet theorem (see 2.8) is due to Ricci (1898).

Theorem. *The functions $g_{ij}(u)$, $h_{ij}^a(u)$, $\Gamma_{ib}^{\perp a}(u)$, $i, j = 1, \ldots, n$, $a, b = 1, \ldots, p$, which satisfy equations (23)–(25) and the condition that the matrix $\|g_{ij}\|$ is positive definite, are realized as the coefficients of the first and second quadratic forms and the coefficients of the normal connection of some surface, which is uniquely determined by them up to a motion.*

In other words, up to a motion a surface is uniquely determined by its two quadratic forms and the normal connection.

We observe that equations (23)–(25) are skew-symmetric in X, Y and so they are automatically satisfied if $Y = fX$. For this reason the Gauss-Peterson-Mainardi-Codazzi system for curves is empty.

3.5. The Geometrical Meaning and Algebraic Properties of the Curvature Tensor. Since the curvature tensor can be expressed in terms of the operators of covariant differentiation, it reflects certain aspects of the intrinsic geometry of the surface M. Let M be a surface in E^3 and $K(x)$ its Gaussian curvature at the point x. Then

$$K(x) = g_x(R_x(\xi, \eta)\xi, \eta),$$

where $\xi, \eta \in T_x M$ is any orthonormal pair. This shows that the curvature tensor is the analogue of the Gaussian curvature in the multidimensional case.

Let $\gamma(\varepsilon)$ be the projection on M of the parallelogram constructed from the vectors $\varepsilon\xi$ and $\varepsilon\eta$, where $\xi, \eta \in T_x M$. Then the operator of parallel transport along $\gamma(\varepsilon)$ has the form $\mathrm{id} + \varepsilon^2 R_x(\xi, \eta) + o(\varepsilon^2)$. This is the geometrical meaning of the curvature operator $R_x(\xi, \eta): T_x \to T_x$. The normal curvature operator $R_x^\perp(\xi, \eta): T_x^\perp \to T_x^\perp$ admits a similar interpretation.

The curvature tensor has remarkable algebraic properties, which we give and comment on below.

Firstly, it follows from the definition that $R(X, Y) = -R(Y, X)$. Hence

$$R_x(\xi, \eta) = -R_x(\eta, \xi) \Leftrightarrow R_{ijk}^l = -R_{jik}^l. \tag{26}$$

Next, the operator $R_x(\xi, \eta)$ is skew-symmetric in the metric g_x on $T_x M$:

$$g_x(R_x(\xi, \eta)\mu, v) + g_x(\mu, R_x(\xi, \eta)v) = 0 \Leftrightarrow R_{ijk}^s g_{sl} + R_{ijl}^s g_{sk} = 0. \tag{27}$$

This follows from the fact that the operator of parallel transport along an infinitesimal curvilinear parallelogram formed from the vectors $\varepsilon\xi$, $\varepsilon\eta$ is equal to $\mathrm{id} + \varepsilon^2 R_x(\xi, \eta) + o(\varepsilon^2)$, and the fact that if the operator $\mathrm{id} + \lambda A + o(\lambda)$ is orthogonal, then the operator A is skew-symmetric.

To discuss other properties of the curvature tensor it is useful to consider the space $\Lambda^2 T_x = \Lambda^2 T_x M$ of tangent bivectors to M at the point x. A *bivector* is a bilinear skew-symmetric function on $T_x^* = T_x^* M$. The pair $\xi, \eta \in T_x$ determines the bivector $\xi \wedge \eta$:

$$(\xi \wedge \eta)(p, q) = p(\xi)q(\eta) - p(\eta)g(\xi), \qquad p, q \in T_x^*.$$

It is said to be *simple*. The simple bivectors generate $\Lambda^2 T_x$. In the presence of metrics a simple bivector $\xi \wedge \eta$ (as well as any other) can be thought of as a skew-symmetric linear operator on T_x:

$$(\xi \wedge \eta)(\zeta) = g_x(\eta, \zeta) \cdot \xi - g_x(\xi, \zeta) \cdot \eta.$$

Bearing this identification in mind, we can rewrite the Gauss equation (23) in the form

$$R_x(\xi, \eta) = \sum_{a=1}^{p} A_a \xi \wedge A_a \eta, \tag{28}$$

where $A_a = A_{m_a}$ is the Peterson operator (see 3.3). It is elementary to verify that for any symmetric operator $A: T_x \to T_x$ we have the identity

$$(A\xi \wedge A\eta)(\zeta) + (A\eta \wedge A\zeta)(\xi) + (A\zeta \wedge A\xi)(\eta) = 0.$$

In view of (28), it leads to the *first Bianchi identity*:

$$R_x(\xi, \eta)\zeta + R_x(\eta, \zeta)\xi + R_x(\zeta, \xi)\eta = 0 \Leftrightarrow R_{ijk}^l + R_{jki}^l + R_{kij}^l = 0. \tag{29}$$

Next, the curvature tensor R_x can be thought of as a linear operator R_x: $\Lambda^2 T_x \to \Lambda^2 T_x$ that acts on simple bivectors according to the rule $R_x(\xi \wedge \eta) = R_x(\xi, \eta)$, where the right-hand side of this equality, in accordance with what we said above, is understood as a bivector. Also, any linear operator $A: T_x \to T_x$ generates a linear operator $A^{[2]}: \Lambda^2 T_x \to \Lambda^2 T_x$ that acts on simple bivectors according to the rule $A^{[2]}(\xi \wedge \eta) = A\xi \wedge A\eta$. Bearing all this in mind, we can write equation (28) in the form

$$R_x = \sum_{a=1}^{p} A_a^{[2]}. \tag{30}$$

A metric g_x on T_x generates a metric $g^{[2]}$ on $\Lambda^2 T_x$:

$$g_x^{[2]}(\xi_1 \wedge \xi_2, \eta_1 \wedge \eta_2) = \det \|g_x(\xi_i, \eta_j)\|.$$

It is not difficult to see that if the operator A is symmetric in the metric g_x, then the operator $A_x^{[2]}$ is symmetric in the metric $g_x^{[2]}$. Since the operators A_a are symmetric, the operator $\sum_a A_a^{[2]}$ is also symmetric, and so in view of (30) the operator $R_x: \Lambda^2 T_x \to \Lambda^2 T_x$ is symmetric. This is equivalent to the fact that

$$g_x(R_x(\xi_1, \xi_2)\eta_1, \eta_2) = g_x(R_x(\eta_1, \eta_2)\xi_1, \xi_2) \Leftrightarrow R_{ijk}^s g_{sl} = R_{kli}^s g_{sj}. \tag{31}$$

The properties (26), (27), (29) and (31) exhaust the algebraic features of the curvature tensor at a fixed point in the sense that any tensor that has these properties can be realized as the curvature tensor of some surface.

3.6. Hypersurfaces. Mean Curvatures. The Formulae of Steiner and Weyl. A hypersurface, that is, an n-dimensional surface in E^{n+1}, has exactly one normal vector at each of its points. Since parallel transport of normal vectors preserves their lengths, it follows that for sufficiently small paths it does not depend on the choice of them. This shows (see the beginning of 3.5) that $R_x^\perp \equiv 0$, that is, the equations (25) hold tautologically for hypersurfaces.

The second quadratic form h of a hypersurface is scalar and, like the Peterson operator $A = A_m$ corresponding to it, it is uniquely determined by the choice of one of the two possible unit normal vector fields m to M. The eigenvalues $k_1, \ldots, k_n$ of the symmetric operator A are called the *principal curvatures* of the hypersurface at the point under consideration. Their geometrical meaning is the same as in the classical theory (see 2.3). The eigenvectors corresponding to them are called the vectors of the *principal directions* at this point. The j-th elementary symmetric function $K_j = \sigma_j(k_1, \ldots, k_n)$ of the principal curvatures is called the j-th *mean curvature* of the hypersurface. The quantity $\dfrac{1}{n} K_1 = \dfrac{1}{n}\Sigma k_i$ is simply called the *mean curvature*, and $K_n = k_1 \cdot \ldots \cdot k_n$ is called the *Gaussian curvature*.

The Peterson operator can be interpreted as the Jacobi matrix of the *spherical map* $x \mapsto m_x$ of the hypersurface M on the unit sphere $S^n \subset E^{n+1}$. Hence $K = \det A$, $d_n\Sigma = K_n d_n v$, where $d_n\Sigma$ and $d_n v$ are the volume elements of S^n and M respectively. Integration of the second equality proves the following generalization of the classical Gauss-Bonnet formula (11) to smooth compact hypersurfaces without boundary:

$$\int_M K_n d_n v = \text{(degree of the spherical map)} \cdot \omega_n, \tag{32}$$

where ω_n is the volume of the unit sphere S^n. If n is even, then the Euler characteristic $\chi(M)$ of M is equal to twice the degree of the spherical map (if n is odd, then $\chi(M) = 0$). Therefore in this case (32) can be rewritten in the form

$$\int_M K_n d_n v = \frac{1}{2}\omega_n \chi(M) \tag{33}$$

(cf. (12)).

There are other remarkable relations in which the integrals of the mean curvatures take part. Suppose, for example, that M is the boundary of a compact domain $D \subset E^{n+1}$ and D_t is the domain consisting of points at a distance t from D (Fig. 13). Then the volume $V(t)$ of D_t can be expressed by the following *Steiner formula*:

$$V(t) = \text{(volume of } D) + \sum_{i=0}^{n} \frac{(-1)^i}{i+1} x_i t^{i+1} \tag{34}$$

where x_0 is the volume of M, and $x_i = \int_M K_i d_n\sigma$, $i > 0$. This formula holds for sufficiently small t.

For small t the volume $v(t)$ of the layer of width $2t$ around M (Fig. 14) is equal to $V(t) - V(-t)$, or by (34)

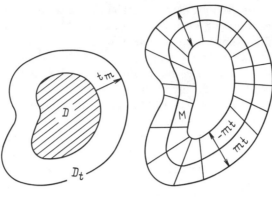

Fig. 13   Fig. 14

$$v(t) = \sum_{s=0}^{[n/2]} \frac{2}{2s+1} x_{2s} t^{2s+1}. \tag{35}$$

It is remarkable that the quantities x_{2s}, and hence $v(t)$, are completely determined by the intrinsic geometry of the hypersurface M. For example, the curvature K_2 is the trace of the operator $A^{[2]}$: $\Lambda^2 T_x \to \Lambda^2 T_x$ (see 3.5). Hence by the Gauss equation (30) it can be expressed in terms of the curvature tensor.

In 1939 H. Weyl found an analogue of (35) for the volume $v(t)$ of a tubular neighbourhood of radius t of an n-dimensional compact surface M without boundary in E^{n+p}:

$$v(t) = \Omega_p \sum_{s=0}^{[n/2]} \frac{x_{2s}}{p(p+2)\ldots(p+2s)} t^{p+2s}. \tag{36}$$

Here Ω_p is the volume of a unit p-dimensional ball, and the quantities x_{2s} are the integrals over M of certain polynomials in the components of the curvature tensor.

The Gauss-Bonnet formula (33) and the Weyl formula (36) have served as the original material for the construction of a differential-geometric theory of characteristic classes (see Ch. 8, § 15).

3.7. Rigidity of Multidimensional Surfaces. If an arbitrarily small part of an n-dimensional surface $M \subset E^{n+p}$ does not admit isometric deformations that change its form (that is, other than a motion), we say that it is *locally rigid*. An everyday test convinces us that surfaces in 3-dimensional space are not locally rigid. Therefore the following result of Beez (1876) and Killing (1885) clearly contradict our intuition. Beez himself regarded it as a proof of the non-existence of multidimensional spaces.

Theorem. *When $n > 2$ a hypersurface $M \subset E^{n+1}$ is locally rigid if the set of points at which the rank of the Peterson operator is greater than two is everywhere dense in M.*

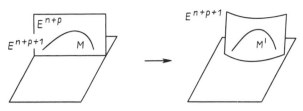

Fig. 15

As simple considerations from linear algebra show, the operator $A^{[2]}$ determines the operator A (and hence the second quadratic form) if the conditions of this theorem are satisfied. From the Gauss equations (30) it follows that the second quadratic form of the hypersurface under consideration can be expressed in terms of the first. Our result now follows from Ricci's theorem (see 3.4).

For sufficiently large p, n-dimensional surfaces in E^{n+p} cease to be locally rigid and begin to admit bendings. Apparently this is always so if $p > \frac{1}{2}n(n-1)$. The elementary reason why n-dimensional surfaces in E^{n+p} are always bendable in E^{n+p+1} is connected with the fact that E^{n+p} itself is bendable in E^{n+p+1} (Fig. 15). On the other hand, for small p surfaces that have a complicated intrinsic geometry are rigid. Simple conditions that guarantee rigidity are given below.

Let T be a linear Euclidean space and let $A_1, \ldots, A_k$ be a set of symmetric operators acting on it. The maximum number of vectors $x_1, \ldots, x_l \in T$ having the property that the vectors $A_i(x_j)$ are linearly independent is called the *type number* of the set $\{A_1, \ldots, A_k\}$. The *type number of an n-dimensional surface* $M \subset E^{n+p}$ at a point $x \in M$ is the type number of the set of its Peterson operators at this point. Let us agree to write

$$h_x(T_x, T_x) = \{h_x(\xi, \eta) | \xi, \eta \in T_x M\} \subset T_x^\perp M.$$

The following theorem of Allendorfer (1939) generalizes the rigidity theorem given above to hypersurfaces.

Theorem. *If the type numbers of an n-dimensional surface $M \in E^{n+p}$ are almost everywhere not less than three and $T_x^\perp M = h_x(T_x, T_x)$, $\forall x \in M$, then this surface is rigid.*

The plan of the proof is as follows. The condition $T_x^\perp M = h_x(T_x, T_x)$ enables us to express by means of (25) the normal connection in terms of the first and second quadratic forms, and the condition on the type numbers enables us to express by means of the Gauss equations (30) the second form in terms of the first. We have thus shown that the intrinsic geometry of the surface under consideration completely determines its extrinsic geometry, that is, its form.

Chapter 3
The Field Approach of Riemann

"In our mind there cannot be any contradiction
when we admit that some forces in nature follow one
particular geometry while others follow different
geometries".

N.I. Lobachevskij

§ 1. From the Intrinsic Geometry of Gauss to Riemannian Geometry

1.1. The Essence of Riemann's Approach. We have already seen that many important aspects of the geometrical structure of multidimensional surfaces in Euclidean spaces have a purely "intrinsic" nature, that is, they can be expressed by means of intrinsic curvilinear coordinates $u^1, \ldots, u^n$ and the first quadratic form $g_{ij} \, du^i \, du^j$. More precisely, intrinsic geometrical quantities are represented by those functions of the components g_{ij} of the first quadratic form and their derivatives that have an objective meaning, that is, not depending to the choice of coordinate system, for example the lengths of curves and the angles between them, the volumes of "subsurfaces", the operation of parallel displacement, the geodesic curvature of curves, the curvature tensor, and so on. Therefore, if such functions are formed from the coefficients of an arbitrary positive definite differential quadratic form, then as before they will have an objective meaning, independently of whether it is the first quadratic form of some surface or not. Thus, fixing a positive definite differential quadratic form leads to the possibility of defining (at least formally) certain "geometrical" quantities and manipulating with them. However, in order to construct a consistent geometrical theory in this way we need to introduce supports of the objects of the geometry in our context of differential quadratic forms. Such objects are naturally represented as "abstract surfaces" that admit the introduction of local curvilinear coordinates but are not to be thought of as situated in some Euclidean space or other.

These are the arguments (in simplified form) that led Riemann [1868] to a new point of view on (metric) geometry, according to which the geometrical objects under investigation are "Riemannian manifolds", that is, abstract surfaces with differential quadratic forms specified on them. Thus, the geometry of Riemann is an abstract form of the intrinsic geometry of Gauss.

(At the time when Riemann was reflecting on the nature of geometry many concepts of the geometry of surfaces were still unknown. Therefore he "had to" discover some of them straight away in abstract form. The most important of them is the curvature tensor.)

1.2 Intrinsic Description of Surfaces. The intuitive idea of an abstract surface that we used above leads after the necessary formalization to the concept of a *smooth manifold*, which is fundamental for differential geometry. A precise definition of it will be given in the next section; now we prepare to introduce it, having considered what precedes it in the intrinsic geometry of Gauss.

In order to obtain a complete representation of the surface of the Earth, it is usual to form geographical atlases from many maps. The topological reason for this is that this surface, which is homeomorphic to a sphere, cannot be covered by a unique one-valued system of curvilinear coordinates. We note that from the metrical point of view only a small part of the Earth's surface can be authentically mapped in practice on a flat sheet of paper.

Similarly, an n-dimensional being can intrinsically describe the world he lives in by forming an atlas of curvilinear coordinate systems covering the separate parts of it. In order to establish the intrinsic metric structure of a surface it is necessary to describe the first quadratic form in each of these coordinate systems.

Let $u_\alpha = (u_\alpha^1, \ldots, u_\alpha^n)$ be one of the curvilinear coordinate systems that occurs in the atlas and covers a domain $M_\alpha \subset M$. Suppose that the first quadratic form in these coordinates is written as $g^{(\alpha)} = g_{ij}^{(\alpha)} \, du_\alpha^i \, du_\alpha^j$. In the domain $M_\alpha \cap M_\beta$ of intersection of the two coordinate systems u_α and u_β the coordinates u_α^j can be expressed in terms of the u_β^j and vice versa. We have thus defined the functions

$$u_\alpha^i = u_\alpha^i(u_\beta^1, \ldots, u_\beta^n), \qquad u_\beta^j = u_\beta^j(u_\alpha^1, \ldots, u_\alpha^n).$$

Since the surface M is smooth, these functions are also smooth. Rewriting the form $g^{(\alpha)}$ in the coordinates u_β, we must obtain the form $g^{(\beta)}$;

$$g^{(\alpha)} = g_{ij}^{(\alpha)} \, du^i \, du^j = g_{ij}^{(\alpha)} \frac{\partial u_\alpha^i}{\partial u_\beta^k} \frac{\partial u_\alpha^j}{\partial u_\beta^l} \, du_\beta^k \, du_\beta^l = g_{kl}^{(\beta)} \, du_\beta^k \, du_\beta^l,$$

Therefore

$$g_{kl}^{(\beta)} = g_{ij}^{(\alpha)} \frac{\partial u_\alpha^i}{\partial u_\beta^k} \frac{\partial u_\alpha^j}{\partial u_\beta^l}. \tag{1}$$

Thus the surface M can be described intrinsically by means of an atlas $\{u_\alpha\}$ of curvilinear coordinate systems, and the intrinsic geometry on it can be described by a family of differential quadratic forms $\{g(\alpha)\}$ in the variables du_α^i, whose coefficients are connected by the relation (1).

1.3. The Field Point of View on Geometry. We can regard the differential form $g_{ij}(u) \, du^i \, du^j$ as a "field" of ordinary quadratic forms on an "abstract surface" M. This means that to each point $x \in M$ there corresponds an ordinary quadratic form $g_{ij}(x)\, du^i\, du^j$ in the variables du^k. Thus geometry according to Riemann is specified by a *field quantity*.

Field quantities are an indispensible attribute of classical (non-quantum) physics. For example, when describing a continuous medium we speak of the temperature field in it, the velocity field of the particles that constitute it, and the field of stresses that arise in it. Moreover, the substances studied by classical physics are fields – electromagnetic, gravitational, and so on.

The mathematical nature of the fields that occur in physics is naturally diverse. If a number is associated with each point of space, we talk about a scalar field (or function). The temperature field, for example, is like this. If a vector is associated in some way with each point of space, then we are concerned with a vector field (for example, a velocity field). An electromagnetic field is the field of the curvature tensor of a connection specified in some fibration on the Minkowski space, and so on.

Geometry from Riemann's point of view appears before us as a special form of a physical field quantity that is able to interact in some way or other with other physical fields. An explicit physical interpretation of this physical quantity as a gravitational field and a concrete definition of the way it interacts with other physical fields led Einstein to the general theory of relativity.

The field nature of fundamental physical quantities, first explicitly realized by Maxwell, is an expression of the fundamental principle of short-range action, according to which the action of one field on another is carried out through space with finite speed, and instantaneous action at a distance is impossible. We see below that all "geometrical quantities" have a field nature. This is one of the reasons why differential geometry plays such a fundamental role in modern physics.

1.4. Two Examples. Of course, the route into Riemannian geometry described above is not the only one. The need to study differential quadratic forms arises very often, and we give two examples of this, one from physics and the other from geometry.

With the non-homogeneous heat equation

$$\sum \frac{\partial}{\partial x^i}\left(a_{ij} \frac{\partial u}{\partial x^j} \right) = h \frac{\partial u}{\partial t}, \tag{2}$$

where $u = u(x_1, \ldots, x_n, t)$ is the temperature function of an n-dimensional body, $a_{ij} = a_{ij}(x)$ are its heat coefficients, and h is the specific heat, we naturally associate the differential quadratic form $A = a_{ij}\, dx^i\, dx^j$. Riemann himself considered the problem of transforming equation (2) to the "homogeneous" form $\sum \dfrac{\partial^2 u}{(\partial x^i)^2} = h \dfrac{\partial u}{\partial t}$ by means of a change of coordinates.

Using the geometrical interpretation of the form A, which was already known to him, Riemann gives a necessary and sufficient condition for a similar change to be possible. It is that the curvature tensor associated with A vanishes. The necessity of this condition is obvious, since the curvature tensor of Euclidean space E^n is zero.

We turn our attention to the fact that if information about the structure of an n-dimensional universe is communicated by means of the processes of heat conduction described by (2), then such a universe is perceived as a Riemannian manifold corresponding to the form A.

Our second example is connected with the description of the natural metric of projective space. We shall think of n-dimensional projective space $\mathbb{R}P^n$ as the manifold of lines passing through the origin in $\mathbb{R}^{n+1}$. It is natural to define the distance between two points of projective space as the smallest angle between the corresponding lines. The fact that the standard non-homogeneous coordinates of a point of projective space are $(x_1, \ldots, x_n)$ means that the direction vector of the corresponding line is $(x_1, \ldots, x_n, 1)$ (Fig. 16). Calculating directly the infinitesimal angle between the vectors $(x_1, \ldots, x_n, 1)$ and $(x_1 + dx_1, \ldots, x_n + dx_n, 1)$, it is not difficult to verify that the square of the length of the infinitesimal segment $(dx_1, \ldots, dx_n)$ at the point $(x_1, \ldots, x_n)$ is equal to

$$ds^2 = \frac{1}{1 + \sum\limits_{i=1}^{n} x_i^2} \left(\sum_{i=1}^{n} dx_i^2 + \sum_{1 \leqslant i < j \leqslant n} (x_i\, dx_j - x_j\, dx_i)^2 \right).$$

This differential form transforms projective space into a Riemannian manifold. We emphasize that this manifold by virtue of its origin is not a part of any Euclidean space.

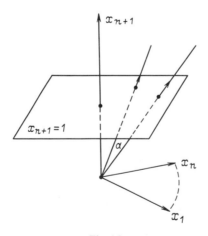

Fig. 16

§2. Manifolds and Bundles (the Basic Concepts)

2.1. Why Do We Need Manifolds? In the previous section we saw that the process of giving an abstract form to the intrinsic geometry of Gauss leads to the necessity of introducing the concept of "abstract surface" or *manifold* (as we shall say, starting from here). Below we mention two other sources that lead to the necessity of considering the totality of objects of one kind or another that admit the introduction of local coordinates.

One of them arises in the study of families of "spatial elements" (lines, planes, spheres, quadrics, and so on) of Euclidean space. Consider, for example, the totality L of all lines in E^3. Almost all lines (namely, those that cut the yz-plane in a point) can be specified by the equations

$$y = ax + b, \qquad z = cx + d,$$

from which it follows that the space L is four-dimensional, since any line of this kind is uniquely described by the four parameters a, b, c, d.

The geometers of the last century enthusiastically studied three-dimensional, two-dimensional and one-dimensional submanifolds of L, which are called complexes, congruences and surfaces of lines, respectively.

The study of totalities of "spatial elements" is motivated not only by purely geometrical aspirations but also, for example, by the desire to interpret visually the integration of differential equations. Let

$$z = f(x, y, a, b) \qquad (a, b \text{ are parameters}) \tag{3}$$

be a "congruence" of surfaces in E^3. Consider together with (3) the relations

$$\frac{\partial z}{\partial x} = f_x(x, y, a, b), \qquad \frac{\partial z}{\partial y} = f_y(x, y, a, b). \tag{4}$$

that follow from it. Eliminating the parameters a and b from (3) and (4), we obtain a relation of the form

$$F\left(x, y, z, \frac{\partial z}{\partial x}, \frac{\partial z}{\partial y}\right) = 0, \tag{5}$$

that is, a differential equation. The family (3) is its *complete integral* in the sense that (almost) any solution of it can be obtained as follows. We choose a one-parameter subfamily of (3), specifying, for example, the curve $(a(\tau), b(\tau))$, $\tau \in \mathbb{R}$, in the parameter space. Then the envelope of the one-parameter family of surfaces

$$z = f(x, y, a(\tau), b(\tau))$$

is a solution of (5).

Spaces of configurations of mechanical systems are another natural source of manifolds. Here the role of "spatial element" is played by any possible position (or configuration) of the mechanical system in question.

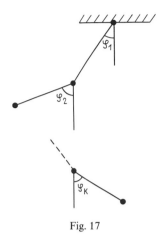

Fig. 17

For example, the position of a k-link plane pendulum is uniquely determined by the angles $\varphi_1, \ldots, \varphi_k$ (Fig. 17). In view of this we can conclude that the space of configurations of this system is a k-dimensional torus

The reader should turn his attention to the fact that the examples considered in this subsection of "totalities" that admit the introduction of local coordinates arise in a natural way in a purely abstract form, not as surfaces of Euclidean space.

2.2. Definition of a Manifold. The definition given below of a smooth (or infinitely differentiable) manifold in the abstract form repeats the way of describing the Earth's surface by means of a geographical atlas.

Let us fix a set M and a non-negative integer n. An (n-dimensional) *chart* on M is a one-to-one map $\varphi: U \to \varphi(U) \subset M$ of some domain $U \subset \mathbb{R}^n$ on a subset $\varphi(U)$ of M. The coordinates $x_1, \ldots, x_n$ of a point $x \in U$ are called local coordinates of the point $p = \varphi(x) \in M$ (in the sense of the chart in question). Therefore, as well as the word chart, in the same sense we use the term *local coordinate system* (on M). The chart is denoted by (φ, U).

Suppose that there are two charts (φ, U) and (ψ, V) on M. Then in the domain of intersection of them, that is, in $\varphi(U) \cap \psi(V)$, there arise on M two local coordinate systems, say x and y. The functions $x_i = f_i(y_1, \ldots, y_n)$ and $y_j = g_j(x_1, \ldots, x_n)$ that express each of them in terms of the other are called *transition functions*. If these functions are smooth (= infinitely differentiable), then the charts are called *smoothly compatible*. The transition functions are often written in the form $x_i = x_i(y_1, \ldots, y_n)$, $y_i = y_i(x_1, \ldots, x_n)$. This is expressive, but may lead to ambiguity (Fig. 18).

Suppose that a system of charts $(\varphi_\alpha, U_\alpha)$ is specified on M. A function $f: M \to \mathbb{R}$ on M is said to be *smooth* with respect to this system if it is infinitely differentiable as a function with respect to any of the local coordinate systems that constitute the system in question.

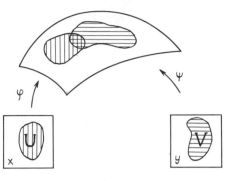

Fig. 18

A system $A = \{(\varphi_\alpha, U_\alpha)\}$ consisting of no more than countably many charts is called a *smooth atlas* if

1) it covers the whole of M, that is, $\bigcup_\alpha \varphi_\alpha(U_\alpha) = M$;

2) any two charts occurring in it are smoothly compatible;

3) for any two distinct points $p, q \in M$ there is a smooth (with respect to A) function f such that $f(p) \neq f(q)$.

The meaning of these conditions is the following. The first means that the whole set M is "cartographized" by means of A. The second guarantees the unambiguity of the property of differentiability: any function that is differentiable with respect to one of the local coordinate systems occurring in A is also differentiable with respect to any other in a domain that overlaps them. Finally the third, which is often called the "Hausdorff property", enables us to distinguish the points of M by means of smooth functions.

Two atlases on a set M are equivalent if the charts of one of them are smoothly compatible with the charts of the other.

We say that a smooth atlas A specifies the *structure of a smooth manifold* on M (or a *smooth structure*); we assume that equivalent atlases specify the same structure. In this case we also say that M is a *smooth manifold of dimension n*.

We denote the set of all smooth (with respect to A) functions on M by $C_A^\infty(M)$. It is not difficult to see that the atlases A and B are equivalent if and only if $C_A^\infty(M) = C_B^\infty(M)$. Therefore the concept of a smooth function on a smooth manifold does not depend on the choice of the defining atlas. The set of all smooth functions on M is denoted by $C^\infty(M)$, that is, $C^\infty(M) = C_A^\infty(M)$. If F is a smooth function of m variables and $f_1, \ldots, f_m \in C^\infty(M)$, then

$$F(f_1, \ldots, f_m) \in C^\infty(M).$$

In particular, $C^\infty(M)$ is an algebra over $\mathbb{R}$ with respect to the natural operations of addition and multiplication of functions.

A smooth structure on a manifold M is a way of distinguishing in the algebra of all functions on this set the subalgebra of those functions that it is proper to call smooth. Of course, this can be done in many different ways.

It is important to note that the algebra $C^\infty(M)$ contains exhaustive information about the set M, since the latter can be reconstructed uniquely from it. Namely, the set M is naturally identified (see Ch. 1, 2.1) with the set of maximal ideals $\mathscr{I}$ of the algebra $C^\infty(M)$ such that $C^\infty(M)/\mathscr{I} = \mathbb{R}$. In view of this we can treat the theory of smooth manifolds as the theory of commutative $\mathbb{R}$-algebras of a special kind. This point of view is more fundamental, but less intuitive (see Ch. 1, 2.1). $\mathbb{R}$-algebras that have the form $C^\infty(M)$ can be described axiomatically, using some concepts of differential calculus in commutative algebras. This is not completely elementary and will be omitted here.

A subset $N \subset M$ is called a *submanifold* of a smooth manifold M if it admits a smooth atlas (ψ_α, V_α) such that the compositions $V_\alpha \xrightarrow{\psi_\alpha} N \hookrightarrow M$ are smooth.

Generalizing the concept of a domain of Euclidean space with a smooth boundary, we can arrive at the concept of a *manifold with boundary*. For this, we need to use as charts parts of the half-space

$$\mathbb{R}^n_+ = \{(x_1, \ldots, x_n) | x_1 \geqslant 0\}.$$

The *boundary* of such a manifold M is an $(n-1)$-dimensional smooth manifold and denoted by ∂M.

A smooth manifold M has a natural *topological structure*. With respect to it closed sets are sets of the form $f = 0$, $f \in C^\infty(M)$. In terms of the defining atlas A the fact that a set W is closed is equivalent in the given sense to the fact that the intersection of W with any chart of this atlas is closed, that is, $\varphi_\alpha^{-1}(W)$ is closed in U_α for all $(\varphi_\alpha, U_\alpha) \in A$.

The existence of a topological structure on a smooth manifold enables us to speak of its *compactness, connectivity* and other topological properties. Condition 3) in the definition of a smooth manifold is equivalent to the requirement that the topology is Hausdorff or separable.

2.3. The Category of Smooth Manifolds. Let M and N be smooth manifolds and $F: M \to N$ a map. With any function f on N we can associate the function $f \circ F$ on M. The map F is said to be *smooth* if the function $f \circ F$ is smooth for any smooth function f on N. In other words, a smooth map F generates the map

$$F^*: C^\infty(N) \to C^\infty(M), \qquad F^*(f) = f \circ F.$$

The following properties of the $*$-operation are obvious:

$$\mathrm{id}^* = \mathrm{id}, \qquad (F \circ G)^* = G^* \circ F^*, \qquad (F^{-1})^* = (F^*)^{-1}.$$

Suppose that a point $x \in M$ lies in the domain of the local coordinates $x_1, \ldots, x_m$, and that the point $y = f(x)$ lies in the domain of the local coordinates $y_1, \ldots, y_n$. In terms of these coordinates the map F is specified locally by a set of functions $y_i = f_i(x_1, \ldots, x_m)$, $i = 1, \ldots, n$, that is,

$$(x_1, \ldots, x_m) \xmapsto{F} (f_1(x), \ldots, f_n(x)), \qquad x = (x_1, \ldots, x_m).$$

The smoothness of the map F is equivalent to the smoothness of the functions f_i with respect to any systems of local coordinates, since $f_i = F^*(y_i)$.

A smooth one-to-one map F for which the inverse map F^{-1} is also smooth is called a *diffeomorphism*. A diffeomorphism of a manifold onto itself is often called a *(smooth) transformation*.

Example. Suppose that $M = N = \mathbb{R}$, and that $F: x \to x^3$; then F is a smooth one-to-one map. It is not a diffeomorphism if we consider the standard smooth structures on M and N, that is, those defined by charts which are the identity maps $\varphi: \mathbb{R} \to M$ and $\psi: \mathbb{R} \to N$. However, if we define a smooth structure in N by the chart $\psi: \mathbb{R} \to N$, $\varphi(x) = x^3$, then F is a diffeomorphism.

In other words, the smooth structure on the line defined by the chart ψ is *equivalent* to the standard structure.

In the general case, two smooth structures on a set M are said to be *equivalent* if there is a map $F: M \to M$ that is a diffeomorphism with respect to these structures.

Example. Milnor showed that on a manifold topologically equivalent to a 7-dimensional sphere S^7 there are exactly 28 different smooth structures. Brieskorn, using singularity theory, showed that the 28 manifolds in C^5 specified by the equations

$$|z_1|^2 + |z_2|^2 + |z_3| + |z_4|^2 + |z_5|^2 = 1, \qquad z_1^2 + z_2^2 + z_3^2 + z_4^3 + z_5^{6k-1} = 0,$$

$k = 1, 2, \ldots, 28$, realize all these smooth structures on S^7.

2.4. Smooth Bundles. The direct (or Cartesian) product $M \times N$ of two smooth manifolds is endowed in a natural way with a smooth structure: if $(\varphi_\alpha, U_\alpha), (\psi_\beta, V_\beta)$ are smooth atlases on M and N respectively, then $(\varphi_\alpha \times \psi_\beta, U_\alpha \times V_\beta)$ is a smooth atlas on $M \times N$. Here $(\varphi_\alpha \times \psi_\beta)(u, v) = (\varphi_\alpha(u), \psi_\beta(v)) \in M \times N$, $u \in U_\alpha$, $v \in V_\beta$. A small complication of this construction leads to one of the most important concepts of differential geometry – a smooth bundle. Namely, a smooth map $\pi: E \to M$ is called a *smooth bundle* over M with fibre N if

1) π is a map on the whole of M;

2) for any point $x \in M$ there is a neighbourhood $\mathcal{O} \ni x$ such that its inverse image $\pi^{-1}(\mathcal{O})$ is fibrewise diffeomorphic to the product $\mathcal{O} \times N$, that is, there is a diffeomorphism $f|_{\mathcal{O}}: \mathcal{O} \times N \to \pi^{-1}(\mathcal{O})$ such that $f|_{\mathcal{O}}(x, y) \in \pi^{-1}(x)$, $\forall x \in \mathcal{O}$, $y \in N$.

The usual notation of a bundle is $\pi: E \to M$ or (E, M, π).

The submanifolds $\pi^{-1}(x) \subset E$, $x \in M$, are called the *fibres* of this bundle. They are all diffeomorphic to N. The space of the bundle E is the union of these fibres that do not intersect in pairs, that is, it is "fibered" by them. The manifold M is called the *base* of the bundle (E, M, π).

Example. The projection $\pi: M \times N \to M$ of a direct product onto its first factor is a smooth bundle, which is called *trivial*.

This example gives the justification for saying that locally bundles are direct products.

Example. The projection of an (open) Möbius band into its central circle (see Fig. 19) is a bundle whose fibre is an open interval. This bundle is not trivial.

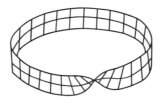

Fig. 19

Example. Associating with each line in E^3 the line parallel to it and passing through the origin, we obtain a non-trivial bundle $\pi: L \to \mathbb{R}P^2$ (for L see 2.1) with fibre $\mathbb{R}^2$.

Let $(\varphi_\alpha, U_\alpha)$ be a smooth atlas on M such that the inverse images $\pi^{-1}(V_\alpha)$, $V_\alpha = \varphi(U_\alpha) \subset M$, admit the "trivializing" maps $f_\alpha = f_{V_\alpha}: V_\alpha \times N \to \pi^{-1}(V_\alpha)$ concerned in condition 2) of the definition of a bundle. The map

$$f_\beta^{-1} \circ f_\alpha: (V_\alpha \cap V_\beta) \times N \to (V_\alpha \cap V_\beta) \times N,$$

maps any "fibre" $x \times N \subset (V_\alpha \cap V_\beta) \times N$, $x \in V_\alpha \cap V_\beta$, diffeomorphically into itself. We denote this diffeomorphism, after the obvious identification of N and $x \times N$, by $h_{\alpha\beta}(x): N \to N$ and call it a *transition function*. It is obvious that

$$h_{\alpha\alpha}(x) = \mathrm{id}, \qquad h_{\beta\alpha}(x) = h_{\alpha\beta}^{-1}(x), \qquad h_{\beta\gamma}(x)h_{\alpha\beta}(x) = h_{\alpha\gamma}(x)$$

and that the diffeomorphisms $h_{\alpha\beta}(x)$ depend smoothly on $x \in U_\alpha \cap U_\beta$. Conversely specifying an atlas $(\varphi_\alpha, U_\alpha)$ on M and a system of diffeomorphisms $h_{\alpha\beta}(x)$ that satisfy the conditions just stated enables us to construct a bundle $\pi: E \to M$ for which the $h_{\alpha\beta}(x)$ are transition functions. Speaking informally, the functions $h_{\alpha\beta}(x)$ show how the manifold E is pasted together from blocks of the form $U_\alpha \times N$.

Example. Let us cover a circle by an atlas of two charts (φ_i, U_i), $i = 1, 2$, in such a way that $V_1 = \{0 < \varphi < \pi + \varepsilon\}$, $V_2 = \{\pi < \varphi < 2\pi + \varepsilon\}$. Then $V_1 \cap V_2$ consists of two arcs $P_\pm$. Let N be the interval $(-1, 1) \subset \mathbb{R}$. The operation of pasting a Möbius band from two rectangles $U_i \times N$, $i = 1, 2$, shown in Fig. 20, corresponds to the transition function $k_{12}(x) = \pm \mathrm{id}$, $x \in P_\pm$.

A map $\sigma: M \to E$ such that $\pi \circ \sigma = \mathrm{id}$ is called a *section* of the bundle (E, M, π). Obviously σ is a section if and only if $\sigma(x) \in \pi^{-1}(x)$ (see Fig. 21). The totality of all smooth sections of the bundle in question is denoted by $\Gamma(\pi)$.

Let (E_i, M_i, π_i) $i = 1, 2$, be certain bundles. The pair of maps $F: E_1 \to E_2$, $f: M_1 \to M_2$ is called a *morphism* of the first into the second if $\pi_2 \circ F = f \circ \pi_1$. In other words, a morphism maps fibres of π_1 into fibres of π_2. The inverse morphism is called an *equivalence* (or *isomorphism*) of the bundles in question.

The bundle π_1 is called a *subbundle* if the morphism (F, f) determines embeddings of submanifolds $F: E_1 \hookrightarrow E_2$, $f: M_1 \hookrightarrow M_2$.

Any smooth map $f: M_1 \to M$ in the base of the bundle $\pi: E \to M$ generates a bundle over M_1, called the *induced bundle* and denoted by $f^*(x): f^*(E) \to M_1$.

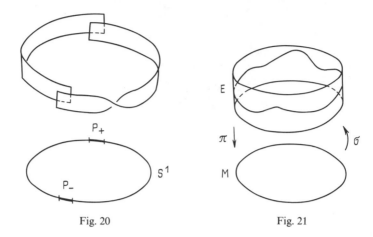

Fig. 20   Fig. 21

A fibre of the induced bundle at a point $y \in M_1$ coincides with the fibre of the original bundle at the point $x = f(y)$. Formally the manifold $f^*(E)$ is defined as the submanifold of $M_1 \times E$ consisting of pairs (y, e) such that $f(y) = \pi(e)$, and the projection $f^*(\pi)$ is defined as the map $(y, e) \to y$.

A bundle (E, M, π) is called a vector bundle if its fibres $\pi^{-1}(x)$ are endowed with the structure of a vector space "smoothly depending" on the point $x \in M$. This means that this bundle can be specified by linear transition functions, that is, $h_{\alpha\beta}(x)$ are linear transformations of the standard fibre N, which is a linear space.

Let S_i, $i = 1, \ldots, k$, be sections of the vector bundle π and let $f_i \in C^\infty(M)$ be functions on the base M. Then there is defined the section $s = \Sigma f_i s_i \colon M \ni x \to s(x) = \Sigma f_i(x) s_i(x)$, since the fibre $\pi^{-1}(x)$ is a vector space. In this sense we say that $\Gamma(\pi)$ is a $C^\infty(M)$-module.

Similarly we can consider bundles whose fibres have some structure or other, for example the structure of an affine space, the structure of a group, and so on. Such bundles are characterized by the fact that the transition functions $h_{\alpha\beta}(x)$ that specify them are transformations of the standard fibre N that preserve the corresponding structure on it.

§3. Tensor Fields and Differential Forms

3.1. Tangent Vectors. The question of carrying over the intuitively obvious concept of a tangent vector to a surface in Euclidean space to abstract manifolds is not completely trival and has its own history. The following three definitions, given in historical sequence, demonstrate the way in which understanding of it has evolved.

Thus, a *tangent* vector to a manifold M at a point a is:

1. The set of numbers $(\alpha_1, \ldots, \alpha_n)$ associated with each local coordinate system $(x_1, \ldots, x_n)$ around the point a, which is transformed according to the rule $(\alpha_i) \mapsto (\beta_i)$, $\beta_i = \sum_k \alpha_k \partial y_i / \partial x_k$, on transition to the coordinates $(y_1, \ldots, y_n)$.

2. The class of smooth curves $\gamma(t)$ on M that touch each other at the point a, where $\gamma(0) = a$.

3. The linear map $\xi \colon C^\infty(M) \to \mathbb{R}$ that satisfies the "*Leibniz rule*" $\xi(fg) = \xi(f)g(a) + \xi(g)f(a)$ (see Ch. 1, §4).

The first of these is a typically bureaucratic substitution of the essence of the matter by an instruction, that is, by the set $(\alpha_1, \ldots, \alpha_n)$. The price for it is non-invariance and "confusion of indices", of which we spoke above (see Ch. 2, §3). The second definition is invariant, but if we start from it, it is not clear how we should put together the tangent vectors. Finally, the third, which is the least intuitive of them, is adequate for the nature of differential geometry, as we see from its motivation, given in Ch. 1, §4. We shall follow this below.

The totality $T_a M$ of all tangent vectors to a manifold M at a point a forms a vector space, which is called the *tangent space* to M at a. The linear structure on $T_a M$ is induced by natural linear operations on the set of linear operators that act from $C^\infty(M)$ into $\mathbb{R}$. In local coordinates $(x_1, \ldots, x_n)$ any tangent vector $\xi \in T_a M$ can be represented in the form $\xi = \sum \alpha_i \dfrac{\partial}{\partial x}\Big|_a$. Here $\left(\dfrac{\partial}{\partial x_i}\Big|_a\right)(f) = \dfrac{\partial f}{\partial x_i}(a)$. This shows that the vectors $\dfrac{\partial}{\partial x_i}\Big|_a$ form a basis of the space $T_a M$ and that $\dim T_a M = n$. The numbers $\alpha_1, \ldots, \alpha_n$ are called the components of the vector ξ in the coordinate system in question (see the first definition above). Intuitively the tangent space $T_a M$ is a linearization of the manifold M close to the point a.

For any point $a \in M$ a smooth map $F \colon M \to N$ induces a linear map $d_a F \colon T_a M \to T_{F(a)} N$, called the *principal linear part* or *differential* of the map F at this point, namely the map $d_a F$ takes a vector $\xi \in T_a M$ into the vector $\eta \in T_{F(a)} N$ that acts on functions $g \in C^\infty(M)$ according to the rule $\eta(g) = \xi(g \circ F)$. In other words, $\eta = \xi \circ F^*$. If in local coordinates the map F is given by functions $y_j = y_j(x)$, $j = 1, \ldots, m$, then its differential is given by the Jacobi matrix $J_a = \left\| \dfrac{\partial y_j}{\partial x_i}(a) \right\|$, that is, if $\xi = \sum \alpha_i \dfrac{\partial}{\partial x_i}\Big|_a$, $\eta = \sum \beta_j \dfrac{\partial}{\partial y_j}\Big|_{F(a)}$, then

$$\begin{pmatrix} \beta_1 \\ \vdots \\ \beta_m \end{pmatrix} = J_a \begin{pmatrix} \alpha_1 \\ \vdots \\ \alpha_n \end{pmatrix}. \tag{6}$$

If $M = N$, $F = \mathrm{id}$, then this formula can be regarded as the rule for transforming the components of the tangent vector under a change of local coordinates.

3.2. The Tangent Bundle and Vector Fields. Let $TM = \bigcup_{x \in M} T_x M$. A map $\tau \colon TM \to M$ that associates a vector $\xi \in T_x M$ with its point of application x is a vector bundle. It is called the tangent bundle of M and has as transition functions the linear transformations of the standard fibre $\mathbb{R}^n$, specified by the Jacobi matrices $\left\| \dfrac{\partial y^i}{\partial x^j} \right\|$. (Here x^i are local coordinates in the domain V_α, and y^j in V_β; see the notation in 2.4.)

Obviously $\tau^{-1}(x) = T_x M$. In view of this, any section $s: M \to TM$ of the tangent bundle can be understood (intuitively) as a field of vectors on M. Conceptually (see Ch. 1, §4) a smooth *vector field* on M is defined as a linear operator

$$X: C^\infty(M) \to C^\infty(M),$$

satisfying the Leibniz rule $X(fg) = X(f)g + fX(g)$. In local coordinates a vector field can be written in the form

$$X = \sum_i \alpha_i(x) \frac{\partial}{\partial x_i}$$

and can be understood as the operation that associates the "vector" $(\alpha_1(x), \ldots, \alpha_n(x))$ with the point x.

The totality $D(M)$ of smooth vector fields on M is a $C^\infty(M)$-module, since $D(M) = \Gamma(\tau)$.

Direct verification shows that the linear operator $[X, Y] = X \circ Y - Y \circ X$ satisfies the Leibniz rule and is therefore a vector field. Obviously

$$[X, Y] = -[Y, X], \qquad [X, [Y, Z]] + [Y, [Z, X]] + [Z, [X, Y]] = 0,$$

that is, $D(M)$ is a Lie algebra with respect to the operation $[\cdot, \cdot]$.

A curve $\gamma: I \to M$, where $I = (a, b)$, is called a *trajectory* of the field $X \in D(M)$ if for any $\tau \in I$ the differential $d_\tau \gamma$ maps the standard "unit" vector $\dfrac{d}{dt}\Big|_\tau \in T_\tau I$ into the vector $X_{\gamma(\tau)}$. This is equivalent to the fact that $X \circ \gamma^* = \gamma^* \circ \dfrac{d}{dt}$. If $\gamma(t) = (x_1(t), \ldots, x_n(t))$ and $X = \sum \alpha_i(x) \dfrac{\partial}{\partial x_i}$, then the condition that $\gamma(t)$ is a trajectory of the field X has the form

$$\frac{dx_i}{dt} = \alpha_i(x(t)), \qquad i = 1, \ldots, n.$$

The theorem on existence and uniqueness for ordinary differential equations shows that through a given point $a \in M$ there passes a unique trajectory $\gamma(t)$ such that $\gamma(\tau) = a$ for a given $\tau \in I$. Fixing the number t, we define a *shift transformation* A_t along the trajectories of the field X by putting $A_t(a) = \gamma(\tau + t)$, where γ is the trajectory mentioned above. We should emphasize that the shift A_t is defined only for points $a \in M$ for which the maximal trajectories passing through them are defined for the values $\tau + t$.

Obviously $A_{t+s} = A_t \circ A_s$, $A_{-t} = A_t^{-1}$ and $A_0 = \text{id}$. For this reason we say that the family of shifts $\{A_t\}$ *forms a one-parameter* (*local*) *group* of transformations of M (or *flow*). A vector field is said to be *complete* if all its maximal trajectories

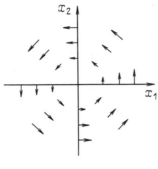

Fig. 22

$\gamma(t)$ are defined for all $t \in \mathbb{R}$. For a complete field the shifts A_t are "real" (not local) diffeomorphisms of M. We note that on a compact manifold any vector field is complete.

Conversely, a (local) one-parameter group of transformations A_t determines a vector field

$$X = \frac{d}{dt} A_t^* \bigg|_{t=0}, \qquad (7)$$

where A_t^* are the "shift" operators in the function space (see 2.3). In other words, X is the *velocity field of the flow* A_t.

Here is the "physical" interpretation: if X is the stationary velocity field of a certain continuous medium that fills a manifold M, then the shifts A_t describe the corresponding flow, that is, $A_t(x)$ is the position at time t of that particle of the medium that is at the point x when $t = 0$.

Example. $X = x_1 \dfrac{\partial}{\partial x_2} - x_2 \dfrac{\partial}{\partial x_1}$ (Fig. 22) is the velocity field of points of the plane that rotate uniformly about the origin counterclockwise with unit speed. Its trajectories are circles with centre at the origin, and the shift transformations A_t are rotations through an angle t.

Generally speaking, a smooth map $F: M \to N$ does not lead to a map of vector fields defined on M, despite the fact that the separate tangent vectors are represented by differentials of it. The reason for this can be seen from an attempt to map the field shown in Fig. 22 onto the x_1-axis by means of the projection $(x_1, x_2) \to x_1$. If F is a diffeomorphism, then such a map exists: a field $X \in D(M)$ is mapped into the field $F(X) = (F^*)^{-1} \circ X \circ F^*$ on N, where $F^*: C^\infty(N) \to C^\infty(M)$ is described in 2.3.

3.3. Covectors, the Cotangent Bundle and Differential Forms of the First Degree. A linear function on $T_a M$ is called a *tangent covector* to M at the point a. In other words, a tangent covector is the element of the space $T_a^* M$ dual to

T_aM, called the *cotangent space*. The value of the covector $v \in T_a^*M$ at the vector $\xi \in TM$ is denoted by $v(\xi)$ or $\langle v, \xi \rangle$.

To any function $f \in C^\infty(M)$ there corresponds a covector $d_a f \in T_a^*M$, $\langle d_a f, \xi \rangle = \xi(f)$, $\xi \in T_aM$, called its *differential* at the point a. Since $\langle d_a x_i, \partial/\partial x_j|_a \rangle = \delta_{ij}$, the differentials $d_a x_i$ form a basis in T_a^*M dual to the basis $\partial/\partial x_j|_a$ in T_aM. Therefore any covector $v \in T_a^*M$ can be uniquely represented in the form $v = \sum v_i d_a x_i$, where $v_i \in \mathbb{R}$ are its *components* in the system of local coordinates $x_1, \ldots, x_n$. If $\xi = \sum \xi_i \partial/\partial x_i|_a$, then $v(\xi) = \sum \xi_i v_i$. Since $\xi(f) = \sum \xi_i \dfrac{\partial f}{\partial x_i}(a)$, the components of the differential $d_a f$ are the derivatives $\dfrac{\partial f}{\partial x_i}(a)$, or $d_a f = \sum \dfrac{\partial f}{\partial x_i}(a) d_a x_i$.

If (y_j) is another coordinate system and $v = \sum v_i d_a x_i = \sum \mu_j d_a y_j$, then

$$\begin{pmatrix} \mu_1 \\ \vdots \\ \mu_n \end{pmatrix} = (J_a^*)^{-1} \begin{pmatrix} v_1 \\ \vdots \\ v_n \end{pmatrix}.$$

Comparison with (6) shows that the components of a covector are transformed in a different way to the components of a vector under transition to new coordinates.

The set $T^*M = \bigcup_{a \in M} T_a^*M$ of all tangent covectors is endowed in a natural way with the structure of a smooth manifold. The projection $\tau^*: T^*M \to M$ that sets up a correspondence between each covector and its point of contact determines a smooth vector bundle (T^*M, M, τ^*) on M, called the *cotangent bundle*. Its transition functions are linear transformations of the standard fibre $\mathbb{R}^n$ specified by the matrix $(J_a^*)^{-1}$ (compare with 3.2).

It is usual to call sections of the cotangent bundle (intuitively these are *covector fields* on M) *differential forms of the first degree*, or *1-forms*. The totality of all 1-forms on M, denoted by $\Lambda^1(M)$, is a $C^\infty(M)$-module: $\Lambda^1(M) = \Gamma(\tau^*)$ (see 2.4). For a 1-form $\omega: M \to T^*M$ let us agree to write ω_a instead of $\omega(a) \in (\tau^*)^{-1}(a) = T_a^*M$.

The *value* $\omega(X)$ *of the 1-form* ω *on the vector field* $X \in D(M)$ is a smooth function on M whose value at the point a is equal to the value of the covector ω_a at the vector X_a:

$$\omega(X)(a) = \omega_a(X_a).$$

In local coordinates a 1-form ω is written in the form $\alpha = \sum \omega_i(x) \, dx_i$. If $X = \sum \alpha_i(x) \dfrac{\partial}{\partial x_i}$, then $\omega(X) = \sum \omega_i \alpha_i$.

The differential 1-form represented by the covector field $a \mapsto d_a f$ is called the differential of the function f. Obviously,

$$df(X) = X(f).$$

For any point $a \in M$ a smooth map $F: M \to N$ induces a linear map $F_a^*: T_{F(a)}^*N \to T_a^*M$, conjugate to the differential $d_a F: T_aM \to T_{F(a)}N$. In local coordinates it is described by the matrix J_a^* (compare (6)).

If $\rho \in \Lambda^1(N)$, then the covector field $a \to F_a^*(\rho_{F(a)})$ on M is a differential form, denoted by $F^*(\rho)$. Thus, a smooth map $F: M \to N$ induces a linear map $F^*: \Lambda^1(N) \to \Lambda^1(M)$ of differential forms. We should draw attention to the essentially different character of the interrelations of vector fields and differential forms with smooth maps of manifolds (see the end of 3.2).

In this subsection 1-forms have been represented as objects dual to vector fields. This point of view is geometrically intuitive, but the approach presented in Ch. 1, §4 is more fundamental.

3.4. Tensors and Tensor Fields. In linear algebra, together with any linear space V it is useful to consider others connected with it in a natural way, for example the conjugate or dual space V^*, the space End V of linear operators acting on V, the space of bilinear functions, and so on. All these spaces are functorially connected with V. This means, in particular, that to any linear transformation $A: V \to V$ there corresponds a linear transformation $\Phi(A): \Phi(V) \to \Phi(V)$ of any such space. Bearing this fact in mind, with any vector bundle $\pi: E \to M$ having V as its standard fibre we can associate a vector bundle $\Phi(\pi): \Phi(E) \to M$ for which the standard fibre is the space $\Phi(V)$. This bundle is determined by the transition functions $H_{\alpha\beta}(x) = \Phi(h_{\alpha\beta}(x))$, where the linear transformations $h_{\alpha\beta}(x): V \to V$ are transition functions of the original bundle. In the previous subsection we defined the cotangent bundle in a similar way.

Application of this general construction to the tangent bundle $\tau: TM \to M$ enables us to obtain all possible *tensor bundles* over M, whose sections are called *tensor fields*. A Riemannian metric is one of the most important examples of a tensor field. Tensor fields are the simplest but also the most important geometrical structures on manifolds. From the point of view of Chapter 1 they represent the intuitive images of those elements of differential calculus that reduce to the consideration of first-order differential operators. Below we give details of this procedure that are necessary for what follows.

Let us begin with the elements of tensor algebra. A multilinear function or *tensor* on V of type (p, q) is a real function $f(v_1^*, \ldots, v_p^*, v_1, \ldots, v_q)$, $v_i \in V, v_j^* \in V^*$, of q vector arguments and p covector arguments that is linear in each argument. The number p is called the *contravariant valency* of the tensor, q the *covariant valency*, and $p + q$ simply the *valency*. A "vector" $v \in V$ is a tensor of type $(1, 0)$, and a "covector" $v^* \in V^*$ is a tensor of type $(0, 1)$.

Natural linear operations on multilinear functions of a given type (p, q) transform the totality of them into a linear space, denoted by $T_p^q(V)$. In particular, $T_1^0(V) = V$, $T_0^1(V) = V^*$. Tensors of type $(p, 0)$ are called *contravariant*, those of type $(0, q)$ are called *covariant*, and if $p \neq 0$, $q \neq 0$ they are called *mixed*; $T_0^0(V) = \mathbb{R}$.

The *tensor product* $f \otimes g \in T_{p+s}^{q+t}(V)$ of tensors $f \in T_p^q(V)$ and $g \in T_s^t(V)$ is defined by the formula

$$(f \otimes g)(v_1^*, \ldots, v_{p+s}^*, v_1, \ldots, v_{q+t}) = f(v_1^*, \ldots, v_p^*, v_1, \ldots, v_q) \cdot$$

$$\cdot g(v_{p+1}^*, \ldots, v_{p+s}^*, v_{q+1}, \ldots, v_{q+t}).$$

In particular, the tensor product $v_1 \otimes \cdots \otimes v_p \otimes v_1^* \otimes \cdots \otimes v_q^*$, $v_i \in V$, $v_j^* \in V^*$ is a tensor of type (p, q). If the vectors $e_1, \ldots, e_n$ form a basis in V, and the covectors $e^1, \ldots, e^n$ form a dual basis in V^*, then tensors of the form $e_{i_1} \otimes \cdots \otimes e_{i_p} \otimes e^{j_1} \otimes \cdots \otimes e^{j_q}$ constitute a basis of the space $T_p^q(V)$. Therefore any tensor $f \in T_p^q(V)$ can be expanded with respect to elements of this basis. The coefficients of this expansion are denoted by $f_{j_1 \ldots j_q}^{i_1 \ldots i_p}$ and called *components of the tensor* f with respect to the basis $e_1, \ldots, e_n$. On transition to another basis $\bar{e}_1, \ldots, \bar{e}_n$ the components of the tensor are transformed according to the rule

$$\bar{f}_{j_1 \ldots j_q}^{i_1 \ldots i_p} = A_{j_1}^{k_1} \ldots A_{j_q}^{k_q} B_{s_1}^{i_1} \ldots B_{s_p}^{i_p} f_{k_1 \ldots k_q}^{s_1 \ldots s_p}. \tag{8}$$

Here $\|A_j^i\|$ is the "transition matrix", that is $\bar{e}_i = A_i^k e_k$, $\|B_s^k\| = \|A_j^i\|^{-1}$, and we are using Einstein's summation rule (Ch. 2).

The operation of tensor multiplication turns the space $T(V) = \sum_{p,q} T_p^q(V)$ into a bigraded non-commutative associative algebra, which is called the *tensor algebra of the space V*. It contains the subalgebras $T^*(V) = \sum_p T_0^p(V)$ and $T_*(V) = \sum_p T_p^0(V)$ of covariant and contravariant tensors, respectively. The group $GL(V)$ of linear transformations of the space V acts in a natural way in $T(V)$ as automorphism group. For example, if $A \in GL(V)$, then

$$A(v_1 \otimes \cdots \otimes v_p) = A(v_1) \otimes \cdots \otimes A(v_p), \qquad v_i \in V.$$

The Lie algebra $gl(V) = \text{End } V$ of the group $GL(V)$ acts in $T(V)$ as the Lie algebra of differentiations. For example,

$$v(v_1 \otimes \cdots \otimes v_p) = v(v_1) \otimes v_2 \otimes \cdots \otimes v_p + \cdots + v_1 \otimes \cdots \otimes v_{p-1} \otimes v(v_p),$$

where $v \in \text{End } V$, $v_i \in V$.

The *contraction operation* $\tau_1^1: T_q^p(V) \to T_{q-1}^{p-1}(V)$ over the first covariant index and first contravariant index is given by

$$\tau_1^1(v_1 \otimes \cdots \otimes v_p \otimes v_1^* \otimes \cdots \otimes v_q^*) = \langle v_1, v_1^* \rangle v_2 \otimes \cdots \otimes v_p \otimes v_2^* \otimes \cdots \otimes v_p^*.$$

In other words, the value of the multilinear function $\tau_1^1 f$, $f \in T_q^p(V)$ of the arguments $v_1, \ldots, v_{q-1}, v_1^*, \ldots, v_{p-1}^*$ is equal to the trace of the automorphism

$$V \ni v \mapsto f(v, v_1, \ldots, v_{q-1}, \cdot, v_1^*, \ldots, v_{p-1}^*).$$

We similarly define the contraction operations τ_j^i with respect to other pairs of indices.

If $f_{j_1 \ldots j_q}^{i_1 \ldots i_p}$ are the components of the original tensor f, then the components of its convolution $\tau_1^1 f$ are $f_{aj_1 \ldots j_{q-1}}^{ai_1 \ldots i_{p-1}}$ (summation over a).

In classical differential geometry the word tensor is understood as a "quantity", which in each coordinate system is characterized by its components, and under a change of coordinates the components are transformed according to (8). The concept of a tensor has been used in this sense up to now in the physics and mechanics literature.

Suppose that $f \in T_0^p(V)$. We put $f^\sigma(v_1, \ldots, v_p) = f(v_{\sigma(1)}, \ldots, v_{\sigma(p)})$, where σ is a permutation of the indices $1, \ldots, p$. The tensor f is said to be *symmetric* (respectively, *skew-symmetric*, or *skew*) if $f^\sigma = f$ for any permutation σ (respectively, $f^\sigma = (-1)^\sigma f$), where $(-1)^\sigma = \pm 1$ depending on the parity of σ. Symmetric (respectively, skew-symmetric) tensors form a subspace of $T_0^p(V)$, denoted by $S^p(V)$ (respectively, $\Lambda^p(V)$).

For any tensor $f \in T_0^p(V)$ we can distinguish its symmetric part $f_{\text{sym}} = \dfrac{1}{p!} \sum_\sigma f^\sigma$

and skew-symmetric part $f_{\text{alt}} = \dfrac{1}{p!} \sum_\sigma (-1)^\sigma f^\sigma$. The operations $f \mapsto f_{\text{sym}}$ and $f \mapsto f_{\text{alt}}$ are usually called *symmetrization* and *alternation*, respectively.

For symmetric tensors $f \in S^p(V)$, $g \in S^q(V)$ we define their *symmetric product*

$$f \circ g = \frac{(p+q)!}{p!q!}(f \otimes g)_{\text{sym}} \in S^{p+q}(V).$$

Similarly, if $f \in \Lambda^p(V)$, $g \in \Lambda^q(V)$, we define the *exterior product*

$$f \wedge g = \frac{(p+q)!}{p!q!}(f \otimes g)_{\text{alt}} \in \Lambda^{p+q}(V).$$

It is not difficult to see that $g \wedge f = (-1)^{pq} f \wedge g$.

The symmetric product turns the direct sum $S^*(V) = \sum_{p \geq 0} S^p(V)$ into a commutative algebra isomorphic to the algebra of polynomials in n variables. Here we assume that $S^0(V) = \mathbb{R}$. Similarly, using exterior multiplication, we obtain the *Grassmannian* or *exterior algebra* $\Lambda^*(V) = \sum_{p=0}^n \Lambda^p(V)$. Here $\Lambda^0(V) = \mathbb{R}$ and we should turn our attention to the fact that $\Lambda^p(V) = 0$ if $p > n = \dim V$. The elements $e^{i_1} \wedge \cdots \wedge e^{i_p}$, $1 \leq i_1 < i_2 < \cdots < i_p \leq n$, form a basis in $\Lambda^p(V)$. Therefore any form $\omega \in \Lambda^p(V)$ can be uniquely respresented in the form $\omega = \sum_{i_1 < \cdots < i_p} \omega_{i_1 \ldots i_p} e^{i_1} \wedge \cdots \wedge e^{i_p}$.

Similar definitions can be given for contravariant tensors $f \in T_p^0(V)$. In this way we arrive at the subspaces $S_p(V) \subset T_p^0(V)$, $\Lambda_p(V) \subset T_p^0(V)$ of symmetric and skew-symmetric tensors, respectively, and the graded algebras $S_*(V) = \sum_{p \geq 0} S_p(V)$, $\Lambda^*(V) = \sum_{p \geq 0} \Lambda_p(V)$. The elements of the space $\Lambda_p(V)$ are usually called *p-vectors*. In particular, bivectors are elements of the space $\Lambda_2(V)$.

Example 1. Billinear forms on V are elements of the space $T_0^2(V)$. We have the direct decomposition $T_0^2(V) = S^2(V) \oplus \Lambda^2(V)$, which corresponds to the well-known representation of a bilinear form as the sum of symmetric and skew-symmetric forms.

Example 2. The linear operators $A: V \to V$ are identified with the tensors $f \in T_1^1(V)$: $A \leftrightarrow f$ if $f(v^*, v) = \langle Av, v^* \rangle$, $v \in V$, $v^* \in V^*$.

We arrive at the concept of a *tensor at a point a of a manifold M* if we assume that $V = T_a M$. Then $V^* = T_a^* M$. As we see, any local coordinate system $x_1, \ldots,$ x_n in a neighbourhood of the point a induces a basis $e_i = \dfrac{\partial}{\partial x_i}\bigg|_a$, $i = 1, \ldots, n$, in $T_a M$ and the conjugate basis $e^i = d_a x_i$ in $T_a^* M$. Therefore any tensor θ_a of type (p, q) at the point a can be written in the form

$$\theta_a = \theta^{i_1\dots i_p}_{j_1\dots j_q} \left.\frac{\partial}{\partial x_{i_1}}\right|_a \otimes \cdots \otimes \left.\frac{\partial}{\partial x_{i_p}}\right|_a \otimes d_a x_{j_1} \otimes \cdots \otimes d_a x_{j_p}. \tag{9}$$

Skew-symmetric tensors ω_a of type $(0, p)$ can be written in the form

$$\omega_a = \sum_{1 \leqslant i_1 < \cdots < i_p \leqslant n} \omega_{i_1\dots i_p} d_a x_{i_1} \wedge \cdots \wedge d_a x_{i_p}. \tag{10}$$

Tensor bundles of one type or other on M are obtained if we put $\Phi = T^q_p$, S^p, S_p, Λ^p or Λ_p in the construction described at the beginning of this subsection. It is appropriate to denote them by $(\tilde{T}^q_p(M), M, T^q_p(\tau)), \dots, (\tilde{\Lambda}_p(M), M, \Lambda_p(\tau))$. In view of (9) a tensor field θ of type (p, q) on M (that is, a section of the bundle $T^q_p(\tau)$) can be written in local coordinates in the form

$$\theta = \theta^{i_1\dots i_p}_{j_1\dots j_q}(x) \frac{\partial}{\partial x_{i_1}} \otimes \cdots \otimes \frac{\partial}{\partial x_{i_p}} \otimes dx_{j_1} \otimes \cdots \otimes dx_{j_q}.$$

Covariant skew-symmetric tensor fields of valency p are called *differential forms* of degree p, or *p-forms* for short. In view of (10) the differential form ω has the following coordinate description:

$$\omega = \sum_{i_1 < \cdots < i_p} \omega_{i_1\dots i_p}(x)\, dx_{i_1} \wedge \cdots \wedge dx_{i_p}.$$

We denote the $C^\infty(M)$-module of differential forms of degree p on M by $\Lambda^p(M)$. Pointwise exterior multiplication induces the operation of exterior multiplication of differential forms, also denoted by the symbol $\wedge : \omega \wedge \rho$. Thus in the space of differential forms

$$\Lambda^*(M) = \sum_{i=0}^{n} \Lambda^i(M)$$

there arises the structure of an associative graded skew-symmetric (or, in other words, supercommutative) algebra. It is called the *algebra of differential forms* (or *Grassmann algebra*) *of the manifold M*. Similarly we define the tensor product and symmetric product of tensor fields.

The value $\omega(X_1, \dots, X_p)$ of a differential form $\omega \in \Lambda^p(M)$ on the vector fields $X_1, \dots, X_p$ is a smooth function on M whose value at the point a is equal to the value of the semilinear function ω_a on the vectors $X_{1,a}, \dots, X_{p,a} \in T_a M$. The function $\omega(X_1, \dots, X_p)$ is skew-symmetric and $C^\infty(M)$ is linear in each argument. Conversely, any differential form ω of degree p on M, like any covariant tensor field, can be understood as a $C^\infty(M)$-valued function $\omega(X_1, \dots, X_p)$ of p arguments $X_i \in D(M)$ that is $C^\infty(M)$-linear in each argument.

Example 3. A Riemannian metric is a symmetric covariant tensor field of valency 2 (see Example 1).

Example 4. The stress state of a continuous medium at a given point $a \in E^3$ is described by stating, for each "elementary" area S at this point, the force F acting on it (see Fig. 23). Since the area S is specified by the bivector $\xi \wedge \eta$, and the force is a covector, we arrive at a linear map $\Lambda^2(V) \to V^*$, $V = T_a(E^3)$, which in the

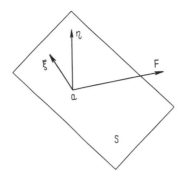

Fig. 23

spirit of Example 2 can be identified with a tensor of type $(0, 3)$. Thus in a domain of the continuous medium we are dealing with there arises the *stress tensor* field.

3.5. The Behaviour of Tensor Fields Under Maps. The Lie Derivative. Any covariant tensor field ω on a manifold N, in particular a differential form, can be pulled back to a manifold M by means of a smooth map $F: M \to N$ as we did in 3.3 for 1-forms. The inverse image of ω is denoted by $F^*(\omega)$. If ω has valency p, then the value of the tensor $F^*(\omega)_a$, $a \in M$, on the vectors $\xi_1, \ldots, \xi_p \in T_a M$ is equal by definition to the value of the tensor $\omega|_a$ on the vectors $\eta_1, \ldots, \eta_p \in T_p N$, where $\eta_i = (d_a F)(\xi_i)$. The map F^* of covariant tensor fields thus defined preserves linear operations, and also the tensor product, the symmetric product and the exterior product.

If M is a submanifold of a smooth manifold N, and $F: M \to N$ is the embedding, then the tensor $F^*(\omega)$ will also be denoted by $\omega|_M$.

As the example of vector fields discussed in 3.3 shows, contravariant tensor fields, generally speaking, are not carried over by smooth maps that are not diffeomorphisms. The image of a contravariant tensor field Θ on M under a diffeomorphism $F: M \to N$ is a field $F_*(\Theta)$ on N which, if it is thought of as a $C^\infty(N)$-valued function of covector fields on N, is defined by

$$F_*(\Theta)(\omega_1, \ldots, \omega_p) = F^{*-1}(\Theta(F^*(\omega_1), \ldots, F^*(\omega_p))),$$

where $\omega_i \in \Lambda^1(N)$.

The flow A_t generated by the vector field X on M (see 3.2) takes with it the covariant and contravariant tensor fields. The rate at which a given tensor field Θ changes is called its *Lie derivative* in the direction of the field X and denoted by $\mathscr{L}_X(\Theta)$. Below we give the three most important explicit formulae for the Lie derivative:

$$\mathscr{L}_X(f) = X(f), \qquad f \in C^\infty(M),$$

$$\mathscr{L}_X(Y) = [Y, X], \qquad Y \in D(M),$$

$$\mathscr{L}_X(\omega)(X_1, \ldots, X_p) = X(\omega(X_1, \ldots, X_p))$$

$$+ \sum_{i=1}^{p} \omega(X_1, \ldots, [X_i, X], \ldots, X_p), \qquad (11)$$

where ω is a covariant field of valency p and $X_i \in D(M)$. The second of these formulae gives a geometrical interpretation of the commutator of two vector fields.

The Lie derivative satisfies the Leibniz rule for differentiating products with respect to all forms of products that occur in tensor algebra. For example,

$$\mathscr{L}_X(\omega \wedge \rho) = \mathscr{L}_X(\omega) \wedge \rho + \omega \wedge \mathscr{L}_X(\rho),$$

where ω and ρ are differential forms.

Having arranged that the covariant and contravariant part of a mixed tensor field are carried over by the current A_t on the same side, we can define the Lie derivative for such fields.

Example. Let X be the velocity field of particles of some continuous medium that fills some domain in E^3. The kinematic characteristic of a flow in this medium – the *rate of deformation tensor* – which is important in mechanics, is (up to a factor $1/2$) the Lie derivative of the standard Euclidean metric $dx^2 + dy^2 + dz^2$ along the field X.

3.6. The Exterior Differential. The de Rham Complex. The operator d: $C^\infty(M) = \Lambda^0(M) \to \Lambda^1(M)$, $f \to df$, is included in a remarkable way in the following sequence of first-order differential operators:

$$\Lambda^0(M) \xrightarrow{d = d_1} \Lambda^1(M) \xrightarrow{d = d_2} \Lambda^2(M) \xrightarrow{d = d_3} \cdots \xrightarrow{d = d_{n-1}} \Lambda^n(M).$$

A conceptual definition of the operators d_p is not simple. They will therefore be characterized by the following properties, which uniquely define them:
1. $\mathbb{R}$-linearity,
2. $d_{p+1} \circ d_p = 0$, or $d^2 = 0$,
3. $d(\omega_1 \wedge \omega_2) = d\omega_1 \wedge \omega_2 + (-1)^p \omega_1 \wedge d\omega_2$ if $\omega_1 \in \Lambda^p(M)$. From these properties it follows immediately that

$$d\omega = \sum_{i_1 < \cdots < i_p} d(\omega_{i_1 \ldots i_p}) \wedge dx_{i_1} \wedge \cdots \wedge dx_{i_p}$$

if $\omega = \sum_{i_1 < \cdots < i_p} \omega_{i_1 \ldots i_p} dx_{i_1} \wedge \cdots \wedge dx_{i_p}$.

The most important property of the operator d, which is called the exterior differential, is its *naturalness*:

$$d \circ F^* = F^* \circ d$$

for any smooth map $F: M \to N$. The infinitesimal analogue of this equality has the form

$$d \circ \mathscr{L}_X = \mathscr{L}_X \circ d.$$

We can give an explicit coordinate-free formula for d:

$$d\omega(X_1, \ldots, X_{p+1}) = \sum_{i=1}^{p+1} (-1)^{i+1} X_i(\omega(X_1, \ldots, \hat{X}_i, \ldots, \hat{X}_{p+1}))$$

$$+ \sum_{i<j} (-1)^{i+j} \omega([X_i, X_j], X_1, \ldots, \hat{X}_i, \ldots, \hat{X}_j, \ldots, X_{p+1});$$

(12)

where $\omega \in \Lambda^p(M)$, $X_i \in D(M)$ and the symbol $\wedge$ denotes omission of the argument over which it stands.

Comparison of (11) and (12) enables us to obtain the *infinitesimal Stokes formula*, which expresses the Lie derivative in terms of exterior differentials:

$$\mathscr{L}_X(\omega) = X \lrcorner \, d\omega + d(X \lrcorner \, \omega).$$

(13)

Here $X \lrcorner: \Lambda^k(M) \to \Lambda^{k-1}(M)$ is the operation of substituting the vector field X in the differential form as the first argument.

If $d\omega = 0$, then the differential form ω is said to be *closed*, and if $\omega = d\rho$, it is said to be *exact*. Since $d^2 = 0$, any exact form is closed. The converse is true only locally (*Poincaré's lemma*).

§ 4. Riemannian Manifolds and Manifolds with a Linear Connection

4.1. Riemannian Metric. The concepts presented in the previous section are a convenient language for an invariant (that is, not using coordinates) presentation of Riemannian geometry.

A *Riemannian metric* on an n-dimensional manifold M is a positive definite symmetric tensor field g of type $(0, 2)$. A manifold M endowed with a Riemannian metric g is said to be *Riemannian* and denoted by (M, g).

If we replace the requirement that g is positive definite by the weaker condition that it is non-degenerate, we arrive at the concept of a *pseudo-Riemannian metric* and a *pseudo-Riemannian manifold*. If at each point $x \in M$ the symmetric form g_x has signature $(k, n - k)$, where k is the number of positive squares in its diagonal representation, we say that the pseudo-Riemannian metric g has *signature* $(k, n - k)$. A metric g of signature $(1, n - 1)$ is said to be *Lorentzian*, and the corresponding manifold (M, g) is called a *Lorentzian manifold*. The main postulate of the general theory of relativity states that physical space-time is a four-dimensional Lorentzian manifold. According to the special theory of relativity, in the absence of gravitation space-time is a flat Lorentzian manifold $M \approx \mathbb{R}^4$ with pseudo-Euclidean *Minkowski metric* $g = c^2 dt^2 - dx_1^2 - dx_2^2 - dx_3^2$, where c is the speed of light in the vacuum.

4.2. Construction of Riemannian Metrics. We give two simple constructions that make it possible, given some (pseudo-)Riemannian manifolds, to construct others.

Firstly, the direct product of two pseudo-Riemannian manifolds of signatures (k, l) and (p, q) respectively obviously turns into a pseudo-Riemannian manifold of signature $(k + p, l + q)$.

Further, just as the metric of a Euclidean space E^n induces a Riemannian metric (first quadratic form) on a surface, the Riemannian metric g of a manifold (M, g) induces a Riemannian metric on an arbitrary submanifold N. Thus, a submanifold of a Riemannian manifold is a Riemannian manifold. We note that for Lorentzian manifolds this is not always so, because the symmetric form h induced on N may turn out to be degenerate.

There arises the question: is there a Riemannian metric on any manifold? The standard technique of "partition of unity" enables us to glue together a global Riemannian metric from local ones and gives an affirmative answer to this question. For Lorentzian manifolds the answer turns out to be more complicated.

A Lorentzian metric exists on any non-compact manifold and also on compact manifolds with Euler characteristic zero.

4.3. Linear Connections. As we showed in Chapter 2, 2.6 and 3.2, the first quadratic form (that is, the Riemannian metric) of a surface determines a connection. Geometrically a connection can be regarded as a law of parallel transport of tangent vectors along curves. Analytically it is convenient to specify a connection by operators of covariant differentiation.

Many fundamental concepts of the intrinsic geometry of surfaces (for example, the curvature tensor) are defined in terms of connections. This led Weyl in 1917 to the idea of defining a connection in the tangent bundle abstractly, independently of the metric, as a field quantity and to regard the geometry corresponding to it as a natural generalization of Riemannian geometry. He also suggested a physical interpretation of this quantity. For an instructive history of the development of these deep ideas see Ch. 6, § 1.

The argument of Ch. 2, 2.6, motivate the following definition. A *linear connection* on a manifold M is a map $X \to \nabla_X$ that associates with each vector field $X \in D(M)$ the operator ∇_X in the space $D(M)$ of vector fields that depends $C^\infty(M)$-linearly on X and satisfies the "Leibniz rule":

$$\nabla_X(fY) = (X(f))Y + f\nabla_X Y, \qquad f \in C^\infty(M), \quad X, Y \in D(M).$$

The operator ∇_X is called the *covariant derivative* in the direction of the field X.

Practically, the specification of a connection ∇ in the domain of definition of a local coordinate system $x^1, \ldots, x^n$ reduces to the choice of an arbitrary set of n^3 smooth functions $\Gamma^i_{jk}(x)$, the *Christoffel symbols*:

$$\nabla_{\partial_i}\partial_j = \Gamma^k_{ij}\partial_k.$$

For $X, Y \in D(M)$ we have

$$\nabla_X Y = (X^i \partial_i Y^j + \Gamma^j_{ik}(x)X^i Y^k)\partial_j$$

or briefly $\nabla_X = X^i\partial_i + \Gamma^j_{ik}X^i$.

The Christoffel symbols $\Gamma^i_{jk}(x)$ and $\Gamma^a_{bc}(y)$ corresponding to systems of local coordinates (x^i) and (y^a) are connected by the relations

$$\Gamma^i_{jk}(x) = \Gamma^a_{bc}(y)\frac{\partial y^b}{\partial x^j}\frac{\partial y^c}{\partial x^k}\frac{\partial x^i}{\partial y^a} + \frac{\partial^2 y^a}{\partial x^j \partial x^k}\frac{\partial x^i}{\partial y^a}.$$

From this it is obvious that a connection is a non-tensor field quantity.

The value $(\nabla_X Y)_x$ of the vector field $\nabla_X Y$ at a point $x \in M$ depends only on the value $\xi = X_x$ of the vector field X at the point x. Therefore (as in Ch. 2, 2.6) we can define the operator $\nabla_\xi \colon D(M) \to T_x M$ of covariant differentiation in the direction of the vector ξ and the operator $\nabla_{\dot{x}(t)}$ of covariant differentiation along the curve $x(t)$, which acts in the space of vector fields defined along this curve. The vector field $Y(t)$ along the curve $x(t)$ is said to be *parallel* if $\nabla_{\dot{x}(t)} Y(t) = 0$, or in local coordinates

$$\dot{Y}^i(t) + \Gamma^i_{jk}(x(t))\dot{x}^j(t) Y^k(t) \equiv 0.$$

Any vector $Y(0) \in T_{x(0)}M$ can be uniquely extended to a parallel vector field $Y(t)$ along the curve $x(t)$, and the map $Y(0) \to Y(t)$ determines a linear isomorphism of tangent spaces $T_{x(0)}M \to T_{x(t)}M$. It is called the *parallel transport of vectors along the curve* $x(t)$ from the point $x(0)$ to the point $x(t)$. A curve $x(t)$ that has a parallel velocity field $\dot{x}(t)$ is called a *geodesic* of the linear connection ∇.

A linear connection ∇ determines a tensor field T of type $(1, 2)$, called the *torsion tensor* and specified by the formula

$$T(X, Y) = \nabla_X Y - \nabla_Y X - [X, Y], \qquad X, Y \in D(M). \tag{14}$$

In local coordinates the components of the torsion tensor have the form

$$T^i_{jk} = \Gamma^i_{jk} - \Gamma^i_{kj}.$$

A connection with zero torsion tensor is said to be *symmetric*.

As in the theory of surfaces, with a Riemannian metric g on a manifold we associate in a natural way a linear connection ∇, called the *Levi-Civita connection*. It is uniquely characterized by the symmetry condition and the fact that the parallel transport determined by it along the curve is an isometry of tangent spaces. The latter condition can be written in the form

$$X \cdot g(Y, Z) = g(\nabla_X Y, Z) + g(Y, \nabla_X Z), \qquad X, Y, Z \in D(M).$$

The Christoffel symbols of the Levi-Civita connection can be expressed in terms of the coefficients g_{ij} of the metric in accordance with formula (16) of Ch. 2.

The *geodesics* of a Riemannian manifold (M, g) are defined as the geodesics of the Levi-Civita connection. They can also be defined as the extremals of the functional $x(t) \mapsto \int_0^{t_0} g(\dot{x}, \dot{x})\, dt$ of the "square of arc length" ("action"), and also, if they are regarded as non-parametrized curves, as extremals of the functional of arc length $x(t) \mapsto \int_0^{t_0} \sqrt{g(\dot{x}, \dot{x})}\, dt$.

The operator ∇_X of covariant differentiation of vector fields can be extended to a differentiation (denoted by the same symbol) of the algebra of tensor fields, which preserves the type of tensors. For example, the action of the operator ∇_X of covariant differentiation in the direction of the vector field X on a covariant tensor field ω of valency k is given by

$$(\nabla_X \omega)(Y_1, \ldots, Y_k) = X \cdot \omega(Y_1, \ldots, Y_k) - \omega(\nabla_X Y_1, Y_2, \ldots, Y_k) - \cdots\cdots$$

$$- \omega(Y_1, \ldots, Y_{k-1}, \nabla_X Y_k), \quad Y_1, \ldots, Y_k \in D(M). \tag{15}$$

In contrast to the Lie derivative, the operator ∇_X is $C^\infty(M)$ linear in X. Therefore formula (15) enables us to define also a covariant tensor field $\nabla\omega$ of valency $k + 1$, called the *covariant differential*

$$(\nabla\omega)(X, Y_1, \ldots, Y_k) = (\nabla_X \omega)(Y_1, \ldots, Y_k).$$

If ω is a k-form and the connection ∇ is symmetric, then the alternation of the covariant differential $\nabla\omega$ coincides with the exterior differential $d\omega$: $(\nabla\omega)_{\text{alt}} = d\omega$. Similarly we can define the covariant differential of any tensor field S of type (k, l) as a tensor field ∇S of type $(k, l + 1)$.

A tensor field S is said to be *parallel* or *covariantly constant* with respect to the connection ∇ if $\nabla S = 0$. In this case parallel transport along any curves preserves the field S.

Example. 1) The metric g is parallel with respect to the Levi-Civita connection.

2) In a simply-connected compact symmetric Riemannian space (see Ch. 4, 4.4) the algebra of parallel differential forms coincides with the algebra of invariant forms and is isomorphic to its de Rham cohomology algebra.

Classical mechanics admits a remarkable interpretation in the language of Riemannian geometry. Namely, the kinetic energy $T = g_{ij}(x^k)\dot{x}^i\dot{x}^j$ of a mechanical system determines a Riemannian metric $ds^2 = g_{ij}(x^k)\,dx^i\,dx^j$ in the configuration space (see 2.1). In the absence of non-holonomic connections and external forces the evolution of the system is described by a motion along a geodesic of this Riemannian metric. For example, for a two-link plane pendulum (see 2.1) the kinetic energy, as a function of the angular coordinates φ_1, φ_2 and the angular speeds $\dot{\varphi}_1$, $\dot{\varphi}_2$ is equal to

$$T = \frac{1}{2}(m_1 V_1^2 + m_2 V_2^2) = \frac{1}{2}r_1^2\dot{\varphi}_1^2 + \frac{1}{2}(r_1^2 + r_2^2 + 2r_1r_2 \cos \varphi_1)(\dot{\varphi}_1^2 + \dot{\varphi}_2^2),$$

where m_1, m_2 are the masses and r_1, r_2 the lengths of the links, and so the corresponding Riemannian metric on the torus (as a quadratic differential form) has the form

$$ds^2 = [2r_1^2 + r_2^2 + 2r_1r_2 \cos \varphi_1]\,d\varphi_1^2 + (r_1^2 + r_2^2 + 2r_1r_2 \cos \varphi_1)\,d\varphi_2^2.$$

In the presence of an external potential field with potential $V(g^i)$ the total energy $T + V$ of the system is preserved, and the trajectories of the system in the configuration space for a fixed value $E = T + V$ of the energy coincide (as non-parametrized curves) with the geodesics of the *Jacobi metric* $ds_E^2 = (E - V(x))\,ds^2$ conformal to the metric ds^2.

4.4. Normal Coordinates. Let M be a manifold with linear connection V; each vector v of the tangent space T_xM uniquely determines a geodesic $\gamma_v(t)$ for which $\gamma_v(0) = x$, $\dot\gamma_v(0) = v$. The map

$$\exp: T_xM \to M, \qquad V \leftarrow \gamma_v(1)$$

is called the *exponential map*. If the connection is not complete, it is defined only in some neighbourhood of the origin in T_xM. The exponential map determines a diffeomorphism of some neighbourhood U of the origin in T_xM onto a neighbourhood $\exp U$ of the point x in M. On the strength of this diffeomorphism the Cartesian cordinates in the tangent space, specified by the basis $e_1, \ldots, e_n$, determine coordinates $x^1, \ldots, x^n$ in the neighbourhood $\exp U$. They are called normal coordinates with centre at the point x and are uniquely defined up to a linear transformation. In the Riemannian case the normal coordinates corresponding to an orthonormal basis $e_1, \ldots, e_n$ are called *normal Riemannian coordinates*.

Normal coordinates are characterized by the fact that in these coordinates the equations of geodesics $\gamma_v(t)$ passing through their origin have the linear form

$$\gamma_v(t) = tv = (tv^1, \ldots, tv^n), \qquad v = v^i e_i.$$

This is equivalent to

$$\Gamma^i_{jk}(x)x^jx^k = 0, \qquad \Gamma^i_{jk}(0) = 0.$$

Similarly, the Riemannian coordinates x^i are characterized by the conditions

$$g_{ij}(x)x^j = x^i, \qquad g_{ij}(0) = \delta_{ij}.$$

Example. Any Riemannian metric of a two-dimensional manifold in normal coordinates x, y has the form

$$\|g_{ij}\| = \begin{pmatrix} 1 & 0 \\ 0 & 1 \end{pmatrix} + f(x, y)\begin{pmatrix} y^2 & -xy \\ -xy & x^2 \end{pmatrix},$$

where $f(x, y)$ is an arbitrary function.

Let us expand the Christoffel symbols $\Gamma^i_{jk}(x^k)$ of the linear connection as a series in the normal coordinates x^i:

$$\Gamma^i_{jk}(x) = \Gamma^i_{jka}x^a + \Gamma^i_{jkab}x^ax^b + \cdots.$$

Since the normal coordinates are defined up to linear transformations, the set $\Gamma^{(s)} = (\Gamma^i_{jka_1 \ldots a_s})$ determines a tensor of type $(1, s + 2)$, called the *s-th normal tensor of the connection* and is expressed in terms of the curvature and torsion tensors R, T and its covariant derivatives of order at most $s - 1$. For example, if $T = 0$

$$\Gamma^i_{jkl} = \frac{1}{3}(R^i_{jkl} + R^i_{kjl}), \qquad R^i_{jkl} = \Gamma^i_{jkl} - \Gamma^i_{jlk}. \tag{16}$$

More generally, expanding a tensor field A of type (k, l) in a series in normal coordinates, we can obtain a set of tensors $A^{(1)}, A^{(2)}, \ldots$ of type $(k, l + 1), (k, l + 2)$,

..., called the 1st, 2nd, and so on, *extension of the tensor A at the point x.* The first extension of the tensor coincides with its covariant differential.

We say that two manifolds M and M' with linear connections are *locally equivalent at the points $x \in M$ and $x' \in M'$* if there is a diffeomorphism $\varphi \colon U \to U'$ of neighbourhoods of these points that takes one connection into the other, and x into x'.

Normal coordinates enable us to solve easily the question of local equivalence of linear connections.

Theorem. *Two analytic manifolds M and M' with linear connections are locally equivalent at points $x \in M$ and $x' \in M'$ if and only if in suitable normal coordinates with centres at these points they have the same curvature tensor and torsion tensor.*

It is sufficient to prove that in suitable normal coordinate systems the Christoffel symbols have the same series expansions, i.e. the same normal tensors at the origin. This follows from the fact that the normal tensors of a connection V are expressed in terms of the curvature and torsion tensors R_x, T_x of the connection V and its extensions at the point x. We do not know whether this theorem is true in the smooth case.

4.5. A Riemannian Manifold as a Metric Space. Completeness. A Riemannian manifold (M, g) can be regarded as a metric space if the distance $d(x, y)$ between points is defined as the lower bound of lengths of curves joining them. If this metric space is complete, we say that the Riemannian manifold (M, g) is complete. According to the Hopf-Rinow theorem, the completeness of the manifold (M, g) is equivalent to the fact that all its geodesics $\gamma(t)$ are complete, that is, they can be defined for any values of the parameter $t \in \mathbb{R}$.

This condition of geodesic completeness can be restated more elegantly if we observe that a uniform motion along geodesics determines a local one-parameter group of transformations $\varphi_t \colon TM \to TM$, called a geodesic flow:

$$\varphi_t \colon v \mapsto \dot{\gamma}_v(t) \quad \text{(see 4.4)}.$$

Geodesics are projections of trajectories of this flow on M. The condition of geodesic completeness means that the geodesic flow is complete, that is, it is a global one-parameter group of transformations of the manifold TM. We note that the concept of geodesic completeness makes sense for an arbitrary manifold with a linear connection.

Any two points x and y of a complete Riemannian manifold are joined by a geodesic of length $d(x, y)$. It is called a *minimal geodesic.* Generally speaking, this is false for complete manifolds with a linear connection, in particular for pseudo-Riemannian manifolds: generally speaking, in such manifolds only two sufficiently close points can be joined by a geodesic (see Example 3 below). In general relativity theory, from a number of physically natural requirements there follows the incompleteness of the Lorentzian metric of space-time. It is interpreted as the existence in space-time of singular points corresponding to "black holes", the initial "Big Bang", and so on.

Example. 1) Rejecting a point from a complete Riemannian manifold (for example, the Euclidean plane), we obtain an incomplete Riemannian manifold, which admits a completion – an embedding in a complete Riemannian manifold.

2) The surface of a cone without its vertex in E^3 is an example of an incomplete locally Euclidean manifold with conical singularity at the vertex that does not admit a completion. The initial singularity of the "Big Bang" in the standard cosmological model of the universe suggested by Fridman has a similar conical character.

3) Let us consider the group of unimodular matrices $SL(2, \mathbb{R})$ with bi-invariant Lorentzian metric given by the Cartan-Killing form in the Lie algebra $sl(2, \mathbb{R})$ (see Ch. 4, 2.1). It is geodesically complete, and its geodesics are trajectories of one-parameter subgroups. The point $\mathrm{diag}(-\lambda, -\lambda^{-1})$, $\lambda > 1$ does not belong to any one-parameter subgroup and so it cannot be joined by a geodesic to the identity of the group.

4.6. Curvature. The curvature of a surface, defined in Ch. 2, shows to what extent a parallel transport of a vector round a small loop differs from the identity transformation. By this property we define the curvature of a manifold M with an arbitrary linear connection V.

We choose two commuting vector fields X and Y in a neighbourhood of a point $a \in M$ and consider the "curvilinear parallelogram" formed by trajectories of these fields (Fig. 24). Let F_t and G_s be shift transformations in time t and s along the trajectories of these fields respectively, and A_t and B_s the parallel transport along these trajectories. Then from the definition of covariant derivative it follows that

$$A_t(Z)(F_t a) = Z(F_t a) + t V_X(Z)(F_t a) + o_2,$$

$$B_s(Z)(G_s a) = Z(G_s a) + s V_Y(Z)(G_s a) + o_2,$$

where o_2 are small quantities of the second order in t and s. Therefore, at the point $b = G_s F_t(a) = F_t G_s(a)$ the vectors $B_s A_t(Z)(b)$ and $A_t B_s(Z)(b)$ will have the form

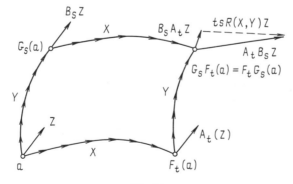

Fig. 24

$$B_s A_t(Z)(b) = Z(b) + t V_X(Z)(b) + s V_Y(Z)(b) + st V_Y V_X(Z)(b) + o_3,$$

$$A_t B_s(Z)(b) = Z(b) + t V_X(Z)(b) + s V_Y(Z)(b) + st V_X V_Y(Z)(b) + o_3,$$

where o_3 is a small quantity of the third order in t and s, and so the amount of deviation under transport round the parallelogram up to small quantities of the third order is characterized by the expression $V_X V_Y(Z) - V_Y V_X(Z)$. For non-commuting fields X and Y we arrive in a similar way at the expression

$$R(X, Y)Z = V_X V_Y(Z) - V_Y V_X(Z) - V_{[X, Y]}(Z), \tag{17}$$

which defines a tensor of type (1, 3). It is called the *curvature tensor of the linear connection* V.

Geometrically, taking account of what we said above, we can interpret the curvature tensor as follows: at each point $a \in M$ it associates with an oriented pair of vectors X, Y the linear curvature operator

$$R(X, Y): T_a M \to T_a M$$

the operator of infinitesimal parallel transport of vectors of $T_a M$ round the parallelogram spanned by the vectors X and Y. The vanishing identically of the curvature tensor is equivalent to the fact that the result of parallel transport of vectors from the point x to the point y is not changed by a deformation of the path joining them (see Ch. 6, 3.3). Hence it follows that any tensor A at the point x is carried by means of parallel transport to a parallel tensor field in any simply-connected neighbourhood of x. In particular, the basis vectors $e_1, \ldots, e_n$ of the tangent space $T_x M$ determine parallel vector fields $X_1, \ldots, X_n$. We suppose that the connection is symmetric. Then, according to (14), the fields $X_1, \ldots, X_n$, and hence the flows $A_t^{(1)}, \ldots, A_t^{(n)}$ generated by them, commute pairwise. Any point y of a sufficiently small neighbourhood of x can be represented uniquely in the form $y = A_{t^1}^{(1)} \ldots A_{t^n}^{(n)} x$, and the parameters $t^1, \ldots, t^n$ can be taken as its local coordinates. It is easy to see that in this coordinate system the Christoffel symbols Γ_{jk}^i of the connection are identically zero. In other words, the connection V is (locally) *flat*. Thus, the vanishing of the curvature tensor of a symmetric linear connection is necessary and sufficient for it to be (locally) flat.

In the Riemannian case (when V is the Levi-Civita connection) the same arguments, applied to the orthonormal basis $e_1, \ldots, e_n$, lead to a local coordinate system $t^1, \ldots, t^n$ in which the metric has the Euclidean form

$$g = \sum_l (dt^l)^2.$$

Thus, the vanishing of the curvature tensor of a Riemannian metric is necessary and sufficient for the Riemannian manifold (M, g) to be *locally Euclidean*.

In the case when the manifold M is simply-connected, and the symmetric connection with zero curvature V is geodesically complete, the fields $X_1, \ldots, X_n$ are defined globally and are complete, and the corresponding coordinates $t^1, \ldots, t^n$ are global. Hence it follows that the manifold (M, V) is isomorphic to the affine space A^n, and in the Riemannian case the Riemannian manifold (M, g) is isometric to Euclidean space.

These results give the simplest example of how the differential-geometric properties of a manifold determine its topology.

We note that, as a rule, a Riemannian metric g is almost uniquely determined by its curvature tensor R. For example, if the curvature operators $R(X, Y)$; $X, Y \in T_x M$, generate an irreducible Lie algebra of endomorphisms of the tangent space, then the value g_x of the metric tensor at x is determined up to a factor as the scalar product with respect to which the curvature operators are skew-symmetric. If this condition is satisfied at any point $x \in M$, then the Riemannian metric g is determined by the curvature tensor up to a constant factor.

4.7. The Algebraic Structure of the Curvature Tensor. The Ricci and Weyl Tensors and Scalar Curvature. From (17) and the Jacobi identity for the commutator of operators it follows that the curvature tensor field of a symmetric linear connection satisfies the following (*first*) *Bianchi identity*:

$$R(X, Y)Z + R(Y, Z)X + R(Z, X)Y = 0. \tag{18}$$

The curvature tensor of a Riemannian manifold (defined as the curvature tensor of the Levi-Civita connection) also satisfies the identities

$$g(R(X, Y)Z, U) = g(R(Z, U)X, Y),$$

$$g(R(X, Y)Z, U) = -g(Z, R(X, Y)U)$$

and, as in Ch. 2, 3.5, for any point $x \in M$ it is identified with the symmetric operator

$$R_x: \sum \xi_i \wedge \eta_i \mapsto \sum R(\xi_i, \eta_i), \qquad \xi_i, \eta_i \in T_x M,$$

of the bivector space $\Lambda^2 T_x M$ that satisfies (18). The set of such operators forms a vector space $\mathcal{R}$, called the *space of curvature tensors*. It is invariant with respect to the natural action of the orthogonal group $SO(T_x M)$. The description of the decomposition of the space $\mathcal{R}$ into irreducible subspaces, given below, enables us to associate with the curvature tensor new tensors that are differential invariants of the metric g.

The contraction determines the projection

$$\text{ric}: \mathcal{R} \to S^2 T_x M,$$

which associates with a tensor $R_x \in \mathcal{R}$ a symmetric bilinear form $\text{ric}(R_x)$, called the *Ricci tensor*. The value $\text{ric}(R_x)(\xi, \eta)$ of this form on vectors $\xi, \eta \in T_x M$ is defined as the trace of the linear operator

$$T_x M \ni \zeta \mapsto R_x(\zeta, \xi)\eta.$$

Obviously, $[\text{ric}(R_x)]_{jk} = R^i_{jki}$.

The symmetric operator $\text{Ric}(R_x)$ corresponding to the form $\text{ric}(R_x)$ with respect to the metric g_x is called the *Ricci operator*, and its trace $\text{Sc}(R_x)$ is called the *scalar curvature*. The Ricci operator is given by the formula

$$\text{Ric}(R_x)\xi = \sum_i R_x(\xi, e_i)e_i, \qquad \xi \in T_x M,$$

where the e_i form an orthonormal basis in $T_x M$. The number $\mathrm{ric}(R_x)(\xi, \xi)$, where ξ is a unit vector, is called the *Ricci curvature* in the direction ξ.

We note that a symmetric operator A of the space $T_x M$ determines in a natural way a symmetric operator $A^{[2]}$ of the space of bivectors belonging to the space $\mathcal{R}$:

$$A^{[2]}\xi \wedge \eta = A\xi \wedge \eta + \xi \wedge A\eta, \qquad \xi, \eta \in T_x M.$$

An arbitrary tensor $R \in \mathcal{R}$ can be represented uniquely in the form

$$R = \frac{\mathrm{Sc}(R)}{n(n-1)}\,\mathrm{id}^{[2]} + \frac{2}{n-2}\,\mathrm{Ric}_0(R)^{[2]} + W(R), \qquad (19)$$

where $\mathrm{Ric}_0(R) = \mathrm{Ric}(R) - \dfrac{\mathrm{Sc}(R)}{n}\,\mathrm{id}$ is the "trace-free" part of the Ricci operator, and $W(R)$ is a tensor belonging to the kernel of the map ric and called the *Weyl tensor*. This decomposition corresponds to the decomposition of the space $\mathcal{R}$ as a sum of three $SO(T_x M)$-invariant subspaces. Thus, the curvature tensor R_x of a Riemannian manifold at an arbitrary point $x \in M$ consists of three terms – the Weyl tensor $W(R_x)$ and two components determined by the scalar curvature $\mathrm{Sc}(R_x)$ and the trace-free part $\mathrm{Ric}_0(R_x)$ of the Ricci operator $\mathrm{Ric}(R_x)$, respectively.

The Weyl tensor $W(R_x)$ is a conformal invariant of the Riemannian metric, that is, it is not changed when the metric is multiplied by a positive function. When $n = \dim M > 3$ its vanishing at all points is necessary and sufficient for the Riemannian manifold to be locally conformally Euclidean (the Weyl-Schouten theorem). When $n = 3$ the Weyl tensor is identically zero and the curvature tensor R_x is completely determined by the Ricci tensor $\mathrm{Ric}(R_x)$ (that is, the map ric is an isomorphism). In the two-dimensional case $\mathrm{Ric}_0(R_x) = W(R_x) = 0$ and the curvature tensor is specified by one number – the scalar curvature, which in the case of a surface reduces to the Gaussian curvature.

All the concepts defined above also make sense for pseudo-Riemannian manifolds. A (pseudo) Riemannian manifold (M, g) is called an *Einstein space* if its Ricci operator at each point is scalar:

$$\mathrm{Ric}(R_x) = \lambda \cdot \mathrm{id}. \qquad (20)$$

In this case, when $n > 2$, $\lambda = \dfrac{1}{n}\mathrm{Sc}(R_x)$ is automatically a constant, called the *cosmological* constant. Equation (20) can be regarded as a system of partial differential equations for the metric g. It is called the Einstein vacuum equation. According to the general theory of relativity the Lorentzian four-dimensional Einstein space describes the geometry of space-time in a domain where matter and fields other than the gravitational field are lacking. The content of the general ("non-vacuum") Einstein equation is that matter and physical (non-gravitational) fields uniquely determine the Ricci tensor (and thereby the first two components of the decomposition of the curvature tensor (19)). Thus, the solution of the Einstein equation reduces to the construction of metrics with a given Ricci tensor.

We note that when $n > 4$ the space of Weyl tensors $r = \mathrm{Ker}(\mathrm{ric})$ is irreducible with respect to the natural action of the orthogonal group. However, when $n = 4$ in the Riemannian case it splits into a sum of two 5-dimensional invariant subspaces. The corresponding components of the Weyl tensor $W(R)$ are called the *self-dual* part W^+ and the *anti-self-dual* part W^- of the Weyl tensor. (The name is determined by the choice of orientation, under a change of which the self-dual part becomes the anti-self-dual part and vice versa). An Einstein space with self-dual Weyl tensor $(W = W^+)$ is said to be *self-dual* or *left-flat*. Complete simply-connected self-dual spaces with positive scalar curvature are exhausted by the sphere S^4 and the complex projective plane $\mathbb{C}P^2$ with the standard metrics (Hitchin's theorem).

4.8. Sectional Curvature. Spaces of Constant Curvature.

Let R_x be the curvature tensor of a Riemannian manifold at the point $x \in M$, regarded as a symmetric operator in the space of bivectors $\Lambda^2 T_x M$. The metric g_x in the tangent space $T_x M$ determines in a natural way the metric $g_x^{(2)}$ in the space of bivectors:

$$g_x^{(2)}(\xi \wedge \eta, \xi \wedge \eta) = g_x(\xi, \eta)g_x(\xi, \eta) - g_x(\xi, \eta)^2, \qquad \xi, \eta \in T_x M.$$

The number

$$K(\sigma) = K(\xi, \eta) = \frac{g_x^{(2)}(R(\xi \wedge \eta), \xi \wedge \eta)}{g_x^{(2)}(\xi \wedge \eta, \xi \wedge \eta)} = \frac{g_x(R(\xi, \eta)\eta, \xi)}{g_x(\xi, \xi)g_x(\eta, \eta) - g_x(\xi, \eta)^2}$$

depends only on the two-dimensional subspace σ spanned by the vectors $\xi, \eta \in T_x M$. It is called the *sectional curvature* in the 2-dimensional direction σ.

This concept was introduced by Riemann. It has a simple geometrical meaning. If $n = \dim M = 2$, then $K(\sigma)$, where $\sigma = T_x M$, is the Gaussian curvature at the point $x \in M$. When $n > 2$, $K(\sigma)$ coincides with the Gaussian curvature of the 2-dimensional submanifold formed by geodesics starting from x and touching the subspace σ. We also note that the Ricci curvature in the direction of the unit vector ξ is equal to

$$\mathrm{ric}(R_x)(\xi, \xi) = \sum_{i=1}^{n-1} K(\xi, e_i), \tag{21}$$

where $e_1, \ldots, e_{n-1}, \xi$ is an orthonormal basis of $T_x M$. If the sectional curvature $K(\sigma)$ in any two-dimensional direction σ is equal to a constant k, we say that the Riemannian manifold is a *space of constant curvature* k. The curvature tensor of such a manifold, regarded as an operator in the space of bivectors $\Lambda^2 T_x M$, is a scalar operator: $R_x = k \cdot \mathrm{id}$. When $n > 2$ the converse is also true: if $R_x = k \cdot \mathrm{id}$ at every point, then $k = \mathrm{const}$ and the Riemannian manifold is a space of constant curvature (Schur's theorem).

Examples. 1) Euclidean space E^n is a space of constant zero curvature.

2) The sphere $S^n \subset E^{n+1}$ of radius r is a space of constant positive curvature $k = 1/r^2$.

3) The ball $L^n = \{x \in \mathbb{R}^n \|x\|^2 < r^2\}$ of radius r with conformal Euclidean metric

$$g = \sum dx_i^2 \bigg/ \left(1 - \frac{1}{r^2}x^2\right)^2$$

is a space of constant negative curvature $k = -1/r^2$. The Riemannian manifold (L^n, g) is a model of n-dimensional Lobachevskij geometry if its straight lines are defined as geodesics in L^n, and the motions as transformations that preserve the metric. This model is said to be *conformal*. Thus, the Riemannian point of view on geometry enables us to give a simple proof of non-contradiction of Lobachevskij geometry, about which we spoke in Ch. 1, 3.3.

It is remarkable that any complete simply-connected space of constant curvature is isometric to one of these spaces, and any simply-connected space of constant curvature is isometric to an open submanifold of one of these spaces. For non-simply-connected spaces this is not so; see Example 2 of 4.5. However, any complete space of constant curvature is obtained from the spaces E^n, S^n, L^n by factorization with respect to some freely acting discrete group of motions (in this connection, see 4.7). We also note that the spaces E^n, S^n L^n are maximally homogeneous: any isometry of tangent spaces $\varphi: T_xM \to T_yM$ can be uniquely extended to a motion of the space $M = E^n$, S^n, L^n that takes x into y.

4.9. The Holonomy Group and the de Rham Decomposition. Let M be a manifold with a linear connection V. A parallel displacement of vectors along a loop $\gamma(t)$ with beginning and end at the point $x = \gamma(0) = \gamma(1) \in M$ is a linear transformation of the tangent space. The set of such transformations forms a linear group Γ_x, called the *holonomy group of the linear connection at the point* x. Its connected component Γ_x^0 consists of parallel displacements along contractible loops and is called the *restricted holonomy group* at the point x. The holonomy groups of different points are isomorphic as linear groups. We can therefore simply talk about the *holonomy group* Γ. We note that a symmetric linear connection V on a manifold M is the Levi-Civita connection of the Riemannian metric if and only if its holonomy group Γ is contained in the orthogonal group $O(n)$. If Γ is irreducible, then the metric g is uniquely defined up to a constant factor as the only parallel field of symmetric tensors of type $(0, 2)$. Any (connected) linear group is the holonomy group of some linear connection. The position is essentially changed if we consider only symmetric linear connections. Henceforth we shall restrict ourselves to a consideration of Riemannian manifolds with Levi-Civita connection V. The holonomy group Γ_x of this connection is called the *holonomy group of the Riemannian manifold*. It is obviously a subgroup of the orthogonal group, but generally speaking it is not closed. Nevertheless, the restricted holonomy group Γ_x^0 is always closed and is a compact Lie group. A Riemannian manifold is said to be *irreducible* if its holonomy group is irreducible as a linear group.

Theorem (de Rham). *A complete simply-connected Riemannian manifold decomposes uniquely into the direct product of a Euclidean space E^k and complete simply-connected irreducible Riemannian manifolds.*

A local version of this theorem, which does not assume completeness, is also true: if the holonomy group of a Riemannian manifold is reducible, then some neighbourhood of any point of it decomposes into a direct product of a Euclidean ball and irreducible Riemannian manifolds.

Thus, knowledge of the holonomy group enables us to reduce the study of arbitrary Riemannian manifolds to irreducible ones. In this connection the following question is interesting: how do we calculate the holonomy group? A connected linear Lie group is uniquely determined by its Lie algebra (see Ch. 4, 2.3). Therefore, a description of the restricted holonomy group Γ_x^0 reduces to a description of its Lie algebra, which is called the *holonomy algebra* and denoted by $\mathring{\Gamma}_x$. According to a theorem of Ambrose and Singer (see Ch. 6, 3.4), the curvature operators $R_x(\xi, \eta)$, $\xi, \eta \in T_xM$, and their successive covariant derivatives $(\nabla_{\zeta_1} \dots \nabla_{\zeta_k} R(\xi, \eta)$, $\zeta_1, \dots, \zeta_k \in T_xM$, generate the linear Lie algebra H_x, the *infinitesimal holonomy algebra*, which is contained in the holonomy algebra $\mathring{\Gamma}_x$. For analytic Riemannian manifolds we have $\mathring{\Gamma}_x = H_x$. In the general case, for an open everywhere dense set of points $x \in M$ the infinitesimal holonomy algebra H_x is the holonomy algebra of some neighbourhood of the point x, regarded as a Riemannian manifold. For holonomy groups of homogeneous Riemannian manifolds see Ch. 4, 4.5.

4.10. The Berger Classification of Holonomy Groups. Kähler and Quaternion Manifolds. There naturally arises the question: which irreducible linear groups $\Gamma \subset O(n)$ are holonomy groups of irreducible Riemannian manifolds? We observe that the curvature tensor R_x at an arbitrary point x of a Riemannian manifold M can be regarded as a 2-form with values in the holonomy algebra $\mathring{\Gamma}_x$ satisfying the first Bianchi identity (18). Similarly the covariant differential ∇R of the curvature tensor at x can be regarded as an element of the space $\mathring{\Gamma}_x \otimes T_x^*M \otimes \Lambda^2 T_x^*M$, satisfying the *second Bianchi identity*

$$(\nabla_X R)(Y, Z) + (\nabla_Y R)(Z, X) + (\nabla_Z R)(X, Y) = 0.$$

Using the theory of representations, Berger found all linear Lie algebras $\mathring{\Gamma}_x$ for which there are non-zero tensors R and ∇R satisfying the conditions described. This leads to the following classification of holonomy groups.

Theorem. *Let M be a simply-connected irreducible Riemannian manifold with non-parallel curvature tensor $(\nabla R \neq 0)$, Then the holonomy group $\Gamma = \Gamma_x$ of this manifold is transitive on the unit sphere in T_xM and is contained in the following list:*

$$SO(n), U(m), SU(m), n = 2m; \quad Sp(1) \cdot Sp(m), Sp(m), n = 4m;$$

$$G_2, n = 7; \quad Spin(7), n = 8.$$

If $\mathrm{ric}(R) \neq 0$, then $\Gamma = SO(n)$, $U(m)$ or $Sp(1) \cdot Sp(m)$. Riemannian manifolds with parallel curvature tensor are said to be *locally symmetric*. Any complete simply-connected locally symmetric manifold is a *symmetric Riemannian space* (see Ch. 4, 4.4) and is essentially uniquely determined, up to transition to the dual space, by its holonomy group.

Riemannian manifolds with any holonomy group from Berger's list are known.

Knowledge of the holonomy group Γ reduces the problem of describing parallel tensor fields to the problem of describing tensors of the tangent space that are invariant under the natural action of the group $\Gamma = \Gamma_x$. In fact, any Γ_x-invariant tensor is carried by means of parallel transports to a parallel tensor field and any parallel field is obtained in this way.

Examples. 1) Let $\Gamma = SO(n)$. All parallel tensor fields are obtained from the metric tensor g (regarded as a map $g: TM \to T^*M$), the inverse tensor $g^{-1} = (g^{ij}): T^*M \to TM$ and the Riemannian volume n-forms $\omega_g = \sqrt{\det\|g_{ij}\|}\, dx^1 \wedge \cdots \wedge dx^n$ by means of the operations of tensor product, contraction and linear combinations.

2) $\Gamma = U(m)$ or $SU(m)$. The unitary group $U(m)$ consists of all orthogonal transformations of Euclidean space $\mathbb{R}^{2m}$ that commute with the operator of the complex structure J ($J^2 = -\mathrm{id}$). This operator, being $U(m)$-invariant, defines on M a parallel field of operators J_x, $J_x^2 = -\mathrm{id}$, a *parallel almost complex structure*. From the condition of parallelism it follows that this almost complex structure is integrable, that is, it defines in M the structure of a complex manifold (see Ch. 7, 5.1). In the situation under consideration all parallel tensor fields in the sense of Example 1 are generated by the tensors g, g^{-1} and J. In particular, on the manifold there is a parallel differential form of degree 2

$$\Omega(X, Y) = g(JX, Y), \qquad X, Y \in D(M).$$

It is called the *Kähler form*.

3) $\Gamma = Sp(m)$. The compact symplectic group $Sp(m)$ is defined as the group of orthogonal transformations of the space $\mathbb{R}^{4m}$ that commute with the three anticommuting complex structures J_α, $\alpha = 1, 2, 3$. The operators J_α define three parallel complex structures on M which together with the tensors g and g^{-1} generate all parallel tensor fields.

A Riemannian manifold M^n with holonomy group Γ is said to be *Kählerian* if $\Gamma \subset U(m)$, $n = 2m$, *hyper-Kählerian* if $\Gamma = Sp(m)$, $n = 4m$, and *quaternion* if $\Gamma \subset Sp(1) \cdot Sp(m)$, $n = 4m$. A hyper-Kählerian manifold is simultaneously Kählerian and quaternion.

A Kählerian manifold can be defined as a Riemannian manifold that has a parallel field of almost complex structures (or a parallel non-degenerate 2-form), and a hyper-Kählerian manifold can be defined as a Riemannian manifold with three anticommuting parallel almost complex structures. Similarly, a quaternion manifold can be characterized in terms of some parallel 4-form or as a Riemannian manifold that has a parallel field $x \mapsto A_x \subset \mathrm{End}\, T_x M$ of algebras of endomorphisms isomorphic to the algebra of quaternions.

Any complete simply-connected Kählerian manifold decomposes into the product of irreducible Kählerian manifolds. Any quaternion manifold M of dimension $4m > 4$ is an Einstein space, and its Ricci tensor is zero if and only if it is hyper-Kählerian. Otherwise M is irreducible.

Example. 1) The complex projective space $\mathbb{C}P^m$ with the standard metric is a Kählerian Einstein manifold (Ric $= \lambda \cdot$ id). Any complex submanifold of a Kählerian manifold (in particular $\mathbb{C}P^m$) is Kählerian.

2) The quaternion projective space $\mathbb{H}P^m$ with the standard metric is a quaternion manifold.

3) Any orientable four-dimensional Riemannian manifold is quaternion, since $SO(4) = Sp(1) \cdot Sp(1)$.

§ 5. The Geometry of Symbols

The history of the discovery of the main concepts of Riemannian geometry is essentially the process of transferring by degrees to the language of differential calculus the regularities to which our intuitive ideas about the surrounding world are subject. For example, the very concept of a Riemannian metric arose as a result of interpreting the infinitesimal nature of the process of measuring lengths. The idea of a straight line as a shortest curve led to the theory of geodesics, understood as extremals of the length functional. The geometrical interpretation of the formal apparatus of covariant differentiation made it possible to bring out the idea of parallel displacement and its mathematical formulation – the theory of connections and so on.

In this section, turning from the historical path, and following the point of view of differential geometry expressed in Ch. 1, we explain by a "picture" which aspect of differential calculus is Riemannian geometry. It turns out that Riemannian metrics are naturally interpreted as symbols of differential operators, covariant differentiation as the operator of displacement of singularities of solutions of the corresponding differential equations, geodesics as trajectories of motion of these singularities, and so on.

The point of view on Riemannian geometry that results from the theory of differential operators is completely analogous to the interpretation of classical mechanics as the limiting case of quantum mechanics, which is based on Bohr's "correspondence principle", and is realized mathematically as the so-called procedure of quasiclassical approximation.

5.1. Differential Operators in Bundles. Let $\alpha: E_\alpha \to M$ and $\beta: E_\beta \to M$ be smooth vector bundles. The spaces $\Gamma(\alpha)$ and $\Gamma(\beta)$ of smooth sections of them are $C^\infty(M)$-modules. The operator of multiplication by a function f that acts in $\Gamma(\alpha)$ or $\Gamma(\beta)$ will also be denoted by f. Following the general algebraic approach mentioned in Ch. 1, § 4, we define a linear *differential operator* of order at most k, acting from sections of α to sections of β, as a linear map $\Delta: \Gamma(\alpha) \to \Gamma(\beta)$ satisfying the condition

$$[f_0, [f_1, \ldots [f_k, \Delta] \ldots] = 0, \qquad \forall f_0, \ldots, f_k \in C^\infty(M).$$

In a local coordinate system $x = (x_1, \ldots, x_n)$ over which the bundles α and β are trivialized simultaneously (see §2.4) their sections can be described by columns of vector-valued functions of x. We can show that the operator Δ is described in the usual way, that is, as a matrix whose elements are coordinate expressions for scalar differential operators.

The operation $\Delta \to f\Delta$, $f \in C^\infty(M)$, converts the totality of all linear differential operators of order at most k, acting from $\Gamma(\alpha)$ to $\Gamma(\beta)$, into a $C^\infty(M)$-module denoted by $\mathrm{Diff}_k(\alpha, \beta)$. Obviously $\mathrm{Diff}_l(\alpha, \beta) \subset \mathrm{Diff}_k(\alpha, \beta)$ if $l < k$, and $\mathrm{Diff}_0(\alpha, \beta)$ coincides with the module $\mathrm{Hom}(\alpha, \beta)$ of all fibrewise linear maps of α into β that cover the identity map $M \to M$ (see 2.4).

5.2. Symbols of Differential Operators. Informally speaking, the *symbol of an operator* $\Delta \in \mathrm{Diff}_k(\alpha, \beta)$ is its leading part with respect to derivatives. More precisely, let

$$\mathrm{Smbl}_k(\alpha, \beta) = \mathrm{Diff}_k(\alpha, \beta)/\mathrm{Diff}_{k-1}(\alpha, \beta)$$

and let $\sigma \colon \mathrm{Diff}_k(\alpha, \beta) \to \mathrm{Smbl}_k(\alpha, \beta)$ be the natural projection. Then $\sigma(\Delta)$ is the symbol of the operator Δ.

Let $\mathbf{1} \colon M \times \mathbb{R} \to M$ be the trivial bundle $\Gamma(\mathbf{1}) = C^\infty(M)$. Then $\mathrm{Diff}_k(\mathbf{1}, \mathbf{1})$ is the module of scalar differential operators of order at most k on M. Writing the operator $\Delta \in \mathrm{Diff}_k(\mathbf{1}, \mathbf{1})$ in local coordinates (see Ch. 1, §4)

$$\Delta = \sum a_\sigma(x) \frac{\partial^{|\sigma|}}{\partial x_\sigma},$$

it is not difficult to verify that its coefficients a_σ, $|\sigma| = k$, are transformed like the components of a symmetric contravariant tensor of order k. Thus, $\sigma(\Delta)$ is a tensor field on M, written in local coordinates in the form

$$\sigma(\Delta) = \sum a_{i_1 \ldots i_k} \frac{\partial}{\partial x_{i_1}} \circ \cdots \circ \frac{\partial}{\partial x_{i_k}}$$

(here the "small circle" denotes symmetric tensor multiplication). In coordinate-free form, the tensor field $\sigma(\Delta)$, understood as a multilinear function on 1-forms (see 3.3) is defined by

$$\sigma(\Delta)(df_1, \ldots, df_k) = [\ldots [\Delta, f_k] \ldots], f_2], f_1]. \tag{22}$$

The fact that it is symmetric is obvious from the identity $[f, [g, \Delta]] = [g, [f, \Delta]]$. If $\Delta \in \mathrm{Diff}_k(\alpha, \beta)$, then (22) defines the symbol $\sigma(\Delta)$ as a symmetric covariant $\mathrm{Hom}(\alpha, \beta)$-valued tensor field on M.

Example. If $\Delta = d \colon \Lambda^i(M) \to \Lambda^{i+1}(M)$ is the operator of outer differentiation, then the value $\sigma(d)(\lambda)$ of its symbol on $\lambda \in \Lambda^1(M)$ is a homomorphism of the module $\Lambda^i(M) (= \Gamma(\alpha))$ into the module

$$\Lambda^{i+1}(M) (= \Gamma(\beta)) \colon \Lambda^i(M) \ni \omega \to \lambda \wedge \omega \in \Lambda^{i+1}(M).$$

In view of what we said above, the symbols of scalar differential operators of the second order are fields of symmetric contravariant tensors of the second order, that is, fields of bilinear forms on the cotangent spaces. If these forms are non-degenerate, then they naturally define a field of bilinear forms on the tangent spaces, that is, a pseudo-Riemannian metric. If $\sigma(\varDelta) = \sum a_{ij} \dfrac{\partial}{\partial x_i} \circ \dfrac{\partial}{\partial x_j}$, $\varDelta \in$ Diff$_2(1, 1)$, then the metric $g = g_\varDelta$ that arises in this way has the form $g_{ij} \, dx_i \, dx_j$, where $\|g_{ij}\| = \|a_{kl}\|^{-1}$. Hence it follows that $\varDelta$ being elliptic is equivalent to $g_\varDelta$ being a Riemannian metric, and $\varDelta$ being hyperbolic is equivalent to the metric $g_\varDelta$ being Lorentzian. Here it is convenient to draw attention to the fact that the most important objects of differential geometry – Riemannian and Lorentzian structures – correspond to the most important equations of mathematical physics.

5.3. Connections and Quantization. Obviously, to one and the same symbol there correspond many differential operators. However, it is remarkable that in the presence of a connection this correspondence can be made one-to-one. For simplicity we describe this effect for scalar operators of the second order. Let Smbl$_2(1, 1) \overset{q}{\to}$ Diff$_2(1, 1)$ be a homomorphism of $C^\infty(M)$-modules that restores an operator from its symbol, that is, $\sigma \circ q = $ id. If $X, Y \in D(M)$, then $q(XY) = X \circ Y$ (symmetric product). Therefore $q(X \circ Y)$ can always be represented in the form

$$q(X \circ Y) = XY - \square_X Y - h(X, Y), \qquad (23)$$

where $\square_X Y \in D(M)$, $h(X, Y) \in C^\infty(M)$. Since q is a homomorphism it follows that the hieroglyphic $\square_X Y$ has the following properties:

$$\square_{fX} Y = f \square_X Y, \qquad \square_X(fY) = X(f)Y + f \square_X Y.$$

Hence (see §4.3) $X \mapsto \square_X$ is a linear connection. Since $q(X \circ Y) = q(Y \circ X)$ it follows that this connection is symmetric. Moreover, it is not difficult to verify that h is a symmetric tensor field of type (2, 0).

Conversely, an arbitrary symmetric connection $\square$ and a form h define by (23) the restoring homomorphism q.

Example. Let g be a pseudo-Riemannian metric on M and g^* the corresponding contravariant symmetric vector field. We define the restoring homomorphism q by means of the Levi-Civita connection of the metric g, assuming that $h = 0$. Then $q(g^*)$ is the *Beltrami-Laplace operator* of the corresponding metric g.

This procedure of restoring an operator from its symbol is completely analogous to the basic procedure of quantum mechanics – quantization. This analogy is connected with the fact that the propagation of singularities of solutions of differential equations is described in the language of symbols. The theory presented above corresponds to the so-called "geometrical singularities". Consideration of singularities of a different type requires a corresponding change in the concept of a symbol. By a suitable choice of this type we can arrive at the situation considered in quantum mechanics.

5.4. Poisson Brackets and Hamiltonian Formalism. Let Smbl $M = \sum_{j \geq 0} \text{Smbl}_i(1, 1)$. The operation of composition of differential operators induces in the space Smbl M the structure of a commutative algebra: $\sigma(\Delta_1) \cdot \sigma(\Delta_2) = \sigma(\Delta_1 \Delta_2)$. The fact that the multiplication thus defined is commutative is equivalent to the fact that if $\Delta_1 \in \text{Diff}_k(1, 1)$, $\Delta_2 \in \text{Diff}_l(1, 1)$, then the operator $[\Delta_1, \Delta_2]$ has order $k + l - 1$. Its symbol is called the *Poisson bracket* $\{H_1, H_2\}$ of the symbols $H_1 = \sigma(\Delta_1)$ and $H_2 = \sigma(\Delta_2)$.

From this definition there immediately follow the properties of the Poisson bracket:

1) $\{H_1, H_2\} = -\{H_2, H_1\}$;
2) $\{\sum \lambda_i H_i, H\} = \sum \lambda_i \{H_i, H\}$, $\lambda_i \in \mathbb{R}$;
3) $\{H_1 H_2, H\} = H_1 \{H_2, H\} + H_2 \{H_1, H\}$;
4) $\{H_1, \{H_2, H_3\}\} + \{H_2, \{H_3, H_1\}\} + \{H_3, \{H_1, H_2\}\} = 0$.

We now interpret these algebraic constructions geometrically, regarding elements of the algebra Smbl M as functions on its real spectrum $\text{Spect}^{\mathbb{R}}(\text{Smbl } M)$ (see Ch. 1, §2). First of all, we mention the following remarkable fact:

$$T^*M = \text{Spect}^{\mathbb{R}}(\text{Smbl } M).$$

Thus, any symbol can be thought of as a function on T^*M. Moreover, by property 3), the operator X_H: Smbl $M \to$ Smbl M for $H \in$ Smbl M defined by $X_H(H_1) = \{H, H_1\}$ is a differentiation of Smbl M, and therefore a vector field on T^*M. In physics and mechanics functions on T^*M are usually called *Hamiltonians*, and the fields X_H are said to be *Hamiltonian*.

The Poisson bracket defines a bivector field Γ on T^*M:

$$\Gamma(dH_1, dH_2) = \{H_1, H_2\}.$$

The field Γ can be interpreted as a map $\Gamma: \Lambda^1(T^*M) \to D(T^*M)$,

$$\Gamma(\Sigma G_i dH_i) = \Sigma G_i X_{H_i}, \qquad G_i, H_i \in C^\infty(T^*M).$$

The map Γ is an isomorphism and therefore enables us to define a differential form Ω on T^*M:

$$\Omega(X, Y) = (\Gamma^{-1}(X))(Y).$$

Property 4) of the Poisson bracket is equivalent to the form Ω being closed. The form Ω defines a *symplectic structure* on T^*M.

Suppose that local coordinates $x_1, \ldots, x_n$ are defined in a domain $U \subset M$. Then in the domain $(\tau^*)^{-1}(U) \subset T^*M$ there arise coordinates $(x_1, \ldots, x_n, p_1, \ldots, p_n)$: if $\rho \in T_a^*M \subset T^*M$, then x_i are the coordinates of the point of contact a, and p_i are the components of the expansion of the covector ρ with respect to the basis $d_a x_1, \ldots, d_a x_n$ in T_a^*M. In these coordinates a symbol, as a function on T^*M, is written in the form

$$H(p, q) = \sigma(\Delta) = \sum a_{i_1 \ldots i_k}(x) p_{i_1} \cdots p_{i_k}.$$

We have

$$\{H_1, H_2\} = \sum_{i=1}^{n} \left(\frac{\partial H_1}{\partial p_i} \frac{\partial H_2}{\partial x_i} - \frac{\partial H_1}{\partial x_i} \frac{\partial H_2}{\partial p_i} \right),$$

$$X_H = \sum_{i=1}^{n} \left(\frac{\partial H}{\partial p_i} \frac{\partial}{\partial x_i} - \frac{\partial H}{\partial x_i} \frac{\partial}{\partial p_i} \right),$$

$$\Gamma = \sum_{i=1}^{n} \frac{\partial}{\partial p_i} \wedge \frac{\partial}{\partial x_i}, \qquad \Omega = \sum_{i=1}^{n} dp_i \wedge dx_i.$$

5.5. Poissonian and Symplectic Structures. Axiomatization of the constructions of the last subsection leads to two important geometrical structures, *Poissonian* and *symplectic*, which are intensively exploited in mathematical physics.

A manifold Φ, in whose algebra of functions there is defined an operation $\{f, g\} \in C^{\infty}(\Phi)$, $f, g \in C^{\infty}(\Phi)$, having the properties 1)–4) given in the previous subsection, is said to be *Poissonian*. A manifold endowed with a closed non-degenerate differential form Ω of degree 2 is said to be *symplectic*, and the form Ω is a *symplectic structure* on it.

A function f on a symplectic manifold determines a *Hamiltonian vector field* X_f:

$$X_f \lrcorner \Omega = -df.$$

A symplectic manifold is automatically Poissonian with respect to the bracket

$$\{f, g\} = \Omega(X_f, X_g\}.$$

Conversely, the mechanism described in the previous subsection enables us to associate a symplectic structure with any *non-degenerate* Poissonian structure. Non-degeneracy means that if $\{H, G\} = 0$ for all G, then $H = \text{const}$.

A Hamiltonian field X_f on a Poissonian manifold Φ, as above, is defined by the condition $X_f(g) = \{f, g\}, f, g \in C^{\infty}(\Phi)$. Functions $f \in C^{\infty}(\Phi)$ for which $X_f = 0$ form a subalgebra $\Pi \subset C^{\infty}(\Phi)$, called the *Poissonian centre*. Any Poissonian manifold is naturally fibered (*Poissonian fibration*) by submanifolds (*Poissonian fibres*), which are uniquely characterized by the fact that on them functions $f \in \Pi$ are constants. Generally speaking, Poissonian fibres can have different dimension. Examples of Poissonian fibrations are given in the next subsection.

5.6. Left-Invariant Hamiltonian Formalism on Lie Groups. An interesting class of Poissonian manifolds is associated with Lie groups (for the necessary information on Lie groups and algebras see Ch. 4, §2).

Let G be a Lie group. A diffeomorphism $L_g: G \to G$, $g \in G$, acting according to the rule $h \to gh$, is called a *left translation*. The action of the group G on scalar differential operators Δ defined on it is determined by the rule

$$\Delta \mapsto L_g(\Delta) = L_g^* \circ \Delta \circ L_{g^{-1}}^*.$$

The operator Δ is said to be *left-invariant* if $L_g(\Delta) = \Delta$ for all $g \in G$.

The symbols of left-invariant operators form a subalgebra $S = S(G)$ of the algebra of symbols Smbl G which is closed with respect to the Poisson bracket. Geometrization of the algebra S (see Ch. 1, §2) gives the following result:

$$\text{Specm}^{\mathbb{R}} S = T_e^* G,$$

where e is the identity of the group G. Since symbols $f \in S$ are interpreted as functions on the manifold $T_e^* G$, the Poisson bracket of them turns the latter into a Poissonian manifold.

This construction admits a purely algebraic description. Namely, let $\mathfrak{G}$ be the Lie algebra of the group G. Its *enveloping algebra* $U(\mathfrak{G})$ is defined as the quotient algebra of the tensor algebra $T(\mathfrak{G})$ (see 3.4) with respect to the ideal generated by elements of the form $X \otimes Y - Y \otimes X - [X, Y]$ where $X, Y \in \mathfrak{G}$, and $[\cdot, \cdot]$ is the Lie bracket in $\mathfrak{G}$. The algebra $U(\mathfrak{G})$ is filtered by subspaces $U_p(\mathfrak{G})$, which are the images of subspaces $T_p^0(\mathfrak{G}) \subset T(\mathfrak{G})$ under the projection $T(\mathfrak{G}) \to U(\mathfrak{G})$. Obviously, $U_l(\mathfrak{G}) \subset U_p(\mathfrak{G})$ if $l < p$. Finally, we consider the graded algebra

$$S(\mathfrak{G}) = \sum_{p \geqslant 0} S_p(\mathfrak{G}), \qquad S_p(\mathfrak{G}) = U_p(\mathfrak{G})/U_{p-1}(\mathfrak{G}),$$

which corresponds to the filtered algebra $U(\mathfrak{G})$. The algebra $S(\mathfrak{G})$ is commutative. In addition, the operation of commutation of elements of $U(\mathfrak{G})$ generates Poisson brackets in $S(\mathfrak{G})$ according to the scheme described in 5.4. The geometrization of $S(\mathfrak{G})$ is

$$\text{Specm}^{\mathbb{R}} S(\mathfrak{G}) = \mathfrak{G}^*.$$

In view of this, the space $\mathfrak{G}^*$ dual to the Lie algebra $\mathfrak{G}$ is turned into a Poissonian manifold. Its Poissonian fibration consisits of orbits of the coadjoint representation of the group G on $\mathfrak{G}^*$.

Example 1. The space $\mathfrak{G}^*$ for the group $SO(3)$ can be identified with the standard Euclidean space $\mathbb{R}^3$, and the coadjoint action of the group $SO(3)$ can be identified with the standard action of $SO(3)$ in $\mathbb{R}^2$. Therefore the orbits of this action (they are Poissonian sheets) are standard spheres in $\mathbb{R}^3$. In the standard coordinates x_1, x_2, x_3 in $\mathbb{R}^3 = \mathfrak{G}^*$ the Poisson bracket has the form

$$\{H, G\} = x_1 \frac{\partial H}{\partial x_2} \cdot \frac{\partial G}{\partial x_3} + \langle\langle \text{c. p}\rangle\rangle - x_1 \frac{\partial G}{\partial x_2} \cdot \frac{\partial H}{\partial x_3} - \langle\langle \text{c. p.}\rangle\rangle,$$

where $H, G \in C^\infty(\mathbb{R}^3)$ and c. p. means "cyclic permutation".

Example 2. The Lie algebra $\mathfrak{G}$ of the Lie group of upper triangular matrices of order 3 with ones on the diagonal is three-dimensional and has a basis e_1, e_2, e_3 in which

$$[e_1, e_2] = e_3, [e_1, e_3] = [e_2, e_3] = 0.$$

In the dual coordinates x_1, x_2, x_3 in $\mathfrak{G}^*$ the Poisson bracket has the form

$$\{H, G\} = x_3 \left(\frac{\partial H}{\partial x_1} \cdot \frac{\partial G}{\partial x_2} - \frac{\partial H}{\partial x_2} \cdot \frac{\partial G}{\partial x_1} \right).$$

In this case the Poissonian fibres are either points in the plane $x_3 = 0$ or planes $x_3 = \text{const} \neq 0$.

Poissonian structures on $\mathfrak{G}^*$ are used to describe mechanical systems whose configuation spaces are Lie groups. For example, for a free rigid body this is the group of motions of the space E^3. They are also widely used in the theory of representations of Lie groups ("the method of orbits") and in the theory of "geometrical quantization".

Chapter 4
The Group Approach of Lie and Klein. The Geometry of Transformation Groups

"... the importance of group theory is not diminished
by the modern development of differential geometry,
on the contrary, this is the only theory that is in a
position to unite the different branches of the latter".
                                        E. Cartan

§ 1. Symmetries in Geometry

The simplicity of an object under investigation is in many cases equivalent to its symmetry. The second of these concepts, in contrast to the first, admits a precise mathematical realization and leads to a consistent and fruitful view of geometry if the homogeneity of space is recognized as an integral part of it. This point of view was first stated by Felix Klein in 1872 in his famous "Erlangen programme". Klein's geometry is a very special but nevertheless very important particular case of Riemann's geometry, in the framework of which significant results are obtained on the basis of a distinctive synthesis of the theorems of group theory and linear algebra. In this chapter we present the philosophy and technology of this approach.

1.1. Symmetries and Groups. The word symmetry in the most general context serves to denote a transformation that superposes an object under consideration on itself. Symmetries of specific mathematical structures usually have special titles. For example, as a rule a symmetry of an algebraic structure is called an automorphism of it. The words homeomorphism and diffeomorphism mean a symmetry (or equivalence) of topological spaces and smooth manifolds, respectively. The symmetries of a given object obviously form a group. The more symmetric an object is, that is, the larger its symmetry group, the simpler its structure is and the more information about it can be deduced from "symmetry arguments". The proof of the theorem of Pythagoras given in Ch. 1, 3.1, is a classic specimen of an argument of this kind, in which similarity transformations occur as symmetry transformations.

In many cases symmetry arguments give the possibility of obtaining necessary results in practice in the complete absence of a deep conceptual understanding of the problem. For this reason the first steps of many theories of the natural sciences consist in the investigation of the relevant symmetric models. Thus, for example, the table of Mendeleev precedes the modern quantum-mechanical theory of the structure of atoms, and the special theory of relativity precedes the general. Apparently this also explains the prominent role played by group theory in modern theories of elementary particles. As for (metrical) geometry, the "supersymmetric" models of Euclid and Lobachevskij and also of spherical geometry precede the conceptual geometry of Riemann.

The study of symmetric models enables us to discover new phenomena and facts, thanks to the possibility of replacing complicated technical tools by simple symmetry arguments. In addition, they are a suitable object for verifying conjectures of general character.

The concept of a group, both genetically and in essence, is an abstract substance that mathematics has extracted from numerous concrete ensembles of symmetries that occur in nature, and above all in geometry. Its interaction with differential calculus has led to rise of the following two fundamental concepts of modern geometry, and in mathematics generally: the concept of a *Lie group*, which is a synthesis of the concepts of a group and a smooth manifold (see 2.1 below), and the concept of an infinitesimal transformation or vector field (see Ch. 1, §4), which enables us to introduce the "group" of infinitesimal symmetry transformations. Formed in a proper way, this line of thought leads to *Lie algebras*, whose fundamental link with Lie groups makes it possible, in a well-known sense, to reduce the theory of the latter to linear algebra. For these important discoveries mathematics is indebted to Sophus Lie.

1.2. Symmetry and Integrability. The simplicity of an algebraic or differential equation means the possibility of solving it in some concrete form. Comparing this with the general remarks made at the beginning of this section, we can arrive at the idea that a proper degree of symmetry implies solubility or integrability, as we usually say when talking about differential equations.

By itself, what we have said appears trivial. We can satisfy ourselves that this is not so if we try, for example, to answer the classical problem of when an algebraic equation

$$a_0 x^n + a_1 x^{n-1} + \cdots + a_{n-1} x + a_n = 0 \tag{1}$$

is soluble in radicals. The answer to this, found by Evariste Galois, is that this is so only when the symmetry group of the equation (called the *Galois group*) has a certain property, called *solubility*. The difficulty we encounter here (and in other similar cases) is connected with the fact that (1) is only an "identity card", and not the concept of an algebraic equation itself. Therefore it is unclear what, properly speaking, we should subject to transformations, in other words, what is an algebraic equation? Only by understanding, following Galois, that in the "identity" (1) there is concealed a certain extension of the algebraic field k generated by the coefficients $a_0, \ldots, a_n$ (see volume 11 of the present series) do we land on the regular path.

Galois theory introduced abstract groups into mathematics. Taking as his aim to carry over the ideas of Galois to differential equations, Lie was forced to construct his theory of "continuous groups". Continuous groups, whose elements are defined by finitely many parameters, are today called Lie groups (see 2.1 below) and play a fundamental role in the realization of Klein's programme, about which we shall speak below. Sophus Lie, as we should expect on the basis of what we said above, was also unable to escape the question of what a differential equation is. Having discovered contact geometry (see Ch. 5, §2) he was able to answer it in the case of equations of the first order in one unknown function. In the general situation the answer to it has been obtained recently (see Ch. 5, §7), but this topic cannot be regarded as exhausted.

1.3. Klein's Erlangen Programme. At the foundation of any geometrical system Klein placed a certain group G, which acts in the space in which it is deployed. The requirement of homogeneity, of which we spoke at the beginning of this section, implies that any two points of this space can be taken into one another by a transformation from the group G. The problem of geometry according to Klein is the study of the invariants of the group G, that is, quantities that do not change under the transformations of this group. According to this point of view, the primary geometrical substance is not a field quantity, as with Riemann, but an a priori given group, which a posteriori can be interpreted as the symmetry group of the resulting geometry.

The following examples, which illustrate this point of view, served Klein as the original material for his general constructions.

Example 1. Euclidean geometry. As the basic group G we consider the group of matrices of order $n + 1$ having the form $\begin{pmatrix} A & \xi \\ 0 & 1 \end{pmatrix}$, where A is an orthogonal $(n \times n)$-matrix and ξ is a column vector of length n, and for the space M on which it acts we take the space G/H of right cosets with respect to the subgroup H consisting of matrices of the form $\begin{pmatrix} A & 0 \\ 0 & 1 \end{pmatrix}$. It is not difficult to see that in each such coset there is exactly one matrix of the form $\begin{pmatrix} E & \xi \\ 0 & 1 \end{pmatrix}$, where E is the unit matrix. Therefore the space G/H can be identified with the n-dimensional space of the columns ξ. A natural transformation of the space $M = G/H$ under the action of an element $g = \begin{pmatrix} A & \eta \\ 0 & 1 \end{pmatrix}$ with the given identification has the form

$$\xi \mapsto A\xi + \eta. \tag{2}$$

There is a unique (up to proportionality) Riemannian metric on M that is invariant with respect to this action. It converts M into "ordinary" Euclidean space of n dimensions whose group of motions (that is, symmetries) is identified with the group G. The Levi-Civita connection with respect to this metric is invariant with respect to G and generates an "ordinary" Euclidean parallel transport.

Example 2. Affine geometry. In this case the group G consists of matrices of the form $\begin{pmatrix} A & \xi \\ 0 & 1 \end{pmatrix}$, where det $A \neq 0$. Suppose, as above, that $H = \left\{ \begin{pmatrix} A & 0 \\ 0 & 1 \end{pmatrix} \right\}$ and $M = G/H$. For the same reasons as above, M can be identified with the space of columns ξ on which the action of G is given by (2). On M there is no invariant Riemannian metric, but there is an invariant torsion-free connection that reduces to an ordinary parallel transport.

Example 3. Projective geometry. As the basic group G we consider the factor group of the group of all non-singular matrices by scalar matrices. Suppose also that $M = G/H$, where the subgroup H is formed by matrices of the form $\begin{pmatrix} \alpha & \beta \\ 0 & 1 \end{pmatrix}$. M can be identified with the manifold of all straight lines in $\mathbb{R}^{n+1}$ passing through the origin. In the situation under consideration, on M there is no Riemannian metric nor a torsion-free connection invariant with respect to the natural action of the group G. Nevertheless, on M there is an invariant *projective connection*, that is, a class of connections that have common (non-parametrized) geodesics.

Example 4. Conformal geometry. Let M be the manifold of generators of the cone $\{x_1^2 + \cdots + x_n^2 - x_{n+1}^2\}$ in $\mathbb{R}^{n+1}$, and let G be the group of linear automorphisms of it, factorized by scalar matrices. Cutting this cone by the plane $x_{n+1} = 1$, we can identify M with the standard n-dimensional sphere S^n. The transformations of G take the standard metric g on S^n into a metric of the form φg, where $\varphi \in C^\infty(S^n)$. In other words, we have defined a *conformal metric* on S^n, that is, a class of metrics of the form φg, invariant with respect to G.

In 2.2 it will become clear why for the spaces M in which the geometries described above are deployed there occur spaces of the form G/H. These and other examples of geometries according to Klein have motivated the examination of the corresponding field theories.

The link between the group approach of Klein and the field approach of Riemann consists in the following. With each geometric field quantity h (for example, a tensor field) on M there is associated the group G of its symmetries (automorphisms). The geometry of the field h is "part" of the geometry of the transformation group G. Conversely, with each transformation group G we can associate distinct field quantities invariant with respect to it. The geometry of these quantities is a specialization of the geometry of G. Of course, this link is meaningful only in the case when the transformation group G is sufficiently large, for example, transitive on M.

§2. Homogeneous Spaces

In this section we describe the basic concepts associated with homogeneous spaces and invariant structures on them. Homogeneous spaces are those objects on which Kleinian geometries are constructed.

2.1. Lie Groups. A manifold G endowed with a compatible group structure is called a *Lie group*. "Compatibility" means that the maps $G \times G \to G$ and $G \to G$ that represent the group operations of multiplication $(g_1, g_2) \to g_1 g_2$ and inversion $g \mapsto g^{-1}$, respectively, are smooth.

Lie groups arise in a natural way as symmetry groups of certain geometrical objects. These are, for example, the symmetry groups of a Riemannian metric or a linear connection. However, generally speaking, the symmetries of a geometrical structure may depend on infinitely many parameters and so do not form a Lie group. For example, the symmetry group of the vector field $\partial/\partial x$ on the (x, y)-plane consists of transformations of the form $(x, y) \to (x + f(y), g(y))$, where $f(y)$ can be an arbitrary smooth function.

Similar situations are considered in Chapter 7.

Example 1. One Lie group is the group of all non-singular linear transformations of an n-dimensional linear space over the field of real or complex numbers. It is called the *full linear group*, denoted by $GL(n, \mathbb{R})$ or $GL(n, \mathbb{C})$, respectively, and as a manifold it is a domain in $\mathbb{R}^{n^2}$ or $\mathbb{C}^{n^2} \approx \mathbb{R}^{2n^2}$.

Example 2. A closed (as a topological subspace) subgroup of a Lie group is also a Lie group. For example, the following classical groups are closed subgroups of the full linear group $GL(n, \mathbb{F})$, where $\mathbb{F} = \mathbb{R}$ or $\mathbb{C}$:

1) $SL(n, \mathbb{F})$, the group of *unimodular* (or volume-preserving) transformations;
2) $O(n, \mathbb{F})$, the group of *orthogonal* transformations;
3) $SO(n, \mathbb{F}) = O(n, \mathbb{F}) \cap SL(n, \mathbb{F})$;
4) $Sp(n, \mathbb{F})$, the group of transformations that preserve a certain nondegenerate skew-symmetric bilinear form on $\mathbb{F}^n$;
5) $U(n)$, the group of transformations that preserve the Hermitian scalar product in $\mathbb{C}^n$.

All these groups are *simple* (except for $O(4, \mathbb{F})$ and $SO(4, \mathbb{F})$), that is, all their normal subgroup are discrete. These are the *classical simple Lie groups*.

The factor group of a Lie group G with respect to a closed normal subgroup is also a Lie group, as well as its universal covering space $\tilde{G}$. Lie groups that have isomorphic universal converings are said to be *locally isomorphic*.

Simple Lie groups admit a full classification (Killing, Cartan). For example, simple complex Lie groups are exhausted (up to a local isomorphism) by the groups $SL(n, \mathbb{C})$, $SO(n, \mathbb{C})$, $Sp(n, \mathbb{C})$ and five *exceptional* groups, denoted by G_2, F_4, E_6, E_7, E_8.

Any Lie group is locally isomorphic to a closed subgroup of the full linear group, but in the large this may not be so. Consider, for example, the universal covering of the group $SL(2, \mathbb{R})$.

2.2. The Action of the Lie Group on a Manifold. If with any element g of a group G we associate a diffeomorphism D_g of a manifold M in such a way that

$$D_{g_1 g_2} = D_{g_1} \cdot D_{g_2}, \qquad D_{g^{-1}} = D_g^{-1},$$

then we say that the left *action* of this group is defined on M.

If G is a Lie group and the map $G \times M \to M$, $(g, x) \to D_g x$, $g \in G$, $x \in M$, is smooth, then the action is said to be *smooth*, and M is called a G-manifold. If M is a vector space, and D_g, $g \in G$, are linear transformations of M, then we say that D is a *linear action* or a *linear representation* of G. The action D is said to be *transitive* if any two points x, $y \in M$ are taken into one another by a transformation of the form D_g. A *homogeneous space* of a Lie group G is a manifold M with a smooth transitive action D of the Lie group G.

Example. 1) A Lie group G can be regarded as a homogeneous space with respect to the left action l_g: $x \mapsto gx$, g, $x \in G$, of G. In addition, G can be regarded as a homogeneous space with respect to the action r_g: $x \mapsto xg^{-1}$ of G on the right, and with respect to the action $p_{(g_1, g_2)}$: $x \mapsto g_1 x g_2^{-1}$ of the group $G \times G$.

2) The linear-fractional action D of the group $GL(2, \mathbb{R})$ on the projective line $\mathbb{R}P^1$ is given by the formula

$$D_g x = \frac{ax + b}{cx + d}, \qquad g = \begin{pmatrix} a & b \\ c & d \end{pmatrix} \in GL(2, \mathbb{R}).$$

3) In the space G/H of right cosets there is canonically defined the structure of a smooth manifold, and the action l_g: $g'H \to gg'H$ takes G/H into a homogeneous space.

It is remarkable that the construction of Example 3 essentially exhausts all homogeneous spaces of the group G. For if M is such a space and $H = \{h \in G | D_h x = x\}$ is the *stabilizer* of a fixed point $x \in M$, then the map $G/H \ni gH \to gx \in M$ defines an isomorphism of the homogeneous spaces G/H and M (that is, diffeomorphism of manifolds compatible with the action of G). The action D of the group G on the manifold M is said to be *effective* (*locally effective*) if the kernel of ineffectiveness $K = \{g \in G : D_g = \text{id}\}$ is trivial (discrete). Factorization of G with respect to the kernel of ineffectiveness enables us to restrict ourselves to the consideration of effective actions.

2.3. Correspondence Between Lie Groups and Lie Algebras. The study of homogeneous spaces can be reduced to linear algebra if we use the fundamental link between Lie groups and Lie algebras. The nature of this link can be understood by considering the effective action D of a Lie group on a G-manifold M. In this case the group G can be identified with the group D_G of transformations of M. A curve $g(t)$ in the group G such that $g(0)$ is the identity of this group determines on M a vector velocity field

$$D_{\dot{g}} = \frac{d}{dt} D_{g(t)}|_{t=0},$$

depending on $\dot{g} = \dot{g}(0)$. The set $\mathfrak{G}(G)$ of such fields forms a *Lie algebra*, that is, it is closed with respect to linear operations and the operation of commutation. It is called the *tangent Lie algebra* of the Lie group G. The shift operators along trajectories of the field $D_{\dot{g}}$ have the form $D_{g(t)}$, where $\{g(t)\}$ is a one-parameter subgroup of G. It generates the whole group G (if the latter is connected) as the group of transformations of M. Thus, to the Lie group G of transformations there corresponds the tangent Lie algebra $\mathfrak{G}(G)$ of vector fields, and G can be restored from $\mathfrak{G}(G)$. The subalgebras (ideals) of $\mathfrak{G}(G)$ correspond to connected (normal) subgroups of G. The space $\mathfrak{G}(G)$ of vector fields is invariant with respect to the action of G induced on it. Thus we have defined a representation of G by automorphisms of the Lie algebra $\mathfrak{G} = \mathfrak{G}(G)$. It is called the *adjoint* action (representation) and denoted by $\mathrm{Ad}_g \colon \mathfrak{G} \to \mathfrak{G}$. Its infinitesimal analogue is the *adjoint representation* ad_X, $X \in \mathfrak{G}(G)$, *of the Lie algebra* $\mathfrak{G}(G)$ *on itself*:

$$\mathrm{ad}_X Y = [X, Y] = \frac{d}{dt}(D_{g(t)} Y)|_{t=0}, \quad X = D_{\dot g(0)}, \quad Y \in \mathfrak{G}(G).$$

This is a homomorphism of the Lie algebra $\mathfrak{G}(G)$ into the Lie algebra of endomorphisms of the linear space $\mathfrak{G}(G)$. Application of this correspondence to the left action of the group on itself (Example 1) enables us to establish a one-to-one correspondence between Lie groups, considered up to a local isomorphism, simply-connected Lie groups, and Lie algebras (considered up to an isomorphism).

Example 1. In example 2) of 2.2, to the curve $g(t) = \begin{pmatrix} a(t) & b(t) \\ c(t) & d(t) \end{pmatrix}$ there corresponds the vector field

$$X = D_{\dot g} = (\alpha x^2 + \beta x + \gamma)\, d/dx, \quad \alpha = c'(0), \quad \beta = a'(0) - d'(0), \quad \gamma = b'(0).$$

Fields of this form X, where $\alpha, \beta, \gamma \in \mathbb{R}$, form the Lie algebra $\mathfrak{G}(G)$ of the group $GL(2, \mathbb{R})$, factorized by the kernel of ineffectiveness, which consists of scalar matrices. By integrating the equation $dx/dt = \alpha x^2 + \beta x + \gamma$ we can restore this group.

Example 2. Let $G \subset GL(n, \mathbb{R})$ be a matrix group, acting naturally in $\mathbb{R}^n$, and $A(t) = \|A_j^i(t)\|$ a curve in G. The velocity vector field corresponding to it has the form $X = D_{\dot A(0)} = \dot A_j^i(0) x^j \partial/\partial x^i$ and is given by the matrix $\dot A(0) = \|\dot A_j^i(0)\|$. To the commutator of fields there corresponds the usual commutator of matrices. Thus, the Lie algebra of the group G is identified with an algebra of matrices. In particular, the Lie algebras of the groups $SO(n)$ and $SL(n)$ are identified with the Lie algebras of skew-symmetric matrices and matrices with zero trace, respectively.

2.4. Infinitesimal Description of Homogeneous Spaces. Let G/H be a homogeneous space with the locally effective action of the group G. By going over to the universal covering we can always assume that G is simply-connected. In this case simply-connectedness of the space G/H is equivalent to connectedness of the stabilizer H. Since a connected subgroup of a Lie group G is uniquely determined by the tangent subalgebra $\mathfrak{h}$ of the Lie algebra $\mathfrak{G}(G)$, the simply-connected homogeneous space G/H is uniquely determined by the pair $(\mathfrak{G}, \mathfrak{H})$, where $\mathfrak{G} = \mathfrak{G}(G)$ is the tangent subalgebra of the simply-connected Lie group G, and $\mathfrak{H}$ is the tangent Lie algebra of the subgroup H. Conversely, a pair $(\mathfrak{G}, \mathfrak{H})$, where $\mathfrak{H}$ is a subalgebra of an arbitrary Lie algebra $\mathfrak{G}$, determines a simply-connected homogeneous space G/H of the simply-connected Lie group G with Lie algebra $\mathfrak{G}$ on condition that to $\mathfrak{H}$ there corresponds a *closed* (connected) subgroup of G. The group G acts locally effectively in G/H if $\mathfrak{H}$ is *effective*, that is, it does not contain non-zero ideals of $\mathfrak{G}$. A multiply-connected homogeneous space M of a locally effective simply-connected Lie groups G is covered by a simply-connected homogeneous space G/H_0, where H_0 is a connected closed subgroup of G and has the form G/H, where H is an arbitrary closed subgroup with the connected component of the identity H_0.

2.5. The Isotropy Representation. Order of a Homogeneous Space.

Any diffeo-morphism $F: M \to M$ that leaves a point $x \in M$ fixed naturally generates a linear map of the tangent space $T_x M$ into itself. Thus, to any element h of the stabilizer H_x of a point $x \in M$ there corresponds a linear transformation $j(h): T_x M \to T_x M$. Hence there arises a representation j of the group H_x by linear transformations of the space $T_x M$. It is called the *isotropy representation* and, as we shall see below, it plays an important role in the theory of homogeneous spaces. The group $\bar{H}_x = j(H_x)$ is called the isotropy group. Let us give an infinitesimal description of it in the spirit of the previous subsection.

Let $M = G/H$, $x = \{H\}$. Then $H_x = H$ and the space $T_x M$ can be identified with $\mathfrak{G}/\mathfrak{H}$. Any linear transformation $L: \mathfrak{G} \to \mathfrak{G}$ that leaves the subalgebra $\mathfrak{H}$ invariant generates a linear transformation $\bar{L}: \mathfrak{G}/\mathfrak{H} \to \mathfrak{G}/\mathfrak{H}$. Under the identification $T_x M \leftrightarrow \mathfrak{G}/\mathfrak{H}$ the transformation $j(h)$, $h \in H$, is identified with $\overline{\mathrm{Ad}\, h}: \mathfrak{G}/\mathfrak{H} \to \mathfrak{G}/\mathfrak{H}$.

The Lie algebra of the isotropy group $j(H)$ is described in these terms as the algebra of linear endomorphisms of the space $\mathfrak{G}/\mathfrak{H}$ having the form $\overline{\mathrm{ad}_Y}$, $Y \in \mathfrak{H}$.

Arguing similarly, we can construct isotropy representations of order $k > 1$ by considering *tangent spaces of order* k at a point $x \in M$. The smallest number k for which an isotropy representation of order k has a discrete kernel is called the *order* of the homogeneous space. There are homogeneous spaces of arbitrarily high order. The possible order of a homogeneous space G/H depends on the structure of the basic group G. For example, if G is semisimple, then the order cannot be greater than two.

2.6. The Principle of Extension. Invariant Tensor Fields on Homogeneous Spaces.

On homogeneous spaces it makes sense to consider homogeneous geo-metrical structures, that is, fields of "geometrical objects" invariant under the action of the basic group G. A field $\theta: x \to \theta_x$ of some geometrical object on $M = G/H$ is G-invariant if $D_g(\theta_x) = \theta_y$, where $y = D_g(x)$. Here the symbol $D_g(\theta_x)$ denotes the image of the geometrical quantity θ_x under the action of the diffeo-morphism D_g. Its exact analytical expression is determined by the law of trans-formation of the geometrical object in question.

If $y = D_{g_1}(x) = D_{g_2}(x)$, then

$$D_{g_1}(\theta_x) = D_{g_2}(\theta_x) \Leftrightarrow D_{g_1^{-1} g_2}(\theta_x) = \theta_x \Leftrightarrow h = g_1^{-1} g_2 \in H_x.$$

Thus, if the field θ is G-invariant, then its value θ_x at an arbitrary point x is invariant under the action of the stabilizer H_x. Conversely, a geometrical quantity θ_x at a point $x \in M = G/H$ that is invariant under the action of the stabilizer H_x can be extended by means of a diffeomorphism D_g, $g \in G$, to a G-invariant field on M. We call it the principle of extension.

We observe that the action of the stabilizer H_x on tensors at a point $x \in M$ reduces to the natural linear action of the isotropy group $\bar{H}_x$ in the space of tensors. Hence we have the following result.

Theorem. *There is a one-to-one correspondence between invariant tensor fields on G/H and tensors on the linear space $\mathfrak{G}/\mathfrak{H}$ that are invariant under the action of the isotropy group.*

One of the typical consequences of this theorem is the following.

If the isotropy group in the space of tensors of type (k, l) over $\mathfrak{G}/\mathfrak{H}$ has an orbit whose closure is compact and lies in an open half-space of this space, then on G/H there is a non-zero invariant tensor field of type (k, l).

In fact, the "centre of gravity" of the closure of such an orbit, which exists in view of its compactness, is invariant with respect to the isotropy group. In view of the second assumption, it is non-zero.

It is well known that the cone of positive definite quadratic forms is convex in the space of all quadratic forms on a given linear space. In addition, the group of automorphisms of a positive definite quadratic form is isomorphic to the orthogonal group, and is thus compact. These two facts together with the previous assumption show that on a homogeneous space G/H there is an invariant Riemannian metric if and only if the closure of the isotropy group in the group of all linear transformations of the quotient space $\mathfrak{G}/\mathfrak{H}$ is compact.

2.7. Primitive and Imprimitive Actions. A smooth action of a Lie group G on a manifold is said to be *imprimitive* if it leaves invariant a nontrivial fibration (or a completely integrable Pfaff system), see Ch. 5, § 3, and *primitive* otherwise. The action of a simply-connected group G in a simply-connected homogeneous space G/H will be primitive if and only if H is a maximal connected subgroup of G. Thus, the description of homogeneous spaces G/H with the primitive action of G reduces to the description of maximal subgroups of the Lie group G or maximal subalgebras of the Lie algebra G. For semisimple Lie groups this description is well known.

Example 1. Let $F(k_1, \ldots, k_p)$ denote a flag manifold of type $(k_1, \ldots, k_p)$, that is, systems of subspaces $0 \subset V_1 \subset V_2 \subset \cdots \subset V_p \subset \mathbb{R}^n$ of dimension $k_1 < \cdots < k_p < n_0$. The group $SL(n, \mathbb{R})$ acts naturally in this manifold, converting it into a homogeneous space. This action is primitive if $p = 1$ and imprimitive if $p > 1$. In any case the order of this homogeneous space is equal to two.

Example 2. Let $\varphi: H \to GL(n, \mathbb{R})$ be an irreducible representation of a Lie group H, where $\varphi(H) \subset SL(n, \mathbb{R})$. If the group $\varphi(H)$ preserves a bilinear form on $\mathbb{R}^n$, then we denote the group of automorphisms of this form by G. Otherwise we put $G = SL(n, \mathbb{R})$. With a few exceptions the action of G in the homogeneous space $G/\varphi(H)$ is primitive.

The general approach presented in this section will be developed and made concrete for the most important geometrical quantities.

§ 3. Invariant Connections on a Homogeneous Space

3.1. A General Description. In this subsection we show how the general principle of extension (see 2.6) is realized in the case of connections. It is convenient to understand a connection on a manifold M as a map $X \to V_X$ that assigns to vector fields $X \in D(M)$ the corresponding covariant derivatives, so that condition (6) of Ch. 2, § 2 is satisfied.

Suppose that V is a connection on M, $X \in D(M)$, and the Lie operator $\mathscr{L}_X$: $D(M) \to D(M)$ acts according to the formula $\mathscr{L}_X Y = [X, Y]$. It follows from (6) of Ch. 2 that the *Nomizu operators*

$$L_X = \mathscr{L}_X - V_X, \tag{3}$$

are $C^\infty(M)$-linear, that is,

$$L_X(\Sigma \varphi_i Y_i) = \Sigma \varphi_i L_X Y_i, \qquad \varphi_i \subset C^\infty(M), \qquad Y_i \in D(M). \tag{4}$$

This means that they are fields of endomorphisms of the tangent spaces to the manifold under consideration. Moreover, in view of (6) of Ch. 2,

$$L_{fX} = f L_X - X \otimes df, \qquad f \in C^\infty(M), \tag{5}$$

where the operator $X \otimes df: D(M) \to D(M)$ acts according to the rule $(X \otimes df)Y = Y(f)X$. In particular, the *Nomizu map* $X \to L_X$ is $\mathbb{R}$-linear. If $X_x = 0$, then the endomorphism $L_X|_x: T_x M \to T_x M$ coincides with the canonical linearization of the field X at the point x. In particular, if M is a homogeneous space, $\mathfrak{H}_x$ is the Lie algebra of the stabilizer of x, and $X \in \mathfrak{H}_x$, then $X_x = 0$ and the endomorphism $L_X|_x$ coincides with the infinitesimal isotropy operator $j(X) = \overline{\mathrm{ad}}\, X$.

Formula (3) shows that the connection can be specified by a family of operators L_X satisfying (4) and (5). If $M = G/H$, then the operators L_Y, $Y \in \mathfrak{G}$, are completely sufficient for this, since in this case any vector field $X \in D(M)$ can be represented in the form $\Sigma \varphi_i Y_i$, $Y_i \in \mathfrak{G}$. If, moreover, the connection is invariant, then by the *principle of extension* it is sufficient to specify endomorphisms $L_Y|_x$, $Y \in \mathfrak{G}$, for some fixed point $x \in M$. The formula

$$D_g \circ L_Y \circ D_g^{-1} = L_{\mathrm{Ad}_g Y}, \qquad g \in G, \quad Y \in \mathfrak{G} \tag{6}$$

expresses the conditions of invariance of a connection in the language of operators L_Y. This means that the image of the endomorphism L_Y under the diffeomorphism D_g is an endomorphism $L_{\mathrm{Ad}_g Y}|_y$, where $y = D_g(X)$.

Thus, an invariant connection on a homogeneous space $M = G/H$ is uniquely determined by a linear map $Y \to L_Y|_x$ of the Lie algebra $\mathfrak{G}$ into the space of endomorphisms End T_xM of the tangent space to M at some fixed point x. Conversely, an a priori chosen linear map $L\colon \mathfrak{G} \to$ End T_xM specifies an invariant connection on M if its restriction to $\mathfrak{H} = \mathfrak{H}_x$ coincides with the infinitesimal isotropy representation $j(\mathfrak{H}) = \overline{\mathrm{ad}}\,\mathfrak{H}$ (see above) and

$$L(\mathrm{Ad}_h X) = j(h)L(X)j(h)^{-1}, \qquad X \in \mathfrak{G}, \quad h \in H = H_x.$$

The first of these conditions guarantees the truth of (5), and the second guarantees the truth of (6).

In terms of Nomizu operators the curvature tensor and torsion tensor of an invariant connection are described by the formulae

$$R(X, Y) = [L_X, L_Y] - L_{[X,Y]},$$

$$T(X, Y) = -(L_X Y - L_Y X - [X, Y]), \qquad X, Y \in \mathfrak{G}, \tag{7}$$

which follow directly from the definitions.

3.2. Reductive Homogeneous Spaces. A homogeneous space $M = G/H$ is said to be *reductive* if the direct decomposition

$$\mathfrak{G} = \mathfrak{H} \oplus \mathfrak{M}$$

holds, in which the subspace $\mathfrak{M} \subset \mathfrak{G}$ is Ad H-invariant. This decomposition is also said to be *reductive*. On any reductive space there is an invariant connection, which can be constructed in the following natural way.

Suppose that $x = \{H\} \in M$. Bearing in mind the identification $\mathfrak{M} \Leftrightarrow \mathfrak{G}/\mathfrak{H} \Leftrightarrow T_xM$, we can understand the map L that specifies an invariant connection on M in the way described in 3.1 as a map $L\colon \mathfrak{G} \to$ End $\mathfrak{M}$ that represents the following composition:

$$\mathfrak{G} = \mathfrak{H} \oplus \mathfrak{M} \xrightarrow{\text{projection}} \mathfrak{H} \xrightarrow{\text{isotropy representation}} \text{End } \mathfrak{M}.$$

The resulting connection on $M = G/H$ is said to be *canonical*. Since $L_X = 0$, if $X \in M$ then the formulae (7) reduce to the following expressions for the curvature tensor and torsion tensor at the point $x = \{H\} \in M$:

$$R(X, Y) = -\mathrm{ad}([X, Y]_{\mathfrak{H}}),$$

$$T(X, Y) = [X, Y]_{\mathfrak{M}}, \qquad X, Y \in \mathfrak{M}. \tag{8}$$

Here the subscripts $\mathfrak{H}$ and $\mathfrak{M}$ denote the projections of the algebra $\mathfrak{G} = \mathfrak{H} \oplus \mathfrak{M}$ onto the subspaces $\mathfrak{H}$ and $\mathfrak{M}$ respectively, and we identify $\mathfrak{M}$ with T_xM.

The following facts show that the field properties of a canonical connection can be expressed in terms of its symmetry group G.

1) The geodesic $\gamma(t)$ starting from the point $x = \{H\}$ in the direction of the vector $X \in \mathfrak{M} = T_xM$ is the trajectory $D_{g(t)}x$ of the one-parameter subgroup $g(t) = \exp tX$, and parallel transport along $\gamma(t)$ is carried out by the diffeomorphisms $D_{g(t)}$. Hence, in particular, it follows that the canonical connection is geodesically complete, and also that parallel transport along any curve $\gamma(t)$ is carried out by diffeomorphisms $D_{g(t)}$, where $g(t)$ is a suitable curve in the group G that satisfies the condition $D_{g(t)}\gamma(0) = \gamma(t)$.

2) The holonomy group is contained in the isotropy group whose Lie algebra is generated by the operators $\mathrm{ad}([X, Y]_{\mathfrak{H}})$, X, $Y \in \mathfrak{M}$. Hence it follows that G-invariant tensor fields on M are parallel. In particular, the curvature tensor field and torsion tensor field are of this kind, that is, $\nabla R = 0$, $\nabla T = 0$.

The following theorem of P.K. Rashevskij shows that reductive homogeneous spaces admit a description in the framework of Riemann's field approach.

Theorem. *Suppose that a connection ∇, defined on a simply-connected manifold M, is geodesically complete and that its curvature tensor and torsion tensor are parallel. Then M is a reductive homogeneous space with respect to the group of automorphisms of ∇, and this connection itself is canonical.*

The result is a consequence of the fact (which follows from (8)) that a simply-connected reductive space as a manifold with a canonical connection is uniquely determined by its curvature tensor and torsion tensor, taken at some fixed point.

The following examples demonstrate the richness of the class of reductive spaces.

Example 1. Left-invariant "absolute parallelism" on Lie groups. A Lie group G, regarded as a homogeneous space with resepct to left action on itself (Example 1 of 2.2), is a reductive space with the trivial reductive decomposition $\mathfrak{G} - \{0\}$ $\oplus$ $\mathfrak{G}$. Its canonical connection is given by the zero map $L: \mathfrak{G} \to \mathrm{End}\ \mathfrak{G}$ and has zero curvature, according to (8). Therefore parallel displacement from a point g_1 to a point g_2 does not depend on the path and is carried out by the left translation l_g, where $g = g_2 g_1^{-1}$. The geodesics of this connection are one-parameter subgroups and left translations of them. In view of (8) its torsion tensor has the form $T(X, Y) = [X, Y]$, X, $Y \in G$, and therefore completely determines the Lie algebra of the group G.

Similarly we can arrive at right-invariant "absolute parallelism" on G.

Example 2. If the group H is semisimple or compact, then the homogeneous space G/H is reductive. This follows from the fact that any representation of a semisimple or compact group is completely reducible. Hence the subalgebra $\mathfrak{H} \subset \mathfrak{G}$, being an $\mathrm{Ad}\ H$-invariant subspace of $\mathfrak{G}$, admits an $\mathrm{Ad}\ H$-invariant complement $\mathfrak{M}$.

Example 3. Homogeneous spaces $M = G/H$ of the group $G = SL(2, \mathbb{R})$. A reductive decomposition, if it exists, reduces to the following form.

1) $H = \left\{ \begin{pmatrix} 1 & a \\ 0 & 1 \end{pmatrix}, a \in \mathbb{R} \right\}$, $M \simeq S^1 \times \mathbb{R}^1$,   (cylinder)

there is no reductive decomposition.

2) $H = \left\{ \begin{pmatrix} a & 0 \\ 0 & a^{-1} \end{pmatrix}, a > 0 \right\}$, $M \simeq S^1 \times \mathbb{R}^1$,   (cylinder)

$\mathfrak{G} = \left\{ \begin{pmatrix} \alpha & 0 \\ 0 & -\alpha \end{pmatrix}, \alpha \in \mathbb{R} \right\} \oplus \left\{ \begin{pmatrix} 0 & \beta \\ \gamma & 0 \end{pmatrix}, \beta, \gamma \in \mathbb{R} \right\}$

3) $H = \left\{ \begin{pmatrix} \cos t & -\sin t \\ \sin t & \cos t \end{pmatrix}, t \in \mathbb{R} \right\}$, $M \simeq \mathbb{R}^2$,   (plane)

$$\mathfrak{G} = \left\{ \begin{pmatrix} 0 & -\alpha \\ \alpha & 0 \end{pmatrix}, \alpha \in \mathbb{R} \right\} \oplus \left\{ \begin{pmatrix} \beta & \gamma \\ \gamma & -\beta \end{pmatrix}, \beta, \gamma \in \mathbb{R} \right\}$$

4) $H = \left\{ \begin{pmatrix} a & b \\ 0 & a^{-1} \end{pmatrix} a > 0, b \in \mathbb{R} \right\}$ (projective line or circle)

$M \simeq \mathbb{R}P^1 \simeq S^1$,

there is no reductive decomposition.

The homogeneous space of Example 2 is identified with a pseudo-Riemannian space of constant curvature, and the homogeneous space of Example 3 is identified with the Lobachevskij plane.

Apart from a canonical connection, a reductive space has a unique invariant torsion-free connection, whose geodesics are trajectories of one-parameter subgroups of the basic group G. This connection is called the *natural connection* and is given by the Nomizu operators

$$L_X Y = \frac{1}{2}[X, Y], \qquad X \in \mathfrak{G}, \quad Y \in \mathfrak{M}.$$

The condition that the canonical and natural connections coincide distinguishes a remarkable subclass of reductive spaces – affine symmetric spaces.

3.3. Affine Symmetric Spaces. We begin with a descriptive field definition of a symmetric space.

A manifold M with a linear connection V is said to be (*affine*) *symmetric* if it admits a *central symmetry* S_x with centre at any point $x \in M$, that is, an *involutive* ($S_x^2 = $ id) *automorphism* of the connection V that has x as an isolated fixed point and $d_x S_x = -1$. A central symmetry S_x preserves any geodesic $\gamma(s)$, $\gamma(0) = x$, and takes $\gamma(s)$ into $\gamma(-s)$, where s is the natural parameter on it. If central symmetries on M can be defined locally, we say that M is a *locally (affine) symmetric space*.

Example 1. Let σ be an involutive automorphism of a Lie group G, and G_0^σ the connected component of the identity of the subgroup G^σ of its fixed points. Let H be a closed sugroup of G such that $G_0^\sigma \subset H \subset G^\sigma$. Then G/H is a reductive space. Endowed with a canonical connection, it becomes an affine symmetric space. To verify this, we observe that the automorphism σ induces an involutive automorphism $\dot\sigma$ of the Lie algebra $\mathfrak{G}$. Let $\mathfrak{H}$ and $\mathfrak{M}$ be the eigensubspaces of the automorphism $\dot\sigma$ corresponding to the eigenvalues 1 and -1, respectively. Then $\mathfrak{H}$ is the tangent Lie algebra of any of the subgroups G_0^σ, H, G^σ and we have the reductive decomposition $\mathfrak{G} = \mathfrak{H} \oplus \mathfrak{M}$ satisfying the additional condition $[\mathfrak{M}, \mathfrak{M}] \subset \mathfrak{H}$. Such a reductive decomposition is said to be *symmetric*. In view of (8), this shows that the canonical connection under consideration is torsion-free and coincides with the natural connection. The central symmetry S_x at the point $x = \{H\} \in G/H$ maps the coset $\{gH\}$ into $\{\sigma(g)H\}$.

Example 2. A Lie group G, regarded as a G-manifold with respect to the left action of G (Example 1 of 3.2) and endowed with the natural connection of the reductive space $G/\{\text{id}\}$, is an affine symmetric space. This example can also be obtained by means of the construction of the previous example. For this we need to consider the group $G \times G$ and the involutive automorphism $\sigma: (g_1, g_2) \to (g_2, g_1)$ of it. Then the symmetric space $(G \times G)/H$, where $H = (G \times G)^\sigma = \{(g, g), g \in G\}$, endowed with the canonical connection, is isomorphic to the homogeneous space $G/\{\text{id}\}$ with the natural connection. This isomorphism is carried out by the map

$$G \times G/H \ni \{(g_1, g_2)H\}| \to g_1 g_2^{-1} \in G.$$

Example 3. Specializing Example 1, we consider the group $G = SL(n, \mathbb{R})$ and the inner involutive automorphism of it $\sigma: g \to aga^{-1}$ generated by the matrix $a = \text{diag}(E_k, -E_l)$, where E_i is the unit $(i \times i)$-matrix. This example is interesting in that the canonical connection of the symmetric space G/G^σ is not the Levi-Civita connection of any pseudo-Riemannian metric.

It is remarkable that the construction described in Example 1 enables us to obtain all affine symmetric spaces. Namely, we have the following result, which characterizes symmetric spaces from the group point of view in an exhaustive way.

Theorem. *Any affine symmetric space is isomorphic to a space of the form G/H, where the closed subgroup $H \subset G$ is such that $G_0^\sigma \subset H \subset G^\sigma$, and σ is an involutive automorphism of the group G.*

The group G whose existence is asserted by this theorem is defined as the group generated by central symmetries, its subgroup H as the stabilizer H_x of some fixed point $x \in M$, and the involution $\sigma: G \to G$ by the rule $g \to S_x g S_x$.

From the infinitesimal point of view (see 2.4) affine symmetric spaces are defined as pairs $(\mathfrak{G}, \mathfrak{H})$ admitting a symmetric reductive decomposition $\mathfrak{G} = \mathfrak{H} \oplus \mathfrak{M}$, that is,

$$[\mathfrak{H}, \mathfrak{H}] \subset [\mathfrak{H}, \mathfrak{M}] \subset \mathfrak{M}, [\mathfrak{M}, \mathfrak{M}] \subset \mathfrak{H}.$$

This decomposition determines an involutive automorphism $\dot\sigma$ of the Lie algebra $\mathscr{G}$: $\dot\sigma(X, Y) = X - Y$, $X \in \mathfrak{H}$, $Y \in \mathfrak{M}$, and consequently an involutive automorphism σ of the simply-connected Lie group G with the Lie algebra $\mathfrak{G}$. The set G^σ of fixed points of σ is the connected closed subgroup of G corresponding to the subalgebra $\mathfrak{H}$. The homogeneous space G/G^σ is a simply-connected symmetric space. This sets up a one-to-one correspondence between symmetric decompositions and simply-connected symmetric spaces.

A very important subclass of affine symmetric spaces – Riemannian symmetric spaces – will be considered in the next section.

§4. Homogeneous Riemannian Manifolds

4.1. Infinitesimal Description. A Riemannian manifold (M, g) that admits a transitive group of motions G is called a *homogeneous Riemannian manifold*. Such a manifold M can be identified with a homogeneous space G/H. An invariant metric g is specified by a scalar product $\langle \cdot, \cdot \rangle$ in the tangent space $T_o G/H = \mathfrak{G}/\mathfrak{H}$ of the point $o = \{H\}$, invariant with respect to the isotropy group $j(H)$. We shall assume that G is a closed subgroup of the full group of motions of M. Then from the fact that the isotropy representation is exact (see 2.5) and the orthogonal group is compact it follows that the stabilizer H is compact. Since the representation of the compact group H is completely reducible, there is a reductive decomposition $\mathfrak{G} = \mathfrak{H} + \mathfrak{M}$ of the Lie algebra $\mathfrak{G}$. As above, we identify $\mathfrak{M}$ with the tangent space $T_o G/H$. Obviously the Levi-Civita connection V of the invariant metric g is G-invariant and by 3.1 it is specified by the Nomizu map $L: \mathfrak{G} \to$ End $\mathfrak{M}$. This map satisfies the identity

$$L_X Y_{\mathfrak{M}} - L_Y X_{\mathfrak{M}} = [X, Y]_{\mathfrak{M}}, \qquad X, Y \in \mathfrak{M}, \tag{9}$$

which expresses the fact that V is torsion-free. In addition, the Nomizu operators L_X are skew-symmetric with respect to the scalar product $\langle \cdot, \cdot \rangle$. This is equivalent to $\mathscr{L}_X(g) = 0$, which expresses the invariance of the metric with respect to infinitesimal motions $X \in \mathfrak{G}$. These properties completely characterize the Nomizu map L of the invariant metric g and lead to the following formula, which expresses the Nomizu operators in terms of the bracket $[\ ,\ ]$ in the Lie algebra $\mathfrak{G}$ and the scalar product $\langle \cdot, \cdot \rangle$:

$$2\langle L_X Y, Z \rangle = \langle [X, Y]_{\mathfrak{M}}, Z \rangle - \langle Y, [X, Z]_{\mathfrak{M}} \rangle$$
$$- \langle X_{\mathfrak{M}}, [Y, Z]_{\mathfrak{M}} \rangle, \qquad X \in \mathfrak{G}, \quad Y, Z \in \mathfrak{M}. \tag{10}$$

This formula enables us to calculate explicitly the covariant derivative at the point $o = \{H\}$ of any G-invariant tensor field A if we know its value A_0 at the point o:

$$(V_X A) = -L_X \cdot A_0, \qquad X \in \mathfrak{M} = T_0 G/H, \tag{11}$$

where the dot denotes the natural infinitesimal action of the operator $L_X \in$ End $\mathfrak{M}$ in the space of tensors. In particular, the formulae (9), (10), (11) enable us to calculate in infinitesimal terms the fundamental differential invariants of the metric g – the curvature tensor, the Ricci tensor, the scalar curvature, the sectional curvatures, the covariant derivatives of the curvature tensor, and the holonomy algebra $\dot{\Gamma}$, which according to the Ambrose-Singer theorem (see Ch. 6, 3.4) is generated by the curvature operators $R(X, Y)$ and their covariant derivatives. From (10) and (11) it follows that the linear Lie algebra Lie$\{L_{\mathfrak{G}}\}$ generated by the Nomizu operators L_X, $X \in \mathfrak{G}$, contains the holonomy algebra $\dot{\Gamma}$ and is contained in its normalizer $N(\dot{\Gamma})$:

$$\dot{\Gamma} \subset \text{Lie}\{L_{\mathfrak{G}}\} \subset N(\dot{\Gamma}). \tag{12}$$

4.2. The Link Between Curvature and the Structure of the Group of Motions.
The infinitesimal formulae for the curvature in combination with the structural
theorems of the theory of Lie groups enable us to obtain a number of results that
link the geometrical properties of an invariant Riemannian metric g in a homogeneous space G/H with the algebraic structure of the group of motions G. Let
us mention some of them.

1) Suppose that G is soluble. Then the scalar curvature $\text{Sc} \leqslant 0$, and the
equality $\text{Sc} = 0$ implies that the metric g is locally Euclidean.

2) If the Ricci curvature $\text{ric}(X, X) \equiv 0$, then the metric g is locally Euclidean.
If $\text{ric}(X, X) \geqslant 0$, $X \in \mathfrak{M}$, then the Riemannian manifold $(G/H, g)$ decomposes
into the direct product of a Euclidean space and a compact homogeneous
Riemannian manifold.

3) If the vector $X \in \mathfrak{M}$ belongs to the centre of the Lie algebra $\mathfrak{G}$, then the
sectional curvature $K(X, Y) \geqslant 0$ for any $Y \in \mathfrak{M}$.

4) If the vector $X \in \mathfrak{M}$ is orthogonal to the derived group $[\mathfrak{G}, \mathfrak{G}] \cap \mathfrak{M}$, then
$\text{ric}(X, X) \leqslant 0$ and $L_X X = 0$. The latter condition is equivalent to the fact that
the orbit of the point o with respect to the one-parameter subgroup generated
by the field X is a geodesic.

4.3. Naturally Reductive Spaces. Let $M = G/H$ be a reductive homogeneous
space with reductive decomposition $\mathfrak{G} = \mathfrak{H} + \mathfrak{M}$. Suppose that the Lie algebra
$\mathfrak{G}$ has an Ad G-invariant non-degenerate symmetric bilinear form whose restriction to $\mathfrak{M}$ is positive definite, and therefore determines a $j(H)$-invariant scalar
product $\langle \cdot, \cdot \rangle$. The homogeneous Riemannian manifold $(G/H, g)$, where g is the
metric corresponding to the scalar product $\langle \cdot, \cdot \rangle$, is called a *naturally reductive
space*. In particular, the homogeneous space G/H of a compact semisimple Lie
group G can be regarded as a naturally reductive space: the invariant metric g is
given by the Ad G-invariant form $(-B)$, where B is the (negative definite) Cartan-
Killing form of the Lie algebra $\mathfrak{G}$.

Example. Let $G = SO(n)$, and let $H = SO(k)$ be a subgroup embedded in the
standard way. The homogeneous space $M = SO(n)/SO(k)$ is identified with the
Stiefel manifold of ordered sets of $n - k$ orthonormal vectors of the space $\mathbb{R}^n$.
The Lie algebra $\mathfrak{G} = SO(n)$ consists of skew-symmetric matrices and has the
reductive decomposition

$$SO(n) \ni X = \overset{k \quad n-k}{\left(\frac{h \mid X_1}{-X'_1 \mid X_2} \right)} = \overset{\mathfrak{H}}{\left(\frac{h \mid 0}{0 \mid 0} \right)} + \overset{\mathfrak{M}}{\left(\frac{0 \mid X_1}{-X'_1 \mid X_2} \right)} \begin{array}{l} h \in SO(k) = \mathfrak{H}, \\ X_2 \in SO(n-k). \end{array}$$

The scalar product in $\mathfrak{M}$ that specifies the naturally reductive metric in M has
the form

$$\langle X, Y \rangle = -\text{tr } XY, \qquad X, Y \in \mathfrak{M},$$

where tr denotes the trace of a matrix.

For a naturally reductive space the operators

$$P_X: \mathfrak{M} \ni Y \to [X, Y]_{\mathfrak{M}} \in \mathfrak{M}, \qquad X \in \mathfrak{M}, \tag{13}$$

are skew-symmetric. Therefore (11) and (10) take the particularly simple form

$$L_X = \frac{1}{2} P_X, \qquad R(X, Y) = \frac{1}{4}[P_X, P_Y] - \frac{1}{2} P_{[X, Y]}, \qquad X, Y \in \mathfrak{M}. \tag{14}$$

In particular, $L_X X = 0$ and the trajectories of the point 0 with respect to the one-parameter subgroups generated by the elements of $X \in \mathfrak{M}$ are geodesics. Thus, in a naturally reductive space the geodesics are described in group terms.

4.4. Symmetric Riemannian Spaces. A remarkable class of naturally reductive spaces that admits a complete classification in the framework of the theory of semisimple Lie groups consists of *symmetric Riemannian spaces*.

Let (M, g) be a Riemannian manifold and V its Levi-Civita connection. The manifold (M, g) is called a *symmetric Riemannian space* if (M, V) is an affine symmetric space. In this case the central symmetries S_x, $x \in M$, generate a transitive group of motions G of the metric g, which contains the connected component of the identity of the full group of motions of the manifold (M, g). The affine symmetric space (M, V) can be regarded as a symmetric Riemannian space (M, g) (where g is the metric with Levi-Civita connection V) if and only if the group G generated by central symmetries has a compact stabilizer H. Together with the results of 3.3 this leads to a group description of symmetric Riemannian spaces as spaces of the form G/H, where H is a compact subgroup satisfying the condition $G_0^\sigma \subset H \subset G^\sigma$ (see 3.3). An invariant metric is defined in such a space almost uniquely. In particular, a connected Lie group G can be regarded as a symmetric Riemannian space if and only if it is the direct product of a vector group $\mathbb{R}^k$ and a compact Lie group. Any simply-connected symmetric Riemannian space decomposes into the direct product of a Euclidean space and irreducible (see Ch. 3, 4.9) symmetric Riemannian spaces. This enables us to restrict ourselves to irreducible spaces.

There is a remarkable duality between compact and non-compact irreducible simply-connected symmetric Riemannian spaces.

To each compact symmetric space $M = G/H$ there corresponds the dual non-compact space $M' = G'/H$ and vice versa. Dual spaces M and M' have identical isotropy groups $j(H)$ (coinciding with the holonomy group), and their curvature tensors at the point $o = \{H\}$ differ in sign. The sectional curvature of a compact (non-compact) space is non-negative (non-positive). Let $\mathfrak{G} = \mathfrak{H} + \mathfrak{M}$ be a symmetric decomposition of the Lie algebra $\mathfrak{G}$ of a compact Lie group G, which determines a simply-connected compact symmetric Riemannian space G/H. Then the change of sign of the commutator of any two elements of $\mathfrak{M}$ determines in $\mathfrak{G}$ the structure of a new Lie algebra $\mathfrak{G}'$ together with a symmetric decomposition $\mathfrak{G}' = \mathfrak{H} + \mathfrak{M}$ of it. To this decomposition there corresponds a dual non-compact symmetric space G'/H.

Examples. 1) The sphere $S^{n-1} = SO(n)/SO(n-1)$ with the standard metric is a compact symmetric Riemannian space. The corresponding symmetric decomposition of the Lie algebra $SO(n+1)$ has the form

$$SO(n+1) \ni \left(\begin{array}{c|c} h & X \\ \hline X' & 0 \end{array} \right) = \underbrace{\left(\begin{array}{c|c} h & 0 \\ \hline 0 & 0 \end{array} \right)}_{\mathfrak{H}} + \underbrace{\left(\begin{array}{c|c} 0 & X \\ \hline -X' & 0 \end{array} \right)}_{\mathfrak{M}} \begin{array}{l} h \in SO(n-1), \\ X \in \mathbb{R}^{n-1}. \end{array} \tag{15}$$

The dual non-compact space is the Lobachevskij space $L^{n-1} = SO(1, n-1)/SO(n-1)$, which is determined by the symmetric decomposition

$$SO(1, n-1) \ni \left(\begin{array}{c|c} h & X \\ \hline X' & 0 \end{array} \right) = \underbrace{\left(\begin{array}{c|c} h & 0 \\ \hline 0 & 0 \end{array} \right)}_{\mathfrak{H}} + \underbrace{\left(\begin{array}{c|c} 0 & X \\ \hline X' & 0 \end{array} \right)}_{\mathfrak{M}} \begin{array}{l} h \in SO(n-1), \\ X \in \mathbb{R}^{n-1}. \end{array}$$

2) The *Grassmann manifold* $SO(n)/SO(k) \times SO(l)$, $n = k + l$, of k-dimensional subspaces of $\mathbb{R}^n$ is a compact symmetric Riemannian space. The dual non-compact space is the space $SO(k, l)/SO(k) \times SO(l)$ of k-dimensional Euclidean subspaces of the pseudo-Euclidean space $\mathbb{R}^{k,l}$ of signature (k, l). The symmetric decomposition corresponding to the Grassmann manifold has the form

$$SO(n) \ni \left(\begin{array}{c|c} \overset{k}{h_1} & \overset{l}{X} \\ \hline -X' & h_2 \end{array} \right) = \underbrace{\left(\begin{array}{c|c} h_1 & 0 \\ \hline 0 & h_2 \end{array} \right)}_{\mathfrak{H}} + \underbrace{\left(\begin{array}{c|c} 0 & X \\ \hline -X' & 0 \end{array} \right)}_{\mathfrak{M}}. \tag{16}$$

The space $\mathfrak{M}$ is identified with the space of matrices of order $k \times l$, and the isotropy representation j of the stationary subalgebra $\mathfrak{H} = so(k) \oplus so(l)$ coincides with the natural representation of this Lie algebra in the space of matrices $\mathfrak{M}$.

$$j(h_1, h_2)X = h_1 X - X h_2, \quad X \in \mathfrak{M}, \quad h_1 \in so(k), \quad h_2 \in so(l).$$

For a symmetric space the operator P_X defined by (13) is equal to zero for all $X \in \mathfrak{M}$ and formula (14) for the curvature operator takes the form

$$R(X, Y) = \frac{1}{2} P_{[X,Y]} = -\frac{1}{2} j([X, Y]), \quad X, Y \in \mathfrak{M};$$

therefore the curvature operator of the Grassmann manifold corresponding to the matrices $X, Y \in \mathfrak{M}$ is equal to

$$R(X, Y) = -\frac{1}{2} j \left(\left[\left(\begin{array}{c|c} 0 & X \\ \hline -X' & 0 \end{array} \right), \left(\begin{array}{c|c} 0 & Y \\ \hline -Y' & 0 \end{array} \right) \right] \right) = \frac{1}{2} j(h_1, h_2),$$

$$h_1 = XY' - YX' \in so(k), \quad h_2 = X'Y - YX' \in so(l).$$

In particular, when $l = 1$ the Grassmann manifold turns into a projective space, the decomposition (16) reduces to (15), and the isotropy representation j of the Lie algebra $\mathfrak{H} = so(n-1)$ is identified with the standard representation of this algebra in the space $\mathfrak{M} = \mathbb{R}^{n-1}$. The curvature tensor is given by the formula

$$R(X, Y) = \frac{1}{2} X \wedge Y : Z \mapsto \frac{1}{2}[\langle Y, Z \rangle X - \langle X, Z \rangle Y], \quad X, Y, Z \in \mathfrak{M}.$$

We thus arrive at a metric of constant curvature in projective space, described in other terms in the example of Ch. 3, 4.8, and also at the standard metric on the sphere, which covers this projective space.

Above we have considered mainly simply-connected symmetric Riemannian spaces, which are completely determined by the corresponding symmetric decompositions. The global theory of Lie groups enables us to obtain a classification of multiply-connected symmetric Riemannian spaces. As a characteristic example we give the following result.

Theorem. *Let $M = E^k \times M^+ \times M^-$ be a decomposition of a simply-connected symmetric Riemannian space into the product of a Euclidean space E^k, a compact symmetric space M^+ and a non-compact symmetric space M^-. Any symmetric space that is covered by M is obtained by factorizing M by a commutative discrete group $H_0 \times H^+$, where H_0 is a lattice in E^k, and H^+ is any finite group of motions of the space M^+ that commutes with its full connected group of motions.*

4.5. Holonomy Groups of Homogeneous Riemannian Manifolds. Kählerian and Quaternion Homogeneous Spaces.
From Berger's classification of holonomy groups (Ch. 3, 4.8) and Assertion 2 of 4.2 we have the following description of holonomy groups of homogeneous Riemannian manifolds.

Theorem. *Let (M^n, g) be a simply-connected homogeneous irreducible Riemannian manifold. Suppose that its holonomy group $\Gamma \neq SO(n)$. Then either (M^n, g) is a symmetric space, or $\Gamma = U(m)$, $n = 2m$, and (M^n, g) is a Kählerian manifold, or $\Gamma = Sp(1) \cdot Sp(m)$, $n = 4m$, and (M^n, g) is a quaternion manifold.*

We observe that in all these cases the holonomy algebra $\dot{\Gamma}$ coincides with its normalizer in the algebra of all endomorphisms. In view of (12) this leads to the following description of the holonomy algebra $\dot{\Gamma}$.

Let (M, g) be a homogeneous Riemannian manifold, and suppose that its universal covering does not contain Euclidean factors. Then the holonomy algebra is generated (as a linear Lie algebra) by the Nomizu operators L_X, $X \in \mathfrak{G}$. In particular, it contains the isotropy algebra $j(\mathfrak{H})$ and any parallel tensor field on M is invariant with respect to the connected group of motions.

Compact quaternion homogeneous Riemannian spaces are simply-connected symmetric Riemannian spaces and have the form $G/N(H_0)$, where G is an arbitrary compact simple Lie group, and $N(H_0)$ is the normalizer of some three-dimensional subgroup H_0 of it (the so-called regular three-dimensional subgroup corresponding to a long root).

Any compact Kählerian homogeneous space M decomposes into the direct product of a complex torus T^{2k} and simply-connected irreducible Kählerian homogeneous spaces of the form $G/C(T)$, where G is a compact simple Lie group, and $C(T)$ is the centralizer of some torus T of the group G.

The homogeneous space $G/C(T)$ is identified with the orbit of the adjoint representation of G, and the invariant Kählerian metrics on it are in a natural one-to-one correspondence with invariant symplectic structures.

§5. Homogeneous Symplectic Manifolds

5.1. Motivation and Definitions. Let us look at symplectic geometry from the point of view of the group approach. The group $\text{Aut}(\Omega)$ of automorphisms of any symplectic manifold (M, Ω) is transitive. To it there corresponds the infinite-dimensional Lie algebra $\text{aut}(\Omega) = \{X \in D(M) | \mathscr{L}_X\Omega = 0\}$ of canonical vector fields. However, an attempt to describe the geometry of the manifold (M, Ω) as the homogeneous space of the group $\text{Aut}(\Omega)$ in the framework of Klein's programme does not lead to success. The reason is that the infinite-dimensional objects $\text{Aut}(\Omega)$ and $\text{aut}(\Omega)$ (which in principle contain the necessary information about the manifold M and the form Ω) do not admit a constructive abstract description that does not appeal to their realization as transformations (global or infinitesimal) of the manifold M, as happens for finite-dimensional Lie groups and Lie algebras. Moreover, the Lie algebra of vector fields $\text{aut}(\Omega)$ is not determined by its restriction to a small neighbourhood of some point. From a theorem of Darboux (see Ch. 5, 4.3) it follows that all such restrictions are essentially isomorphic and do not contain any global information. A reasonable modification of the group approach in the case under consideration consists in replacing the group of automorphisms by the pseudogroup of local automorphisms and interpreting the symplectic structure as a pseudogroup structure in the framework of the theory of pseudogroups (see Ch. 7, §7). Nevertheless, it is useful to investigate the structure of symplectic manifolds that admit a transitive Lie group G of automorphisms, that is, diffeomorphisms $F: M \to M$ such that $F^*(\Omega) = \Omega$. We shall call a triad (M, Ω, G) of this kind a *homogeneous symplectic manifold*. The Lie algebra $\mathfrak{G}$ of vector fields that corresponds to G consists of symplectic fields. If the Hamiltonians f_X of fields $X \in \mathfrak{G}$ (defined locally and up to a constant) can be chosen globally and so that the map $\mathfrak{G} \to C^\infty(M)$, $X \to f_X$, is a homomorphism of Lie algebras (where $C^\infty(M)$ is regarded as a Lie algebra with respect to the Poisson bracket), then we say that the group of automorphisms G is *Hamiltonian*. A homogeneous symplectic manifold (M, Ω, G) with a Hamiltonian group G is called a *homogeneous Hamiltonian manifold*.

5.2. Examples. 1) Suppose that the symplectic form Ω of a manifold M is exact, $\Omega = d\rho$, where ρ is a 1-form. Then the Lie group G of transformations of M that preserve ρ is Hamiltonian. In particular, the Lie group G of transformations of an arbitrary manifold Q acts on the cotangent bundle $M = T^*Q$ with the standard symplectic structure $d\theta$ (see Ch. 3, §5) as a Hamiltonian group.

2) Let $(\mathbb{R}^{2n}, \sigma)$ be a symplectic vector space with the standard symplectic form $\Omega = \sum dp_i \wedge dq_i$. The group $Sp(n)$ of its linear automorphisms and the group $\mathbb{R}^{2n}$ of parallel displacements are Hamiltonian. However, the group $\mathbb{R}^{2n}$ is a non-Hamiltonian transitive group of automorphisms of the symplectic torus $(\mathbb{R}^{2n}/\mathbb{Z}^{2n}, \Omega)$.

3) Let G be a Lie group with Lie algebra $\mathfrak{G}$ and let Ad^*G be the coadjoint action of G in the conjugate space $\mathfrak{G}^*$:

$$(\text{Ad}_g^*\xi, X) = (\xi, \text{Ad}_g^{-1}X), \qquad \xi \in \mathfrak{G}^*, \quad X \in \mathfrak{G}.$$

The orbit M of any point $\xi \in \mathfrak{G}^*$ with respect to this action is endowed with a canonical G-invariant symplectic form Ω:

$$\Omega_\eta(\text{ad}_X^*\eta, \text{ad}_Y^*\eta) = \eta([X, Y]) \qquad X, Y \in \mathfrak{G}, \quad \eta \in M,$$

where $\mathfrak{G}^* \ni \eta \mapsto \text{ad}_X^*\eta = -\eta \circ \text{ad}_X$ is the coadjoint representation of $\mathfrak{G}$, and the tangent space $T_\eta M$ is identified with the subspace $\text{ad}_{\mathfrak{G}}^*\eta \subset \mathfrak{G}^*$. Although, generally speaking, the form Ω is not exact, the action of G in M is Hamiltonian: the field ad_X^* has Hamiltonian

$$f_X: M \ni \xi \mapsto f_X(\xi) = \xi(X) \quad \text{and} \quad \{f_X, f_Y\} = f_{[X,Y]}.$$

Suppose, for example, that $G = SO(3)$. The space $\mathfrak{G}$ is identified with the space $so(3) = \mathbb{R}^3$ of skew-symmetric matrices of order 3, the orbits of G in $\mathfrak{G}^*$ are two-dimensional spheres, and the form Ω is identified with the volume form.

5.3. Homogeneous Hamiltonian Manifolds

Theorem. *The orbits of the coadjoint action (see Example 3 above) and coverings of them exhaust all homogeneous Hamiltonian manifolds.*

The main interest is in compact symplectic manifolds. Let (M, Ω) be such a manifold, and G the Hamiltonian group of transformations of it. Then its Lie algebra $\mathfrak{G}$ has an Ad G-invariant positive definite form

$$\langle X, Y \rangle = \int_M f_X f_Y \Omega^m, \qquad \dim M = 2m,$$

where f_X is the Hamiltonian of the field $X \in \mathfrak{G}$, and $\Omega^m = \Omega \wedge \cdots \wedge \Omega$ is the volume form. Hence it follows that the universal covering of G is isomorphic to the group $\mathbb{R}^k \times G'$, where G' is a compact semisimple Lie group. It is easy to show that the transitive Hamiltonian group G of transformations of a symplectic manifold has discrete centre. These arguments lead to the following result.

Theorem. *Let (M, Ω, G) be a compact homogeneous Hamiltonian manifold. Then the group G is compact and semisimple, and the symplectic manifold (M, Ω) is simply-connected and isomorphic to the orbit of the coadjoint action of the group G.*

5.4. Homogeneous Symplectic Manifolds and Affine Actions. Now let (M, Ω, G) be an arbitrary homogeneous symplectic manifold. As usual, we identify M with the quotient space G/H. The inverse image $\pi^*\Omega$ of the form Ω under the map $\pi: G \to G/H$ is a left-invariant closed 2-form on G. The form $\pi^*\Omega$ is determined by its value ω on $T_e G \approx \mathfrak{G}$ (where e is the identity of G). The 2-form ω on G satisfies the closure condition

$$d\omega(X, Y, Z) = \omega([X, Y], Z) + \omega([Y, Z], X) + \omega([Z, X], Y)$$

$$= 0, \qquad X, Y, Z \in \mathfrak{G}.$$

The kernel ker $\omega = \{X \in \mathfrak{G} | \omega\{X, Y\} = 0, \forall Y \in \mathfrak{G}\}$ of the form ω coincides with the stationary subalgebra $\mathfrak{H}$. Hence it is obvious that ω determines a symplectic homogeneous manifold $(G/H, \Omega)$ up to coverings.

The following question arises: does any closed 2-form ω on the Lie algebra $\mathfrak{G}$ have a symplectic manifold corresponding to it? An affirmative answer to this comes from the following result. A closed 2-form ω in $\mathfrak{G}$ determines the affine action A^ω of the Lie algebra $\mathfrak{G}$ in the space $\mathfrak{G}^*$:

$$\mathfrak{G} \ni X \mapsto A^\omega_X \xi = -\xi \circ \mathrm{ad}_X + \omega_X, \quad \xi \in \mathfrak{G}^*, \quad X \in \mathfrak{G}, \quad \omega_X = \omega(X, \cdot).$$

This action can be extended to the affine action of a simply-connected Lie group G with Lie algebra $\mathfrak{G}$, and the stabilizer of the origin $0 \in \mathfrak{G}$ is a closed subgroup H_ω with Lie algebra $\mathfrak{h}_\omega = \ker \omega$. The form ω naturally generates a homogeneous symplectic structure on the homogeneous manifold G/H^0_ω (where H^0_ω is the connected component of the identity of the stabilizer H_ω), which is the universal covering of the orbit G/H_ω of the point 0.

Theorem. *Manifolds of the form G/H^0_ω exhaust all simply-connected homogeneous symplectic manifolds.*

We note that in contrast to the Hamiltonian case the Lie group G of automorphisms of a compact homogeneous symplectic manifold (M, Ω, G) is not necessarily compact, but its maximal semisimple subgroup S is compact and is a normal subgroup.

Chapter 5
The Geometry of Differential Equations

> "... differential equations represent an outstanding
> object of research in differential geometry; we
> can talk about both the geometrical meaning of a
> differential equation itself and, in particular, the
> geometrical meaning of integration of a differential
> equation".
>
> F. Klein

The words "the geometry of differential equations" are used to describe two very dissimilar topics in mathematics. They are often used when we have in mind the geometrical aspects of the theory of dynamical systems with finitely or infinitely many degrees of freedom. Here we are essentially concerned with the geometry of vector fields, more precisely, with problems of constructing and investigating the corresponding phase portraits. In view of the theorem on the rectification of a vector field, the local theory in the case under consideration is trivial if we leave aside singular points. For similar reasons the terminology in this topic consists mainly of morphological concepts, as happens in sciences of a descriptive character.

Things are quite different if we turn to partial differential equations. Here there is a very rich and meaningful local theory if we disregard certain exceptions, the most important of which are scalar equations of the first order. We should say that the simplest question of formal solubility (see § 5 below) may turn out to be very complicated if we need to use complex apparatus (for example, Spencer cohomology). Generally speaking, the solutions of partial differential equations have a rather complicated and interesting internal structure, which can be described by the corresponding differential invariants. Riemannian geometry is one of the uncountable geometrical systems that arise in this way (see Ch. 3, § 5, and also Ch. 7, § 1).

In this chapter we shall give a first idea of this topic of differential geometry, whose foundation was laid by Monge in his "Analyse appliqué à la géométrie" (Monge [1850]).

§ 1. Elementary Geometry of a First-Order Differential Equation

It is convenient to trace the process of geometrization of partial differential equations by first-order scalar equations, since they are particularly simple.

1.1. Ordinary Differential Equations. It is well known that a direct geometrical interpretation of the equation

$$F(x, y, y') = 0 \tag{1}$$

consists in the following. With each point (x_0, y_0) in the plane Π of the variables (x, y) we associate the family $K(x_0, y_0)$ of lines passing through it, whose gradient p satisfies the equation $F(x_0, y_0, p) = 0$. Thus, generally speaking, in Π there arises a multivalued field of directions. The function $y = f(x)$ is a solution of (1) if the tangent to its graph at any point $(x, f(x)) \in \Pi$ is a line of $K(x, f(x))$.

We can get rid of the multivaluedness as follows. We call a pair consisting of a point $a \in \Pi$ and a line L passing through it and not parallel to Oy a *contact element* of the function, or in the modern nomenclature its *1-jet* (Fig. 25). Any 1-jet is specified by a triple (x_0, y_0, p_0), where $a = (x_0, y_0)$, and p_0 is the gradient of L. The specification of a 1-jet is equivalent to the specification of the value $y_0 = f(x_0)$ of the function $y = f(x)$ at the point x_0 and its derivative $p_0 = f'(x_0)$. Hence equation (1) can be treated as a hypersurface $\mathscr{E} = \{F(x, y, p) = 0\}$ in the space $J^1 \approx \mathbb{R}^3$ of all 1-jets (x, y, p), and a solution of it as a curve of the form $y = f(x), p = f'(x)$, that lies entirely on $\mathscr{E}$. The tangents to such curves determine

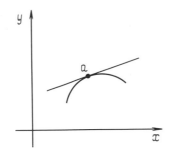

Fig. 25

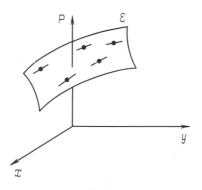

Fig. 26

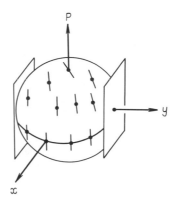

Fig. 27

a field of directions C on $\mathscr{E}$ (Fig. 26). Conversely, the projections on the plane Π of curves that touch this field of directions are local solutions of (1).

Example. If $F(x, y, y') = x^2 + y^2 + y'^2 - 1$, then the surface $\mathscr{E}$ is the sphere $x^2 + y^2 + p^2 = 1$ in $J^1 = \mathbb{R}^3$, and the field of directions C is defined everywhere on it except for the equator $x^2 + y^2 = 1$, $p = 0$. In fact, by vertical tangents it can be extended by continuity to the equator, except for the points $A_{\pm} = \{x = p = 0, y = \pm 1\}$ (Fig. 27). Thus, integral curves of the field C fill the domain $\mathscr{E}\backslash(A_+ \cup A_-)$. Their projections on Π are the graphs of solutions of the equation under consideration. At points of the circle $x^2 + y^2 = 1$ on the plane Π these graphs have a singularity of "beak" type.

1.2. The General Case.

The geometrical picture of the previous subsection is enriched by new details if we consider the equation

$$F\left(x_1, \ldots, x_n, u, \frac{\partial u}{\partial x_1}, \ldots, \frac{\partial u}{\partial x_n}\right) = 0, \tag{2}$$

assuming that $n > 1$. As above, with each function $f(x_1, \ldots, x_n)$ we associate its graph L_f, an n-dimensional hypersurface in the space J^0 with coordinates $(x_1, \ldots, x_n, u)$ specified by the equation $u = f(x)$. A pair (a, P), where $a \in J^0$, and P is a hyperplane passing through the point a, is called a *contact element* or *1-jet*. The pair $(a, T_a(L_f))$, where $a = (\bar{x}_1, \ldots, \bar{x}_n, f(\bar{x}_1, \ldots, \bar{x}_n)) \in L_f$, is a 1-jet of the function f, or equivalently a tangent element of the hypersurface L_f at the point a. A 1-jet is specified by a set of $2n + 1$ numbers $(\bar{x}_1, \ldots, \bar{x}_n, \bar{u}, \bar{p}_1, \ldots, \bar{p}_n)$, where $a = (\bar{x}_1, \ldots, \bar{x}_n, \bar{u}) \in J^0$, and the equation of P has the form

$$u - \bar{u} = \bar{p}_1(x - \bar{x}_1) + \cdots + \bar{p}_n(x - \bar{x}_n). \tag{3}$$

The set specifies a 1-jet of f at the point $(\bar{x}_1, \ldots, \bar{x}_n)$ if

$$\bar{u} = f(\bar{x}), \qquad \bar{p}_i = \frac{\partial f}{\partial x_i}(\bar{x}).$$

Let J^1 be the space of all 1-jets $(x_1, \ldots, x_n, u, p_1, \ldots, p_n)$. Equation (2) can now be interpreted as a certain condition on 1-jets. With it we associate the hypersurface $\mathscr{E} = \{F(x_1, \ldots, x_n, u, p_1, \ldots, p_n) = 0\}$ in J^1. The function $f(x)$ is a solution of it if and only if $L_f^1 \subset \mathscr{E}$, where L_f^1 is the manifold of 1-jets of the function f.

The family of tangent elements of solutions of (2) that pass through the point $a = (\bar{x}, \bar{u}) \in J^0$ is identical to the set of 1-jets satisfying the equation $F(\bar{x}, \bar{u}, p) = 0$. The envelope $K(\bar{x}, \bar{u})$ of this family of hyperplanes is called the *Monge cone* of equation (2) at the point $(\bar{x}, \bar{u})$. The directions of the generator of this cone, called *characteristics*, have the form

$$\left(\frac{\partial F}{\partial p_1}, \ldots, \frac{\partial F}{\partial p_n}, \sum_i p_i \frac{\partial F}{\partial p_i}\right),$$

where the values of the derivatives $\partial F/\partial p_i$ and the coordinates p_i are calculated at the point $(\bar{x}, \bar{u}, p)$ on condition that $F(\bar{x}, \bar{u}, p) = 0$. Hence, in particular, it is

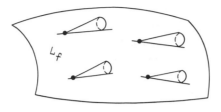

Fig. 28

obvious that for quasilinear equations the Monge cone degenerates to a line—the *Monge axis*—and conversely.

The field of Monge cones $(x, u) \to K(x, u)$ on J^0 when $n > 1$ replaces the multivalued field of directions considered in the previous subsection. In particular, the function $f(x)$ is the solution of (2) if and only if the surface L_f touches the Monge cone at each point of it (Fig. 28). The lines along which this tangency takes place form the *characteristic field* of directions on L_f. The integral curves of this field are called *characteristics*, or *characteristic curves*. Thus, the hypersurface L_f is woven from characteristics, and the graphs of different solutions can touch each other only along characteristics. The characteristics are projections on J^0 of integral curves, lying on $\mathscr{E}$, of the vector field

$$X_F = -\sum_i \frac{\partial F}{\partial p_i} \frac{\partial}{\partial x_i} + \left(F - \sum_i p_i \frac{\partial F}{\partial p_i}\right) \frac{\partial}{\partial u} + \sum_i \left(\frac{\partial F}{\partial x_i} + p_i \frac{\partial F}{\partial u}\right) \frac{\partial}{\partial p_i}, \qquad (4)$$

which touches $\mathscr{E}$.

1.3. Geometrical Integration. The family of solutions $V(x, \lambda)$, which depends on the parameters $\lambda = (\lambda_1, \ldots, \lambda_p)$, is said to be *complete* if any 1-jet $\theta \in \mathscr{E}$ can be represented as a 1-jet of the function $V(x, \bar\lambda)$ at the corresponding point for a suitable choice of the parameter $\lambda = \bar\lambda$. It is not difficult to see that the smallest number of parameters necessary for this is n. In this case $V(x, \lambda)$ is called the *complete integral* of (2).

Let $f(x)$ be a solution of (2). From the definition of the complete integral it follows that any 1-jet of the function f can be represented as a 1-jet of the function $V(x, \lambda)$ for a suitable choice of the parameters λ. These values of the parameters are (locally) unique and are constant along characteristic curves lying on L_f, since the graphs of solutions of (2), if they touch each other, do so necesarily along a whole characteristic (see 1.2). We express this fact by the functional dependence $\lambda_i = \lambda_i(\tau_1, \ldots, \tau_{n-1})$ (briefly, $\lambda = \lambda(\tau)$), where $\tau = (\tau_1, \ldots, \tau_{n-1})$ are the values of the parameters that distinguish the characteristic in question inside the $(n - 1)$-parameter family of them that fills L_f.

From what we have said it follows that L_f is the envelope of an $(n - 1)$-parameter family of surfaces $L_{V(x, \tau)}$, where $V(x, \tau) = V(x, \lambda(\tau))$. It is obvious from 1.2 that the converse is also true. Thus, by constructing the envelopes of different $(n - 1)$-parameter families of solutions that form the complete integral we can obtain all non-singular solutions of (2). These arguments do not cover singular cases when the characteristics are not defined (see the example in 1.1).

These geometrical constructions, set up analytically, constitute *Lagrange's method* of integrating equations of the form (2).

§2. Contact Geometry and Lie's Theory of First-Order Equations

With any function f we can associate an n-dimensional submanifold L^1_f of J^1 consisting of all its 1-jets, or equivalently all tangent elements of the surface L_f. We emphasize that not every n-dimensional submanifold of J^1 has the form L^1_f. For example, when $n = 1$ among its curves $y = f(x), p = g(x)$ there are only those for which $g(x) = f'(x)$ (see §1.1).

The equivalence:

f is a solution of (2) $\Leftrightarrow L^1_f \subset \mathscr{E}$

leads to the following reformulations:

an equation of the form (2) $\Leftrightarrow$ the hypersurface $\mathscr{E} \subset J^1$,

$$\text{a solution of it} \Leftrightarrow \begin{cases} \text{an } n\text{-dimensional submanifold of the form} \\ L^1_f \text{ lying on } \mathscr{E}. \end{cases}$$

Thus, in order to obtain a consistent geometrical treatment of first-order equations in terms of the space J^1 it is necessary to equip it with an additional geometrical structure that gives the possibility of distinguishing submanifolds of the form L^1_f from the others. Such a structure is the distribution, described below, of $2n$-dimensional subspaces or contact structure on J^1.

2.1. Contact Structure on J^1. The tangent vector to the curve $L^1_f = \{u = f(x),$ $p = f'(x)\}$ at the point $\theta = (x, u, p) \in J^1$ is equal to $(1, p, f''(x))$. All such vectors lie in a plane $C_\theta \subset T_\theta(J^1)$. Thus on J^1 there arises a field of planes or a distribution $C: \theta \to C_\theta$ (Fig. 29), called the *contact structure*. By construction the tangent lines

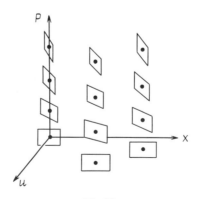

Fig. 29

to the curves L_f^1 lie in the planes C_θ, that is, these curves are *integral curves* for the distribution C.

The plane C_θ is annihilated by the covector $d_\theta u - p\, d_\theta x \in T_\theta^* J^1$. Hence it is obvious that the distribution C is specified by the *Pfaff equation* $\omega = 0$, where $\omega = du - p\, dx$. In particular, $\omega|_{L_f^1} = 0$. Conversely, the restriction of the form ω to the curve $\Gamma = \{u = f(x), p = g(x)\}$ is equal to $(f' - g)\, dx$. Hence $\omega|_\Gamma = 0 \Leftrightarrow \Gamma = L_f^1$. Thus, an integral curve of the distribution C has the form L_f^1 if we can take the coordinate x as a parameter along it.

Thus, when $n = 1$ the contact structure on J^1 enables us to identify the curves L_f^1 among the others.

These constructions can easily be generalized to arbitrary dimensions n. In this case the contact plane $C_\theta \subset T_\theta J^1$ is defined as the linear hull of those vectors $\xi \in T_\theta J^1$ that touch all possible submanifolds of the form L_f^1 passing through the point θ. The form ω that annihilates the resulting distribution (field) of $2n$-dimensional subspaces on J^1 has the form

$$\omega = du - p_1\, dx_1 - \cdots - p_n\, dx_n.$$

The submanifolds L_f^1 are *integral submanifolds* for the distribution C, that is, $T_\theta(L_f^1) \subset C_\theta$, $\forall \theta \in L_f^1$ or equivalently $\omega|_{L_f^1} = 0$. Conversely, n-dimensional integral submanifolds $L \subset J^1$, projected without degeneracy on the x-space, have the form L_f^1. We note that integral manifolds cannot have dimension greater than n.

2.2. Generalized Solutions and Integral Manifolds of the Contact Structure.
Equipping the manifold J^1 with a contact structure gives the possibility of saying geometrically that such a solution of the equation $\mathcal{E}$ belongs to J^1 and hence of successively developing the whole theory of scalar first-order differential equations geometrically.

Bearing in mind the characterization, given in the previous subsection, of submanifolds L_f^1, it is appropriate to call an n-dimensional integral manifold $L \subset J^1$ an (ordinary) solution of the equation $\mathcal{E}$ if 1) $L \subset \mathcal{E}$ and 2) L is projected without degeneracy onto the x-space. The latter condition, being associated with a map into the x-space which is "extraneous" for J^1, seems superfluous from the point of view of the intrinsic contact geometry of the space J^1. It is therefore natural to omit it. This was first done by Lie, who thereby arrived at a generalization of the concept of a solution for equations of the form (2). In this connection it is interesting to know what integral manifolds of the contact structure on J^1 look like. The examples given below answer this question to a certain extent.

Examples. 1) The fibres of the projection $J^1 \to J^0$, $(x, u, p) \to (x, u)$, specified by the equations $x_i = $ const, $u = $ const.

2) The manifold L given by the equations $x_i = c_i$; $i = k + 1, \ldots, n$,

$$p_j = c_j, \quad j = 1, \ldots, k,$$

$$u = c + c_1 x_1 + \cdots + c_k x_k, \qquad \text{where } c, c_i \in \mathbb{R}, \quad i = 1, \ldots, n.$$

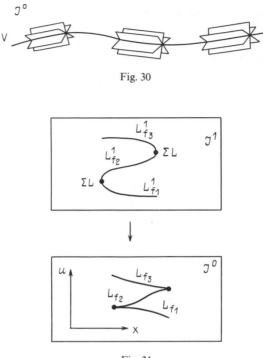

Fig. 30

Fig. 31

Example 1 is a special case of this ($k = 0$). The projection of L into J^0 is a k-dimensional plane.

3) Let $V \subset J^0$ be a k-dimensional submanifold, where $0 \leqslant k \leqslant n$. The manifold $L \subset J^1$ composed of all contact elements (a, P) such that $a \in V$, $P \supset T_a V$ (see Fig. 30) is an integral manifold. V is its image under the projection $J^1 \to J^0$. If $V = \{u = c + c_1 x_1 + \cdots + c_k x_k, x_{k+1} = c_{k+1}, \ldots, x_n = c_n | c_i \in \mathbb{R}\}$, then this construction gives the manifold L of Example 2), and if $k =. n$ it gives a manifold of the form L_f^1.

Locally any integral manifold, if we discard an "inessential" subset of it, is structured like the integral manifold described in Example 3).

Let ΣL denote the subset of singular points of the projection on the x-space of an integral manifold $L \subset J^1$. As a rule, this subset has dimension less than n (this can always be achieved by an arbitrarily small deformation of L). In this case the n-dimensional integral manifold $L - \Sigma L$ is locally projected diffeomorphically onto the x-space and is therefore locally representable in the form L_f^1. The function f occurring here can vary depending on the domain of the manifold $L - \Sigma L$. In other words, it is multivalued (Fig. 31). Thus, an integral manifold L, on condition that $\dim(\Sigma L) < n$, can be interpreted as a manifold of the form L_f^1, where the function f is multivalued. In this connection it is appropriate to call the generalized solutions of (2) defined above *multivalued*.

Example. The Clairaut equation $u = \sum\limits_{i=1}^{n} x_i \dfrac{\partial u}{\partial x_i} + \varphi\left(\dfrac{\partial u}{\partial x_1}, \ldots, \dfrac{\partial u}{\partial x_n}\right)$ has gener-

alized solution L given by the equations

$$x_i = -\frac{\partial \varphi}{\partial p_i}(p_1, \ldots, p_n), \qquad i = 1, \ldots, n,$$

$$u = \varphi(p_1, \ldots, p_n) - \sum_{i=1}^{n} p_i \frac{\partial \varphi}{\partial p_i}(p_1, \ldots, p_n).$$

The set ΣL on L is distinguished by the additional equation

$$\det \left\| \frac{\partial^2 \varphi}{\partial p_i \partial p_j} \right\| = 0.$$

2.3. Contact Transformations. A diffeomorphism $F: J^1 \to J^1$ that preserves the distribution C is called a *contact transformation*. This means that $d_\theta F(C_\theta) = C_{F(\theta)}$, or equivalently that $F^*(\omega) = \lambda\omega$, $\lambda \in C^\infty(J^1)$. In coordinates the latter condition is written in the form

$$dU - P\,dX = \lambda(du \quad p\,dx), \tag{5}$$

where $X = F^*(x)$, $U = F^*(u)$, $P = F^*(p)$ and we use the abbreviated notation $x = (x_1, \ldots, x_n)$, $p = (p_1, \ldots, p_n)$.

Obviously, contact transformations map integral manifolds into integral manifolds. The converse is also true, since the distribution C, as its construction shows, is uniquely determined by its integral manifolds (for example, those having the form L_f^1).

Examples. 1) An arbitrary transformation of the space J^0 $f: (x, u) \mapsto (X(x, u), U(x, u))$ can be extended to a contact transformation $F: (x, u, p) \mapsto (X, U, P)$. The function $P = P(x, u, p)$ is found from (5):

$$P = \left[\frac{\partial X}{\partial x} + p\frac{\partial X}{\partial u}\right]^{-1}\left(\frac{\partial U}{\partial x} + \frac{\partial U}{\partial u}p\right).$$

Here we have used the obvious matrix notation. For example, $\dfrac{\partial X}{\partial x} = \left\|\dfrac{\partial X_i}{\partial x_j}\right\|$.

Transformations of this kind are called *point transformations*.

2) The *Legendre transformation*

$$(x, u, p) \mapsto (p, u - px, -x)$$

is remarkable in that it interchanges the "coordinates" x and the "impulses" p.

3) The *Euler transformation* or *partial Legendre transformation*

$$(x_1, \ldots, x_n, u, p_1, \ldots, p_n) \mapsto (p_1, \ldots, p_k, x_{k+1}, \ldots$$

$$\ldots, x_n, u - p_1 x_1 - \cdots - p_k x_k, -x_1, \ldots, -x_k, p_{k+1}, \ldots, p_n).$$

For any two integral manifolds L_1 and L_2 and arbitrary points $a_i \in L_i$ on them there is a contact transformation $F: J^1 \to J^1$ that maps a certain neighbourhood of the point a_1 on L_1 into a neighbourhood of the point a_2 on L_2, and the point a_1 into a_2. It is easy to see that by point transformations (Example 1) corresponding to diffeomorphisms of the form $(x, u) \mapsto (x + c, u)$ and $(x, u) \mapsto (x, u + f(x))$ of the space J^0 we can achieve this for manifolds L_i of the form L_f^1. Moreover, a sufficiently small neighbourhood of a point $a \in \Sigma L$ on an integral manifold L can be mapped by an Euler transformation (Example 3) onto a manifold of the form L_f^1. This reduces the general case to that already considered. Thus, from the point of view of contact geometry all integral manifolds are locally structured in the same way.

It is remarkable that any contact transformation can be locally specified by one function of $2n + 1$ variables, for example, $v = \Phi(X, x, u)$. In fact, using the relations

$$U = \Phi, \qquad P = \partial\Phi/\partial X, \qquad p = -\frac{\partial\Phi}{\partial x}\bigg/\frac{\partial\Phi}{\partial u},$$

the last two of which follow from (5), we can find X, U, P as functions of x, u, p if

$$\det\left\|\frac{\partial}{\partial X}\left(\frac{\partial\Phi}{\partial x}\bigg/\frac{\partial\Phi}{\partial u}\right)\right\| \neq 0.$$

The function $\Phi(X, u, p)$ is called the *generating function of the contact transformation.*

Example. The function $\Phi = u - \sum x_i X_i$ is the generating function of the Legendre transformation.

2.4. Contact Vector Fields. A vector field on J^1 is called a *contact vector field* if the current $A_t: J^1 \to J^1$ generated by it consists of contact transformations. The manifold $A_t(L_h^1)$ is an integral manifold and (locally) for sufficiently small t it is projected without degeneracy onto the x-space (see Fig. 32). It therefore has the form $L_{h_t}^1$ for some function $h_t = h_t(x)$ (see 2.1). Hence the field X generates the (local) current $\{h_t\}$ in the space of functions $h = h(x)$. It is easy to see that the "velocity vector" of this current $\dot h = \dfrac{dh_t}{dt}\bigg|_{t=0}$ depends only on the 1-jet of the

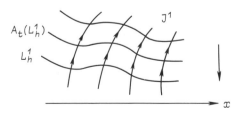

Fig. 32

function h. More precisely, $\dot{h}_1(x) = \dot{h}_2(x)$ if $[h_1]_x^1 = [h_2]_x^1$. This gives the possibility of defining a function $f \in C^\infty(J^1)$ whose value at the point $\theta = [h]_x^1$ is equal to $\dot{h}(x)$. Since $[h]_x^1 = \left(x, h(x), \dfrac{\partial h}{\partial x} \right)$, what we have said can be expressed by the formula

$$\left. \frac{dh_t}{dt} \right|_{t=0} = f\left(x, h(x), \frac{\partial h}{\partial x}(x) \right).$$

Thus, any contact vector field X generates a current in the space of functions of the variables x, whose velocity vector is described by a function $f \in C^\infty(J^1)$. It is remarkable that conversely the field X itself is uniquely determined by this function. For this reason it is denoted by X_f:

$$X_f = -\sum_i \frac{\partial f}{\partial x_i} \frac{\partial}{\partial p_i} + \left(f - \sum_i p_i \frac{\partial f}{\partial p_i} \right) \frac{\partial}{\partial u} + \sum_i \left(\frac{\partial f}{\partial x_i} + p_i \frac{\partial f}{\partial u} \right) \frac{\partial}{\partial p_i}.$$

In this expression we recognize the field of characteristics (4).

A function f that describes the field X in this way is called a *generating function*, or its *contact Hamiltonian*, and is defined by the formula

$$f = \omega(X), \quad \omega = du - \sum_i p_i \, dx_i.$$

Contact vector fields form a Lie algebra whose structure carries over to the space of functions $C^\infty(J^1)$ by means of the one-to-one correspondence $X \leftrightarrow f$ established above. The resulting Lie operation on $C^\infty(J^1)$ is denoted by $\{\cdot, \cdot\}$ and called the *Lagrange bracket*:

$$(f, g) \mapsto \{f, g\} = \omega([X_f, X_g]), \qquad f, g \in C^\infty \cdot (J^1).$$

In accordance with this definition,

$$[X_f, X_g] = X_{\{f,g\}}.$$

In coordinates we have

$$\{f, g\} = \sum_i \left(\frac{df}{dx_i} \frac{\partial g}{\partial p_i} - \frac{dg}{dx_i} \frac{\partial f}{\partial p_i} \right) + f \frac{\partial g}{\partial u} - g \frac{\partial f}{\partial u},$$

where $d/dx_i = \partial/\partial x_i + p_i \dfrac{\partial}{\partial u}$. For functions f and g that do not depend on the variable u this formula turns into a formula for Poisson brackets (see Ch. 5, §5.4) on the space $T^*(\mathbb{R}^n)$, which is understood as the space of the variables (x, p).

2.5. The Cauchy Problem. The geometrical solution of the Cauchy problem for equation (2) goes as follows. The hypersurface N in the x-space on which the initial conditions are specified, together with the initial conditions themselves, that is, the function φ on N, specify an $(n - 1)$-dimensional submanifold $N_\varphi = \{(x, u) | x \in N, u = \varphi(x)\}$ in J^0. The totality of those elements of tangency at a point $y \in N_\varphi$ that contain the subspace $T_y(N_\varphi)$ form a straight line L_y that lies in

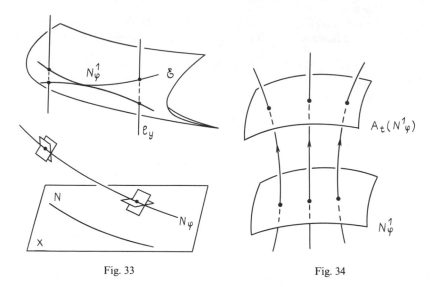

Fig. 33   Fig. 34

the fibre of the projection $\pi_{1,0}: J^1 \to J^0$ over the point y (see Fig. 33). The line L_y may intersect the equation $\mathscr{E} \subset J^1$ in certain points (the non-characteristic case), and it may touch it (the characteristic case) (see Fig. 33). These points of intersection form an $(n-1)$-dimensional integral manifold (possibly with singularities) $N_\varphi^1 \subset \mathscr{E}$, called the *Cauchy data* (in the geometrical sense). In the non-characteristic case it is non-singular and the field X_f, where $\mathscr{E} = \{f = 0\}$, does not touch it. Therefore the manifold $L = \bigcup_t A_t(N_\varphi^1)$ obtained by carrying the Cauchy data by the current A_t generated by the field X_f (see Fig. 34) has dimension n. It is an integral manifold, and since $0 = f = \omega(X_f)$ on the hypersurface $\mathscr{E}$ it is a solution of (2). If we adopt the abstract intrinsic point of view, we can take for the Cauchy data any $(n-1)$-dimensional submanifold of $\mathscr{E}$ with which we can proceed in exactly the same way to find an n-dimensional integral manifold passing through it. The construction of the manifold N_φ^1 is connected with discovering the points of intersection of the lines L_y with $\mathscr{E}$ and therefore reduces to the solution of "algebraic" equations. The current A_t is found by integrating the system of ordinary differential equations corresponding to the field X_f.

2.6. Symmetries. Local Equivalence. A *symmetry of the equation* $\mathscr{E} \subset J^1$ is a contact transformation $F: J^1 \to J^1$ such that $F(\mathscr{E}) = \mathscr{E}$. Obviously, symmetry transformations map integral manifolds lying in $\mathscr{E}$ into manifolds of the same kind, that is, it maps solutions of the equation $\mathscr{E}$ into solutions of $\mathscr{E}$, thereby generat transformation of its space of solutions.

A contact vector field X_g touching the equation $\mathscr{E} \subset J^1$ is called an *infinitesimal symmetry* of it. A necessary and sufficient condition for this is

$$\{f, g\} = 0 \qquad \text{on} \qquad \mathscr{E} = \{f = 0\}.$$

The current A_t generated by the field X_f consists of contact transformations and therefore in turn it generates a current $\hat{A}_t$ on the space of solutions of the equation $\mathscr{E}$. If $\hat{A}_t = $ id, then the field X_g is said to be *characteristic* with respect to $\mathscr{E}$. This is equivalent to the fact that $g|_{\mathscr{E}} = 0$. In particular, the field X_f is characteristic. Any solution of the equation $\mathscr{E}$ is woven from trajectories of the characteristic vector field; this was used above in the solution of the Cauchy problem.

Equations $\mathscr{E} \subset J^1$, $\mathscr{E}' \subset J^1$ are said to be *equivalent* if they are mapped into each other by some contact transformation. We can also talk about their *local equivalence* close to given points $a \in \mathscr{E}$, $a' \in \mathscr{E}'$.

Equivalent equations have the same stock of generalized (or multivalued) solutions (see 2.2), but generally speaking different stocks of ordinary solutions. The latter follows from the fact that a contact transformation may map integral manifolds of the form L_h^1 into integral manifolds of a different type (see 2.2).

A point $a \in \mathscr{E}$ is said to be *regular* if $T_a\mathscr{E} \neq C_a$, and *singular* otherwise.

Theorem (Lie). *Any two equations $\mathscr{E} \subset J^1$, $\mathscr{E}^1 \subset J^1$ are locally equivalent close to their regular points.*

Example. The Legendre transformation maps the equation $u = \sum x_i \partial u/\partial x_i$ into the equation $u = 0$. From Lie's theorem it follows that locally any equation (2) can be reduced to the equation $u = 0$ close to any regular point of it.

The problem of local classification of equations $\mathscr{E} \subset J^1$ close to singular points admits a solution, but it is substantially more complicated.

§ 3. The Geometry of Distributions

In the previous section we interpreted differential equations of the first order as hypersurfaces in some universal manifold J^1, endowed with a contact structure. This is completely analogous to what happens in the theory of surfaces:

$$J^1 \leftrightarrow E^n$$

contact structure $\leftrightarrow$ Euclidean metric.

Going over to the intrinsic point of view in the latter, we can arrive at Riemannian geometry, as we have already seen. If we carry out a similar transiton in the situation under consideration, we arrive at Cartan's theory of differential equations as "differential systems" (see § 6 below).

The intrinsic point of view in the geometrical theory of first-order differential equations consists in restricting to the hypersurface $\mathscr{E}$ a contact structure with ambient manifold J^1, and after this forgetting the existence of the latter. This is quite sufficient, since the integral manifolds of this restricted contact structure are solutions of the original equation $\mathscr{E}$, and conversely.

A restricted contact structure is a field of subspaces on $\mathscr{E}$. In this section we give initial information connected with this concept, which is fundamental for the geometry of differential equations.

3.1. Distributions. A *distribution* $\mathcal{F}$ *on a smooth manifold* M is a field of subspaces $\mathcal{F}_x \subset T_x M$, $x \in M$, of dimenson m, that depend smoothly on a point $x \in M$. The number m (respectively, $n - m$) is called the *dimension (codimension)* of the distribution $\mathcal{F}$. The following (local) methods of describing distributions are the most widely used:

1) by a choice of m independent vector fields belonging to $\mathcal{F}$, that is, vector fields $X_1, \ldots, X_m$ such that $X_i(a) \in \mathcal{F}_a$ for all points a of the neighbourhood in question;

2) by a choice of $n - m$ independent differential 1-forms $\omega_1, \ldots, \omega_{n-m}$ that annihilate $\mathcal{F}$, that is, $\omega_i(\xi) = 0$ for all $\xi \in \mathcal{F}_a$.

In the first case the distribution $\mathcal{F}$ will be denoted by $\mathcal{F}(X_1, \ldots, X_m)$, and in the second case by $\mathcal{F}(\omega_1, \ldots, \omega_{n-m})$.

Examples. 1) A distribution C that specifies a contact structure on J^1 (see 2.1) can be represented either in the form $\mathcal{F}(\omega)$, where $\omega = du - \sum p_i \, dx_i$, or in the form

$$\mathcal{F}\left(\frac{\partial}{\partial x_1} + p_1 \frac{\partial}{\partial u}, \ldots, \frac{\partial}{\partial x_n} + p_n \frac{\partial}{\partial u}, \frac{\partial}{\partial p_1}, \ldots, \frac{\partial}{\partial p_n}\right).$$

2) Any distribution of codimension 1 locally has the form $\mathcal{F}(\omega)$ for some 1-form ω. In the large, on the whole manifold, this is false, generally speaking. Thus a distribution on $\mathbb{R}^{n+1} \times \mathbb{R}P^n$, which in coordinates $(x_1, \ldots, x_{n+1}, [y_1 : \ldots : y_{n+1}])$, where $[y_1 : \ldots : y_{n+1}]$ are projective coordinates in $\mathbb{R}P^n$, is described by the equation $\sum y_i \, dx_i = 0$, cannot be specified by a unique differential form.

A submanifold $L \subset M$ is called an *integral submanifold* for the distribution $\mathcal{F}$ if $T_x L \subset \mathcal{F}_x$ for all $x \in L$. For distributions of the form $\mathcal{F} = \mathcal{F}(\omega_1, \ldots, \omega_{n-m})$ this is equivalent to the fact that

$$\omega_1|_L = \cdots = \omega_{n-m}|_L = 0.$$

Examples. 1) A distribution of dimension 1 is a field of directions. The integral curves are its one-dimensional integral manifolds.

2) In the space $\mathbb{R}^3$ with coordinates (x, u, p) we consider two forms $\omega_1 = du$ and $\omega_2 = du - p \, dx$. The integral manifolds of the distribution $\mathcal{F}(\omega_1)$ are either planes $u = \text{const}$ or arbitrary curves lying in them. The distribution $\mathcal{F}(\omega_2)$ does not have 2-dimensional integral manifolds (see 3.2). Curves of the form L_f^1 are examples of 1-dimensional integral manifolds of this distribution.

An integral manifold L of a distribution $\mathcal{F}$ is said to be *locally maximal* if no domain $U \subset L$ can be embedded in an integral manifold of higher dimension.

Examples. 3) In Example 2 integral manifolds of the form $u = \text{const}$ of the distribution $\mathcal{F}(\omega_1)$ (and only they) are locally maximal.

4) In the space $\mathbb{R}^8$ with coordinates $(x_1, x_2, u, p_1, p_2, p_{11}, p_{12}, p_{22})$ the forms

$$\begin{cases} \omega_0 = du - p_1 \, dx_1 - p_2 \, dx_2, \\ \omega_1 = dp_1 - p_{11} \, dx_1 - p_{12} \, dx_2, \\ \omega_2 = dp_2 - p_{12} \, dx_1 - p_{22} \, dx_2 \end{cases}$$

specify a 5-dimensional distribution. Surfaces of the form $L_f^2 = \left\{ u = f(x_1, x_2), \right.$ $p_i = \dfrac{\partial f}{\partial x_i}, p_{ij} = \dfrac{\partial^2 f}{\partial x_i \partial x_j} \right\}$ are locally maximal 2-dimensional integral manifolds, and the subspaces $x_1 = a_1$, $x_2 = a_2$, $u = c$, $p_1 = b_1$, $p_2 = b_2$ are 3-dimensional. There are no integral manifolds of dimension 4 or more in this distribution (see 2.2).

A distribution $\mathscr{F}$ on a manifold M is said to be *completely integrable* (or involutive) if through each point $x \in M$ there passes an integral manifold of dimension equal to dim $\mathscr{F}$.

In Example 2 the distribution $\mathscr{F}(\omega_1)$ is competely integrable, but the distribution $\mathscr{F}(\omega_2)$ is not. The distribution of Example 4 is not completely integrable.

Let $L \subset M$ be an integral manifold of a completely integrable distribution $\mathscr{F} = \mathscr{F}(\omega_1, \ldots, \omega_{n-m})$. Then, since the operator of exterior differentiation is natural, we have $(d\omega_i)|_L = d((\omega_i)|_L) = 0$. Hence the differential 2-forms $d\omega_i$ at each point $x \in M$ vanish identically on the subspace $\mathscr{F}_x$.

From elementary algebraic facts of the calculus of exterior forms it follows that this is equivalent to

$$d\omega_i \quad \sum \lambda_{ij} \wedge \omega_j \tag{6}$$

for some 1-forms λ_{ij}.

Condition (6) can be written in another way:

$$d\omega_i \wedge \omega_1 \wedge \cdots \wedge \omega_{n-m} = 0 \tag{7}$$

for all $i = 1, \ldots, n - m$.

Thus, conditions (7) are necessary in order that the distribution $\mathscr{F}(\omega_1, \ldots, \omega_{n-m})$ should be completely integrable. A fundamental theorem of Frobenius asserts that these conditions are sufficient.

Example. Finding integral manifolds of the distribution of codimension 1 in the space $\mathbb{R}^{n+1}$ given by the 1-form $\omega = du - f_1(q) \, dq_1 - \cdots f_n(q) \, dq_n$ is equivalent to solving the system of differential equations $p_1 = f_1(q), \ldots, p_n = f_n(q)$. The conditions of the theorem of Frobenius $\omega \wedge d\omega = 0$ in the given case are equivalent to the compatibility conditions $\dfrac{\partial f_i}{\partial q_j} - \dfrac{\partial f_j}{\partial q_i} = 0$, $i, j = 1, \ldots, n$, of this system.

In the case when the distribution $\mathscr{F}$ is given by a set of vector fields $X_1, \ldots, X_m$, $\mathscr{F} = \mathscr{F}(X_1, \ldots, X_m)$, formula (12) of Ch. 3,

$$d\omega_s(X_i, X_j) = X_i(\omega_s(X_j)) - X_j\omega_s(X_i) - \omega_s([X_i, X_j])$$

shows that the conditions (6) are equivalent to

$$[X_i, X_j] = \sum_k A_{ij}^k X_k, \tag{8}$$

where the A_{ij}^k are smooth functions.

We owe the conditions (8) and (6) of complete integrability to Lie and Frobenius, respectively. When forming them Frobenius used the "bilinear co-

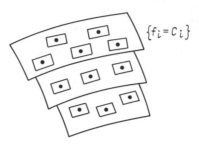

Fig. 35

variant" of a form (of the first degree)—the inverse image of its modern exterior differential. This is one of the sources of the differential calculus of forms (see Ch. 3, § 3).

The maximal integral manifolds of a completely integrable distribution (locally) fibre the manifold M as shown in Fig. 35. Hence, again locally, we can choose $n - m$ independent functions $f_1, \ldots, f_{n-m}$ whose joint level surfaces coincide with maximal integral manifolds. This is equivalent to the fact that (locally) a completely integrable distribution is given by complete differentials: $df_1 = 0, \ldots, df_{n-m} = 0$.

3.2. A Distribution of Codimension 1. The Theorem of Darboux. The conditions of the theorem of Frobenius for distributions $\mathscr{F}(\omega)$ of codimension 1 have the form $\omega \wedge d\omega = 0$. Since, as a rule, $\omega \wedge d\omega \neq 0$ when $n > 2$, a typical distribution of codimension 1 is not completely integrable, that is, it does not have integral manifolds of codimension 1. To estimate the possible dimension of integral manifolds of the distribution $\mathscr{F}(\omega)$ we use the implication

$$\omega|_L = 0 \Rightarrow (d\omega)|_L = 0.$$

This means that the subspace $T_a L$ tangent to an integral manifold L at a point $a \in L$ is *isotropic* for the restriction of the 2-form $d_a \omega$ to the subspace F_a, that is, $d_a \omega(\zeta, \eta) = 0$ for all $\zeta, \eta \in T_a L$. If the rank of the restriction Ω of the 2-form $d_a \omega$ to the subspace F_a is equal to $2r$ (it is always an even number), then it is easy to see that the dimension of an isotropic subspace of F_a cannot exceed $n - r - 1$.

The number r, which is an invariant of the distribution at the point $a \in M$, can be calculated as follows. The rank of the form Ω (see above) is equal to $2r$ if $\Omega^r \neq 0$ but $\Omega^{r+1} = 0$. In terms of the form ω these conditions have the form

$$\omega_a \wedge (d_a \omega)^r \neq 0, \qquad \text{but} \qquad \omega_a \wedge (d_a \omega)^{r+1} = 0,$$

which gives the possibility of finding the number $r = r(a)$.

The *class of the distribution* $\mathscr{F} = \mathscr{F}(\omega)$ *at the point* $a \in M$ is the number $2r + 1$. If this number does not depend on the point $a \in M$, it is called the *class of the distribution* $\mathscr{F}$ *on the manifold* M. From this definition it follows that a distribution $\mathscr{F}$ of class $2r + 1$ does not have integral manifolds of dimension greater than

$$n - r - 1 = \dim \mathscr{F} - r.$$

Examples. 1) The class of a completely integrable distribution is equal to 1.

2) The contact distribution C on the space of 1-jets J^1 has class $2n + 1 = \dim J^1$.

The product $\mathscr{F} \times \mathscr{G}$ of distributions $\mathscr{F}$ and $\mathscr{G}$ defined on the manifolds M and N respectively is a distribution on $M \times N$:

$$(\mathscr{F} \times \mathscr{G})_{(x,y)} = \mathscr{F}_x \oplus \mathscr{G}_y \subset T_x M \oplus T_y N = T_{(x,y)}(M \times N).$$

Obviously, $\dim(\mathscr{F} \times \mathscr{G}) = \dim \mathscr{F} + \dim \mathscr{G}$.

The distribution $\mathscr{T}_N \colon x \to T_x N$ on the manifold N is said to be *trivial*. We observe that $\operatorname{codim}(\mathscr{F} \times \mathscr{T}_N) = \operatorname{codim} \mathscr{F}$ and the class of a distribution of codimension 1 does not change on multiplication by a trivial distribution.

Theorem of Darboux (for distributions). *Any distribution of codimension 1 that has constant class is locally the product of a contact distribution and a trivial distribution.*

Among contact distributions we must also include the zero-dimensional distribution $x \to 0 \in T_x \mathbb{R}$ on the line. Then a completely integrable distribution (it is locally representable as the product of this distribution and a trivial distribution) turns out to be included by this formulation. We also observe that contact distributions are indecomposable, that is, they cannot be represented as the product of distributions on manifolds of lower dimension.

The parallel result also holds for 1-forms.

Theorem of Darboux (for 1-forms). *Suppose that a differential form ω of degree 1 determines a distribution of class $2r + 1$ on a manifold M. Then in a neighbourhood of each point $a \in M$ we can choose a system of local coordinates $x_1, \ldots, x_{2r+1}, \ldots, x_n$, $x_i(a) = 0$, in which the form is written in one of the following two ways:*

$$\omega = x_1 \, dx_{r+1} + \cdots + x_r \, dx_{2r} + dx_{2r+1},$$

if

$$\omega \wedge (d\omega)^r \neq 0, \qquad \text{but} \qquad (d\omega)^{r+1} = 0,$$

or

$$\omega = (1 + x_1) \, dx_{r+2} + \cdots + x_{r+1} \, dx_{2r+2},$$

if

$$\omega \wedge (d\omega)^r \neq 0 \qquad \text{and} \qquad (d\omega)^{r+1} \neq 0.$$

The theorem of Darboux makes clear the meaning of the concept of the class of a distribution—it is the smallest number of variables on which a differential 1-form that specifies it can depend.

According to the theorem of Darboux, a trivial distribution $\mathscr{F}_0 \subset \mathscr{F}$ occurring as a factor in the distribution $\mathscr{F}$ can be distinguished in an intrinsically invariant way. It is said to be *characteristic*, and its integral manifolds are called *charac-*

teristics. If $\mathscr{F} = \mathscr{F}(\omega)$, then a characteristic distribution has the form

$$\mathscr{F}_{0,a} = \{\xi \in \mathscr{F}_a | \xi \lrcorner d_a \omega = 0\}.$$

Maximal integral manifolds of the distribution $\mathscr{F}$ are "woven" from characteristics. Hence locally they are products of characteristics and integral manifolds of the contact structure (see 2.2).

3.3. Involutive Systems of Equations. We use the description (given above) of integral manifolds of distributions of codimension 1 in the investigation of general, that is, generally speaking, overdetermined, systems of first-order differential equations. Let $\mathscr{E} \subset J^1$ be such a system and suppose that codim $\mathscr{E} = k > 1$. The restriction of a contact distribution C with space J^1 to the manifold $\mathscr{E}$ determines on it a family of subspaces

$$C(\mathscr{E}): \mathscr{E} \ni \theta \mapsto C_\theta(\mathscr{E}) = C_\theta \cap T_\theta \mathscr{E}.$$

The dimension of the space $C_0(\mathscr{E})$ is equal to $2n - k$ if $C_\theta \not\supset T_\theta \mathscr{E}$, or $2n - k + 1$ if $C_\theta \supset T_\theta \mathscr{E}$. In the first case the point $\theta \in \mathscr{E}$ is said to be *regular*, and in the second case *singular*.

Henceforth we shall assume that the equation $\mathscr{E}$ is *regular*, that is, all its points are regular. Then the family of subspaces $C_0(\mathscr{E})$ is a distribution $C(\mathscr{E})$ on a manifold $\mathscr{E}$ of dimension $2n - k$. Obviously, $C(\mathscr{E}) = \mathscr{F}(\bar{\omega})$, where $\bar{\omega}$ is the restriction of the form $\omega = du - p\,dx$ to the manifold $\mathscr{E}$.

We say that an equation $\mathscr{E}$ is *soluble at a point* $\theta \in \mathscr{E}$ if there is an n-dimensional integral manifold $L \subset \mathscr{E}$ of the distribution $C(\mathscr{E})$ that passes through this point. An equation $\mathscr{E}$ is said to be *involutive* if it is soluble at each point of it.

The solubility condition can be expressed in terns of the class $2r + 1$ of the distribution $C(\mathscr{E})$. In fact, in this case maximal integral manifolds of the distribution $C(\mathscr{E})$ have dimension equal to dim $C(\mathscr{E}) - r = 2n - k - r$. Hence for the solubility of the equation $\mathscr{E}$ it is necessary and sufficient that $n = 2n - k - r$, that is, $r = n - k$, or that

$$\bar{\omega} \wedge (d\bar{\omega})^{n-k+1} = 0.$$

If $\mathscr{E} = \{f_1 = 0, \ldots, f_k = 0\}$, where the functions $f_i \in C^\infty(J^1)$ are independent, then the condition for involutivity can be described by means of the Lagrange bracket:

$$\{f_i, f_j\} = 0 \quad \text{on} \quad \mathscr{E}.$$

The characteristic distribution $C_0(\mathscr{E})$ of the distribution $C(\mathscr{E})$ for involutory equations is spanned by the contact vector fields $X_{f_1}, \ldots, X_{f_k}$: $C_0(\mathscr{E}) = \mathscr{F}(X_{f_3}, \ldots, X_{f_k})$. Hence every maximal integral manifold $L \subset \mathscr{E}$ is invariant with respect to these fields. This enables us, as we did in 2.5 for definite equations, to construct its solution geometrically.

We define the *Cauchy data* for an involutive differential equation $\mathscr{E} \subset J^1$, codim $\mathscr{E} = k$, as an $(n - k)$-dimensional integral manifold $L_0 \subset \mathscr{E}$ of the distribution $C(\mathscr{E})$ that does not touch the characteristic distribution $C_0(\mathscr{E})$. The solution of the Cauchy problem consists in restoring the n-dimensional integral manifold $L \subset \mathscr{E}$ containing L_0. For this it is sufficient to draw a k-dimensional characteristic through each point $\theta \in L_0$.

3.4. The Intrinsic and Extrinsic Geometry of First-Order Differential Equations.

The question of the relation between the intrinsic and extrinsic geometries of a system of equations $\mathscr{E} \subset J^1$ according to Klein (see Klein [1872]) is equivalent to the question of the link between the inner and outer automorphism (that is, symmetries) of this equation. Let us consider this in more detail.

An *outer symmetry of a differential equation* $\mathscr{E} \subset J^1$ is a contact transformation $F\colon J^1 \to J^1$ that preserves $\mathscr{E}$.

An *inner symmetry of a differential equation* $\mathscr{E}$ is a transformation $F\colon \mathscr{E} \to \mathscr{E}$ that preserves the distribution $C(\mathscr{E})$: $d_\theta F(C_\theta(\mathscr{E})) = C_{F(\theta)}(\mathscr{E})$.

We denote by $\mathrm{Sym}_o(\mathscr{E})$ (respectively, $\mathrm{Sym}_i(\mathscr{E})$) the group of outer (respectively, inner) symmetries of the differential equation $\mathscr{E}$. The operation of restricting an outer symmetry to $\mathscr{E}$ is generated by the homomorphism $\kappa\colon \mathrm{Sym}_o(\mathscr{E}) \to \mathrm{Sym}_i(\mathscr{E})$. We say that the intrinsic geometry is determined by the extrinsic if κ is an epimorphism. A differential equation is said to be *rigid* if κ is an isomorphism.

Example. The simplest ordinary differential equation $y' = 0$ determines a plane $\mathscr{E}$ in the 3-dimensional space J^1, $\mathscr{E} = \{(x, u, p) \in J^1 | p = 0\}$. The distribution $C(\mathscr{E})$ is the distribution of tangents to the curves $u = \mathrm{const}$. Hence every inner symmetry of the equation $\mathscr{E}$ has the form $F\colon (x, u) \mapsto (Q(x, u), U(u))$. This transformation determines a pointwise contact transformation $\tilde{F}\colon (x, u, p) \mapsto (Q(x, u), U(u), P(x, u, p))$, where $P = pU_u/Q_x + pQ_u$ (see 2.3). Obviously, $F(\mathscr{E}) = \mathscr{E}$ and $\tilde{F}|_{\mathscr{E}} = F$. Thus, the map κ is an epimorphism. The one-parameter group of shifts along the contact vector field X_f, where $f = p^2$, induces identity transformations of the submanifold $\mathscr{E}$, so the map κ has a nontrivial kernel.

The situation described in this example is general for involutive systems of scalar differential equations of the first order. This assertion is a consequence of a theorem of Lie for involutive differential equations that generalizes the similar theorem in 2.6.

Theorem (Lie). *Involutive systems of first-order differential equations of the same codimension are locally equivalent at regular points.*

In particular, every involutive system of differential equations of codimension k is locally equivalent to the system $p_1 = \cdots = p_k = 0$. Arguments similar to those given in the previous example show that for involutive systems of the first order the map κ is (locally) an epimorphism with non-trivial kernel. Hence in the case under consideration the intrinsic geometry is determined by the extrinsic, although the link between them is not rigid. In the next section we discuss these questions for general differential equations of arbitrary order. As a rule, such equations are rigid. The exceptions are the first-order differential equations considered here and ordinary differential equations.

§4. Spaces of Jets and Differential Equations

The first steps in geometrizing the theory of differential equations of arbitrary order can be carried out along the same lines as for first-order equations. Let us take these steps, paying attention to those additional details that arise. They are important in that they emphasize the sharp difference between the geometry of equations of higher orders and the geometry of first-order equations considered above.

In the case under consideration, manifolds of jets have the same foundation on which the basic constructions of the geometrical theory of equations of higher order are carried out. In this chapter they are described locally for simplicity. This deficiency is made good in Ch. 6, §4.

4.1. Jets. A general system of partial differential equations of order k in the vector-valued functions $h(x) = (h^1(x), \ldots, h^m(x))$ defined on the space $\mathbb{R}^n \ni x$ has the form

$$F_s\left(x, h(x), \ldots, \frac{\partial^{|\sigma|}h}{\partial x^\sigma}(x), \ldots\right) = 0, \tag{9}$$

where $\sigma = (\sigma_1, \ldots, \sigma_n)$ is a multi-index of length $|\sigma| = \sigma_1 + \cdots + \sigma_n \leqslant k$.

The functions F_i that occur in this system can be regarded as functions on a set of rows of the form $\left(h(a), \ldots, \dfrac{\partial^{|\sigma|}h}{\partial x^\sigma}(a), \ldots\right)$, $a \in \mathbb{R}^n$, $|\sigma| \leqslant k$. These rows are denoted by $[h]_a^k$ and are called k-jets of the function $h(x)$ at points $a \in \mathbb{R}^n$. A k-jet can be identified with a segment of the Taylor series of the function $h(x)$ at the point $a \in \mathbb{R}^n$.

We say that functions $f(x)$ and $g(x)$ are k-equivalent at the point $a \in \mathbb{R}^n$ if their difference $f(x) - g(x)$ has order of smallness at least $k + 1$ at this point. Geometrically the condition of k-equivalence implies tangency of the graphs of $f(x)$ and $g(x)$ at the point $x = a$ with order at least k. The k-equivalence class of a function at a point $a \in \mathbb{R}^n$ is called a k-jet. Obviously, this invariant definition of a jet coincides with that given above.

Let $J^k(n, m)$ denote the set of all k-jets of vector-valued functions defined on the arithmetic space $\mathbb{R}^n$ and taking values in $\mathbb{R}^m$. As coordinates in this space, which we call the *space of k-jets*, we choose the functions

$$(x_1, \ldots, x_n, u^1, \ldots, u^m, p^i_\sigma, 1 \leqslant i \leqslant m, |\sigma| \leqslant k),$$

where

$$u^i([h]_a^k) = h^i(a), \qquad p^i_\sigma([h]_a^k) = \frac{\partial^{|\sigma|}h}{\partial x^\sigma}(a).$$

Thus, the set $J^k(n, m)$ forms an arithmetic space of dimension

$$n + m + nm + \cdots + m\binom{n + k - 1}{k}.$$

There are the natural maps

$$\pi_{k,l}: J^k(n, m) \to J^l(n, m), \quad [h]_x^k \mapsto [h]_x^l \quad \text{(reduction of jets)}, \quad k > l,$$

$$\pi_k: J^k(n, m) \to \mathbb{R}^n, \quad [h]_x^k \mapsto x.$$

The system of differential equations (9) defines a closed subset $\mathscr{E} = \{F_s(q, u^1, \ldots, u^m, p_\sigma^i) = 0\}$ in $J^k(n, m)$, which as before we call a *differential equation*. The fact that the vector-valued function $h(x)$ is a solution of (9) can be reformulated by saying that the n-dimensional submanifold L_h^k defined by the relations

$$u^i = h^i(x), \quad p_\sigma^i = \frac{\partial^{|\sigma|}h}{\partial x^\sigma}(x), \quad 1 \leqslant i \leqslant m, \quad |\sigma \leqslant k,$$

lies in the set $\mathscr{E}$.

4.2. The Cartan Distribution. Thus, from the geometrical point of view, solving the system of differential equations (9) is equivalent to constructing n-dimensional submanifolds of the form L_h^k lying in the set $\mathscr{E}$. Having in mind the intrinsic description of such submanifolds of the space $J^k(n, m)$, with each point $\theta \in J^k(n, m)$ we associate a linear subspace $C_\theta \subset T_\theta(J^k(n, m))$ spanned by the union of the tangent subspaces to all possible submanifolds of the form L_h^k passing through this point.

A distribution $C: \theta \mapsto C(\theta)$ on the space $J^k(n, m)$ is called a *Cartan distribution*. Differential 1-forms that annihilate this distribution have the form

$$\omega_\sigma^i = dp_\sigma^i - \sum_{j=1}^n p_{\sigma+1_j}^i \, dx_j. \tag{10}$$

Here $1 \leqslant i \leqslant m$, $|\sigma| \leqslant k - 1$, and $\sigma + 1_j = (\sigma_1, \ldots, \sigma_j + 1, \ldots, \sigma_n)$. Hence it follows that the dimension of the Cartan distribution is equal to $n + m \binom{n + k - 1}{k}$.

Obviously, submanifolds of the form L_h^k are integral manifolds of the Cartan distribution. From the explicit form of the 1-forms (10) that specify this distribution it is easy to derive the converse: every integral manifold $L \subset J^k(n, m)$ of this distribution, where $\dim L = n$, on which the functions $x_1, \ldots, x_n$ can be chosen as coordinates (or equivalently, that is diffeomorphically mapped onto $\mathbb{R}^n$ by the projection $\pi_k: J^k(n, m) \to \mathbb{R}^n$) has the form $L = L_h^k$ for some smooth vector-valued function h. Bearing this in mind, we can reformulate the notion of a solution as follows. An ("ordinary") *solution of a system of differential equations* $\mathscr{E} \subset J^k(n, m)$ is an arbitrary n-dimensional integral manifold of the Cartan distribution lying in $\mathscr{E}$ and projected diffeomorphically onto $\mathbb{R}^n$. As in 2.2, we generalize the concept of a solution. A *multivalued solution of a system of differential equations* $\mathscr{E}$ is an arbitrary n-dimensional maximal integral manifold of the Cartan distribution lying in $\mathscr{E}$.

Example. Riemann surfaces can be regarded as multivalued solutions of the Cauchy-Riemann system of differential equations in $J^1(2, 2)$.

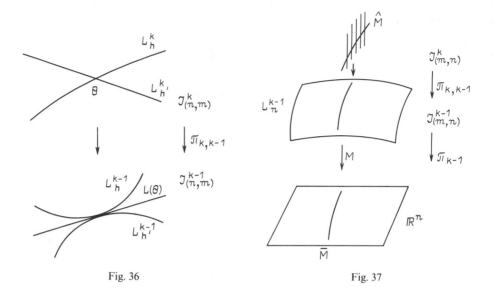

Fig. 36   Fig. 37

The fact that derivatives of order k are obtained by differentiating derivatives of order $k-1$ lies at the basis of the following useful construction in spaces of jets. With each element $\theta = [h]_a^k \in J^k(n, m)$ we associate the subspace $L(\theta) \subset T_{\theta'}J^{k-1}(n, m)$ tangent to the submanifold L_h^{k-1} at the point $\theta' = \pi_{k,k-1}(\theta) = [h]_a^{k-1} \in J^{k-1}(n, m)$ (see Fig. 36). This subspace is determined by the element θ and does not depend on the choice of the function h that represents this element. It is called an *R-plane*. This construction enables us to describe integral manifolds of the Cartan distribution in the same way as we did in 2.2 for a contact structure in the space $J^1 = J^1(n, 1)$. To this end we choose an arbitrary submanifold $M \subset L_h^{k-1} \subset J^{k-1}(n, m)$ that is projected diffeomorphically onto the submanifold $\overline{M} = \pi_{k-1}(M) \subset \mathbb{R}^n$ (see Fig. 37).

Let $\hat{M}$ denote the set of elements $\theta \in J^k(n, m)$ such that $L(\theta) \supset T_{\theta'}(M)$, $\theta' = \pi_{k,k-1}(\theta)$. It is not difficult to verify that $\hat{M}$ is a smooth submanifold of dimension $n - l + m \binom{l + k - 1}{k}$, where $l = \operatorname{codim} \overline{M}$, of the manifold $J^k(n, m)$ on which the 1-forms ω_σ^i, $|\sigma| < k$, vanish. Thus, it is an integral manifold of the Cartan distribution. Moreover, it is locally maximal. The number l is called the *type of the integral manifold $\hat{M}$*. It turns out that any maximal integral manifold locally has the form $\hat{M}$ for some manifold M if we discard an "inessential" set of its points. This enables us to speak about the type of an arbitrary maximal integral manifold.

Maximal integral manifolds of a contact distribution C in the space $J^1 = J^1(n, 1)$ have dimension n (see 2.2). In the remaining cases, when $km \geqslant 2$, $n > 1$, the dimension of locally maximal integral manifolds depends monotonically on their type l. Thus this dimension is equal to n when $l = 0$, and is equal to

$m\binom{n+k-1}{k} > n$ when $l = n$. Maximal integral manifolds corresponding to these extreme cases locally have the form L_h^k and $\pi_{k,k-1}^{-1}(\theta')$, $\theta' \in J^{k-1}(n, m)$ respectively. Simple calculation of the dimensions shows that the fibres of the map $\pi_{k,k-1}$ have the greatest dimension among locally maximal integral manifolds of the Cartan distribution if $km \geqslant 2$, $n \geqslant 2$. As we shall see below, this fact plays an important role in the geometry of spaces of jets.

4.3. Lie Transformations. We now describe the group of transformations of the spaces $J^k(n, m)$ that preserve the Cartan distribution. They are called *Lie transformations*, and the corresponding vector fields are called *Lie fields*.

Examples. 1) The transformations $F: J^0(n, m) \to J^0(n, m)$, $J^0(n, m) = \mathbb{R}^n \times \mathbb{R}^m$, are called *point transformations*. They can naturally be extended to transformations $F^{(k)}: J^k(n, m) \to J^k(n, m)$ if we put $F^{(k)}(h)_a^k = [F(h)]_a^k$, where $F(h)$ denotes the vector-valued function whose graph is obtained from the graph of h by using the transformation F. Of course, the transformation $F^{(k)}$ is not defined at those points $\theta = [h]_a^k$ of the space $J^k(n, m)$ for which the submanifold $F(L_h)$ does not have the form of the graph of some function. Obviously, $F^{(k)}$, where it is defined, preserves the class of submanifolds of the form L_h^k and hence preserves the Cartan distribution.

2) In a similar way contact transformations $F: J^1(n, 1) \to J^1(n, 1)$ are lifted to transformations $F^{(k)}: J^k(n, 1) \to J^k(n, 1)$, $F^{(k)}([h]_a^k) = [F(h)]_a^k$, where $F(h)$ denotes a function for which $F(L_h^1) = L_{F(h)}^1$. As above, the transformation $F^{(k)}$ is defined, generally speaking, on an open everywhere dense domain in $J^k(n, 1)$, and where it is defined it preserves the Cartan distribution.

3) Similarly, any Lie transformation $F: J^k(n, m) \to J^k(n, m)$ is lifted to a Lie transformation in $J^l(n, m)$, where $l \geqslant k$.

Any Lie field X on $J^k(n, m)$ can be lifted to a field $\tilde{X}$ on $J^l(n, m)$ if $l \geqslant k$. In fact, for this it is sufficient to lift transformations A_t of the current generated by the field X, and then put $\tilde{X} = \dfrac{dA_t^*}{dt}\bigg|_{t=0}$.

It turns out that these examples exhaust all Lie transformations.

Theorem (Lie and Backlund). *Any Lie transformation (field) of some domain of the space $J^k(n, m)$ is either*

i) *the lift of a contact transformation (field) if $m = 1$, or*

ii) *the lift of a point transformation (field) if $m \geqslant 2$.*

The proof of this theorem follows from the description of locally maximal integral manifolds of the Cartan distribution given at the end of 4.2. Namely the Lie transformations $A: J^k(n, m) \to J^k(n, m)$ preserve the class of maximal integral manifolds of highest dimension whatever the projections $\pi_{k,k-1}: J^k(n, m) \to J^{k-1}(n, m)$ if $km \geqslant 2$. Hence A determines a transformation $\bar{A}$ of the space $J^{k-1}(n, m)$, which is obviously also a Lie transformation. This method can be used as long as the condition $km \geqslant 2$ is satisfied. Hence, acting in this way we finally arrive at either contact transformations if $m = 1$, or point transformations if $m \geqslant 2$.

4.4. Intrinsic and Extrinsic Geometries. Integration of general systems of differential equations $\mathscr{E} \subset J^k(n, m)$, just like first-order differential equations (see § 2), can be considered either from the extrinsic point of view, as the construction of integral manifolds $L \subset J^k(n, m)$, dim $L = n$, lying on the manifold $\mathscr{E}$, or from the intrinsic point of view, as the construction of integral manifolds $L \subset \mathscr{E}$ of the distribution $C(\mathscr{E})$ that are the restriction of the Cartan distribution to $\mathscr{E}$, that is, $C(\mathscr{E})$: $\mathscr{E} \ni \theta \to C_\theta(\mathscr{E}) = C_\theta \cap T_\theta \mathscr{E}$. Correspondingly we can talk about the extrinsic geometry of the system $\mathscr{E}$, its outer symmetries, that is, Lie transformations that preserve the submanifold $\mathscr{E}$, and about the intrinsic geometry, inner symmetries, that is, diffeomorphisms of $\mathscr{E}$ that preserve the distribution $C(\mathscr{E})$, and so on.

In the theory of surfaces we have already observed the effect of rigidity, that is, how the intrinsic geometry completely determines the extrinsic (see Ch. 2, 3.7). In the geometry of differential equations rigidity is also a common phenomenon. More precisely, an equation $\mathscr{E}$ is said to be *rigid* if the distribution $C(\mathscr{E})$ enables us to reconstruct the embedding of $\mathscr{E}$ in J^k together with the Cartan distribution on J^k. It turns out that with the exception of first-order scalar equations and ordinary differential equations, all systems of differential equations that are not too overdetermined are rigid.

As above, let $\mathrm{Sym}_o(\mathscr{E})$ denote the group of outer symmetries of the system of differential equations $\mathscr{E}$, and let $\mathrm{Sym}_i(\mathscr{E})$ denote the group of inner symmetries. The operation of restricting the outer symmetries to $\mathscr{E}$ determines a homomorphism κ: $\mathrm{Sym}_o \mathscr{E} \to \mathrm{Sym}_i \mathscr{E}$. Obviously, κ is an isomorphism if $\mathscr{E}$ is a rigid system of equations.

The character of the link between extrinsic and intrinsic geometries of differential equations conditions the specific nature of the corresponding equivalence problem. The latter, in turn, also has intrinsic and extrinsic aspects.

We say that systems of differential equations $\mathscr{E}_1$ and $\mathscr{E}_2$ in $J^k(n, m)$ are *locally equivalent at points* $\theta_1 \in \mathscr{E}_1$ and $\theta_2 \in \mathscr{E}_2$ if there is a (local) Lie transformation $F: J^k(n, m) \to J^k(n, m)$ such that $F(\mathscr{E}_1) = \mathscr{E}_2$, $F(\theta_1) = \theta_2$. Similarly, such systems are *intrinsically equivalent at points* $\theta_1 \in \mathscr{E}_1$ and $\theta_2 \in \mathscr{E}_2$ if there is a local diffeomorphism $F: \mathscr{E}_1 \to \mathscr{E}_2$, $F(\theta_1) = \theta_2$, that takes the distribution $C(\mathscr{E}_1)$ into the distribution $C(\mathscr{E}_2)$.

The problem of intrinsic equivalence is actually *Pfaff's problem* on local classification of distributions in its classical version.

In Ch. 6, 4.5 in the bundles $\pi_{k, k-1}$: $J^k(n, m) \to J^{k-1}(n, m)$ we shall construct an affine structure that is invariant with respect to the group of Lie transformations if $k \geqslant 2$, $m \geqslant 2$ or $k \geqslant 3$, $m = 1$. Hence it follows that the problem of classifying general systems of differential equations ($m \geqslant 2$) of order $k \geqslant 2$ or the classification of general scalar differential equations ($m = 1$) of order $k \geqslant 3$ contains the problem of classifying submanifolds of affine space with respect to the group of affine transformations, and consequently it has no special meaning.

§5. The Theory of Compatibility and Formal Integrability

Differentiating the equations of the system under consideration and carrying out algebraic manipulations on them we can attempt to determine whether we arrive at a contradiction in this way. Formal integrability of a system of differential equations implies, roughly speaking, its consistency in this sense. The role of geometry here is determined by the fact that it enables us to order this process of successive differentiation and elimination, showing its intrinsic structure.

The extrinsic and intrinsic points of view of the geometry of differential equations find their expression in the theory of compatibility. The extrinsic, which is based on the concept of prolonging a system of differential equations, will be considered in this section. The intrinsic, which is Cartan's theory of systems in involution, will be considered in the next section.

5.1. Prolongations of Differential Equations. We begin with the simplest example. Consider a regular overdetermined system of differential equations $\mathscr{E} \subset J^1(n, 1)$ specified by functions $f_1, \ldots, f_k \in C^\infty(J^1)$. A necessary and sufficient condition for this system to be soluble is that the equation $\mathscr{E}$ is involutive (see 3.4). This condition, $\{f_s, f_t\}|_{\mathscr{E}} = 0$, from the geometrical point of view denotes the possibility of the submanifold $\mathscr{E}$ being touched by integral manifolds of the form L_h^1. Obviously, an analogous condition is also necessary for an arbitrary system of differential equations $\mathscr{E} \subset J^k(n, m)$ to be soluble. Having this in mind, we consider a submanifold $\mathscr{E}^{(1)} \subset J^{k+1}(n, m)$ consisting of elements $\theta \in J^{k+1}(n, m)$ for which the R-planes $L(\theta)$ touch the submanifold $\mathscr{E}$. In accordance with the general point of view on differential equations as subsets in spaces of jets, $\mathscr{E}^{(1)}$ should be regarded as a differential equation of order $k + 1$. It is called the *first prolongation of the system of differential equations $\mathscr{E}$.*

The first prolongation $\mathscr{E}^{(1)}$ is given by the equations

$$F_s = 0, \qquad D_j F_s = 0, \tag{11}$$

where $F_s = 0$ are the equations of the required system $\mathscr{E}$, and $D_j = \partial/\partial x_j + \sum_{i,\sigma} p^i_{\sigma+1_j} \dfrac{\partial}{\partial p^i_\sigma}$ is the *total derivative operator*.

Example. Let us obtain the conditions for a system of differential equations $\mathscr{E} \subset J^1(n, 1)$ to be involutive, using its first extension $\mathscr{E}^{(1)} \subset J^2(n, 1)$. In view of (11) it is described by the equations

$$f_s(x, u, p) = 0, \qquad \frac{\partial f_s}{\partial x_j} + p_j \frac{\partial f_s}{\partial u} + \sum_i p_{ij} \frac{\partial f_s}{\partial p_i} = 0,$$

where $p_{ij}([f]_a^2) = \dfrac{\partial^2 f}{\partial x_i \partial x_j}(a)$.

Let R denote the symmetric matrix $\|p_{ij}\|$. Then we can write the prolonged system in vector form

$$f_s = 0, \qquad R(a_s) + b_s = 0,$$

where a_s is the vector with components $\partial f_s/\partial p_j$, and b_s is the vector with components $\dfrac{\partial f_s}{\partial x_j} + p_j\dfrac{\partial f_s}{\partial u}$. The conditions for the existence of a symmetric matrix R that satisfies these equations are equivalent to the following:

$$\langle R(a_s), a_t\rangle - \langle a_s, R(a_t)\rangle = \langle a_s, b_t\rangle - \langle a_t, b_s\rangle = 0,$$

that is, the conditions for the equation $\mathscr{E}, \{f_s, f_t\}|_{\mathscr{E}} = 0$ to be involutive.

In this example the conditions for involutivity turn out to be sufficient for the differential equation to be soluble. In the general case this is not so. Hence we need to consider the following prolongations of the original system. More precisely, for each natural number $l \geq 1$ we define the subset $\mathscr{E}^{(l)} \subset J^{k+l}(n, m)$ formed by $(k + l)$-jets $[h]_a^{k+l}$ for which the integral manifolds L_h^k touch the equation $\mathscr{E}$ at the point $[h]_a^k \in \mathscr{E}$ with order at least l. The system of differential equations $\mathscr{E}^{(l)}$ of order $k + l$ is called the l-th *prolongation of the system $\mathscr{E}$.* It is written in the form

$$F_s = 0, \qquad D^{\tau}(F_s) = 0, \tag{12}$$

where $\tau = (\tau_1, \ldots, \tau_n)$ are multi-indices of length at most l, and $D^{\tau} = D_1^{\tau_1} \circ \cdots \circ D_n^{\tau_n}$.

5.2. Formal Integrability. The projection $\pi_{k+l+1,k+l} \colon J^{k+l+1}(n, m) \to J^{k+l}(n, m)$ generates a map $\mathscr{E}^{(l+1)} \to \mathscr{E}^{(l)}$. Generally speaking, the sets $\mathscr{E}^{(l)}$ are not smooth submanifolds of $J^{k+l}(n, m)$. We especially distinguish the class of *regular differential equations*, understanding the latter to be those whose l-prolongations are smooth manifolds for all $l \geq 1$.

A regular system of differential equations $\mathscr{E} \subset J^k(n, m)$ is said to be *formally integrable* if all the projections $\pi_{k+l,k+l-1} \colon \mathscr{E}^{(l)} \to \mathscr{E}^{(l-1)}$ are smooth bundles when $l \geq 1$. For a formally integrable system of differential equations, at each point $\theta_k \in \mathscr{E}$ we can find an infinite sequence of elements $\theta_{k+l} \in \mathscr{E}^{(l)}$ that project into one another, $\pi_{k+l+1,k+l}(\theta_{l+1}) = \theta_l$. By a theorem of Borel there is a smooth function $h(x)$ such that $[h]_a^{k+l} = \theta_l$ for all values of l. The integral manifold L_h^k corresponding to this function touches the manifold $\mathscr{E}$ at a point $\theta_k \in \mathscr{E}$ with infinite order, and the sequence $\{\theta_{k+l}\}$ can be represented as the formal solution of the system of differential equations $\mathscr{E}$.

Thus, formal integrability of the equation $\mathscr{E}$ means that every "solution of order l" $\theta_{k+l} \in \mathscr{E}^{(l)}$ can be prolonged in a regular way to a formal solution of this equation.

Examples. 1) A regular involutive system of differential equations $\mathscr{E} \subset J^1(n, 1)$ is formally integrable.

2) The classical equations of mathematical physics (the wave equation, the Laplace equation, and so on) are formally integrable.

5.3. Symbols. Formal integrability of a differential equation $\mathscr{E}$ is equivalent to the (smooth) solubility of the system (12) with respect to derivatives of higher

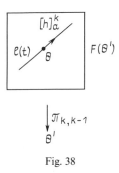

Fig. 38

order. We observe that in this system higher derivatives occur linearly with coefficients $\dfrac{\partial F_s}{\partial p_\sigma^i}$, $|\sigma| = k$. Consequently, to solubility and hence formal integrability there corresponds the "highest part" of the differential equation $\mathscr{E}$, that is, its symbol (see Ch. 3, § 5). More precisely, the *symbol of a system of differential equations* $\mathscr{E} \subset J^k(n, m)$ at a point $\theta = [f]_a^k \in \mathscr{E}$ is a subspace $g(\theta)$ of the tangent space $T_\theta(F(\theta'))$ to the fibre $F(\theta') = \pi_{k,k-1}^{-1}(\theta')$, $\theta' = \pi_{k,k-1}(\theta)$:

$$g(\theta) = T_\theta(\mathscr{E}) \cap T_\theta(F(\theta')).$$

The tangent space $T_\theta(F(\theta'))$ can be identified with the space of k-jets $[h]_a^k$ of the vector-valued functions h that have a zero of order k at the point $a \in \mathbb{R}^n$. For this it is sufficient to observe that each point of the function determines a line $l: t \mapsto [f + th]_a^k$ in $F(\theta')$ such that the tangent vector to it at the point θ coincides with $[h]_a^k$ (see Fig. 38).

In addition, we can regard the "leading part" $[h]_a^k$ of a vector-valued function h such that $[h]_a^{k-1} = 0$ as a symmetric function on the tangent space $T_a \mathbb{R}^n$ with values in $\mathbb{R}^m$:

$$\frac{\partial}{\partial x_{i_1}} \times \cdots \times \frac{\partial}{\partial x_{i_k}} \mapsto \frac{\partial^k h}{\partial x_{i_1} \ldots \partial x_{i_k}}(a) \in \mathbb{R}^m.$$

In the dual basis this function, which we shall also denote by $[h]_a^k \in \mathbb{R}^m \otimes S^k(\mathbb{R}^n)$, is written in the form

$$[h]_a^k = \sum_{|\sigma|=k} \frac{1}{\sigma!} \frac{\partial^k h}{\partial x^\sigma}(a)\, dx^\sigma,$$

where dx^σ denotes the symmetric product of differentials of the coordinates: $dx^\sigma = dx_1^{\sigma_1} \circ \cdots \circ dx_n^{\sigma_n}$.

In this interpretation, with the vector $\sum\limits_{i, |\sigma|=k} \xi_\sigma^i \dfrac{\partial}{\partial p_\sigma^i}$ tangent to the fibre $F(\theta')$ we associate the homogeneous polynomial

$$\sum_{i, |\sigma|=k} \frac{1}{\sigma!} \xi_\sigma^i e_i \otimes dx^\sigma,$$

where $e_1, \ldots, e_m$ is the standard basis in $\mathbb{R}^m$.

Hence the symbol $g(\theta)$ can be described either in the form

$$g(\theta) = \left\{ \sum_{i, |\sigma|=k} \xi_\sigma^i \frac{\partial}{\partial p_\sigma^i} \,\middle|\, \sum_{i, |\sigma|=k} \xi_\sigma^i \frac{\partial F_s}{\partial p_\sigma^i}(\theta) = 0 \right\},$$

or as the intersection of the kernels of the maps

$$\sigma_s \colon \mathbb{R}^m \otimes S^k(\mathbb{R}^n) \to \mathbb{R},$$

where

$$\sigma_s = \sum_{i, |\tau|=k} \frac{\partial F_s(\theta)}{\partial p_\tau^i} e_i^* \otimes \frac{\partial^{|\tau|}}{\partial x^\tau} \in (\mathbb{R}^m)^* \otimes S_k(\mathbb{R}^n),$$

and $e_1^*, \ldots, e_m^*$ is the dual basis in $(\mathbb{R}^m)^*$.

The symbol $g^{(l)}(\theta_{k+l}) \subset \mathbb{R}^m \otimes S^{k+l}(\mathbb{R}^n)$ of the l-th prolongation at the point $\theta_{k+l} \in E^{(l)}$ is completely determined by the symbol $g(\theta)$ of the equation E at the point $\theta = \pi_{k+l,k}(\theta_{k+l})$. Namely,

$$g^{(l)}(\theta_{k+l}) = \left\{ [h]_a^{k+l} \in \mathbb{R}^m \otimes S^{k+l}(\mathbb{R}^n) \,\middle|\, \left[\frac{\partial^l h}{\partial x_{i_1} \ldots \partial x_{i_l}} \right]_a^k \in g(\theta) \right\},$$

for all sets of numbers $i_1, \ldots, i_l$, $1 \le i_j \le n$.

The subspace $g^{(l)}(\theta_{k+l})$ is called the l-th *prolongation of the symbol* $g(\theta)$ and denoted by $g^{(l)}(\theta)$.

5.4. The Spencer δ-Cohomology. Formal integrability of a system of differential equations means that the families of subspaces $\mathscr{E} \ni \theta \to g^{(l)}(\theta)$ must form vector bundles, $l = 0, 1, \ldots$ The apparatus of Spencer δ-cohomology enables us to reduce the infinite number of these conditions to a finite number.

The elements of the space $\mathbb{R}^m \otimes S^r(\mathbb{R}^n) \otimes \Lambda^t(\mathbb{R}^n)$ will be regarded as differential t-forms on the manifold $\mathbb{R}^n$ whose coefficients are homogeneous polynomials of degree r with values in the vector space $\mathbb{R}^m$. The exterior differential induces the map

$$\delta \colon \mathbb{R}^m \otimes S^r(\mathbb{R}^n) \otimes \Lambda^t(\mathbb{R}^n) \to \mathbb{R}^m \otimes S^{r-1}(\mathbb{R}^n) \otimes \Lambda^{t+1}(\mathbb{R}^n),$$

called the *Spencer δ-operator*.

Then $\delta^2 = 0$, and the sequence

$$0 \to \mathbb{R}^m \otimes S^r(\mathbb{R}^n) \xrightarrow{\delta} \mathbb{R}^m \otimes S^{r-1}(\mathbb{R}^n) \otimes (\mathbb{R}^n)^* \xrightarrow{\delta} \mathbb{R}^m \otimes S^{r-2}(\mathbb{R}^n) \otimes \Lambda^2(\mathbb{R}^n) \xrightarrow{\delta} \cdots$$

$$(13)$$

forms the *Spencer δ-complex*.

From the definition of the symbol $g^{(l)}(\theta)$ it follows that

$$\delta(g^{(l)}(\theta)) \subset g^{(l-1)}(\theta) \otimes (\mathbb{R}^n)^*.$$

Hence

$$\delta(g^{(l)}(\theta) \otimes \Lambda^t(\mathbb{R}^n)) \subset g^{(l-1)} \otimes \Lambda^{t+1}(\mathbb{R}^n),$$

and we can consider the subcomplex of the complex (13)

$$0 \to g^{(l)}(\theta) \xrightarrow{\delta} g^{(l-1)} \otimes \Lambda^1(\mathbb{R}^n) \xrightarrow{\delta} g^{(l-2)} \otimes \Lambda^2(\mathbb{R}^n) \to \cdots, \qquad (14)$$

called the *Spencer δ-complex of the system of differential equations $\mathcal{E}$ at the point* $\theta \in \mathcal{E}$.

We denote the cohomology of this complex at the term $g^{(r)} \otimes \Lambda^i(\mathbb{R}^n)$ by $H^{r,i}(\mathcal{E}, \theta)$:

$$H^{r,i}(\mathcal{E}, \theta) = \frac{\operatorname{Ker}(\delta \colon g^{(r)}(\theta) \otimes \Lambda^i(\mathbb{R}^n) \to g^{(r-1)}(\theta) \otimes \Lambda^{i+1}(\mathbb{R}^n))}{\operatorname{Im}(\delta \colon g^{(r+1)}(\theta) \otimes \Lambda^{i-1}(\mathbb{R}^n) \to g^{(r)}(\theta) \otimes \Lambda^i(\mathbb{R}^n))}$$

and call it the *Spencer δ-cohomology of the system of differential equations $\mathcal{E}$ at the point* $\theta \in \mathcal{E}$.

We note that $H^{r,0}(\mathcal{E}, \theta) = 0$, $H^{r,1}(\mathcal{E}, \theta) = 0$ and the first non-trivial Spencer δ-cohomology can be in dimension 2. It corresponds to formal integrability.

We say that a system of differential equations $\mathcal{E}$ is *2-acyclic* if $H^{r,2}(\mathcal{E}, \theta) = 0$ for all $r \geqslant 0$, $\theta \in \mathcal{E}$.

The next theorem, due to Goldschmidt, gives a criterion for formal integrability.

Theorem. *Let $\mathcal{E}$ be a system of differential equations such that*
1) *it is 2-acyclic;*
2) *the family of vector spaces $\mathcal{E} \ni \theta \mapsto g(\theta)$ forms a vector bundle over the manifold $\mathcal{E}$;*
3) *the map $\pi_{k+1,k} \colon \mathcal{E}^{(1)} \to \mathcal{E}$ is a smooth bundle.*
Then $\mathcal{E}$ is formally integrable.

5.5. Involutivity. The Spencer δ-cohomology gives algebraic obstructions to formal integrability. If this cohomology is trivial, then the system of differential equations is said to be (Spencer) *involutive*. The next result, the so-called Poincaré δ-lemma, shows that involutivity can be achieved by finitely many prolongations.

Theorem. *For an arbitrary system of differential equations $\mathcal{E} \subset J^k(n, m)$ there is a number $l_0 = l_0(m, n, k)$ such that $H^{l,i}(\mathcal{E}, \theta) = 0$ for all $l \geqslant l_0$, $i \geqslant 0$, $\theta \in \mathcal{E}$.*

The number l_0 itself can be calculated from the following formulae:

$$l_0(0, n, 1) = 0, \qquad l_0(m, n, 1) = m \binom{n + a}{n - 1} + a + 1,$$

$$l_0(m, n, k) = l_0(b, n, 1),$$

where

$$a = l_0(m, n - 1, 1), \qquad b = \sum_{i=0}^{k} m \binom{n + i - 1}{i}.$$

This theorem shows that for formally integrable systems of differential equations their prolongations of order l_0 are involutive equations. This corresponds to an assertion of E. Cartan that "if a given system is not in involution, there is a regular way of obtaining from it a sequence of new systems admitting the same solutions as the given system. We can prove that under certain conditions, which are not easy to formulate, everything stops when we arrive at a system in involution". Cartan's formulation of the conditions for involutivity is given in the next section.

In the case when $\mathscr{E} \subset J^k(n, m)$ is an *analytic system of differential equations*, formal integrability implies integrability in analytic functions. This result (the Cartan-Kähler theorem) is a generalization of the well-known theorem of Cauchy and Kovalevskaya.

Theorem. *Let $\mathscr{E} \subset J^k(n, m)$ be a formally integrable analytic system of differential equations. Then for any point $\theta \in \mathscr{E}^{(l)}$ there is an analytic vector-valued function h that is a solution of this system $\mathscr{E}$ such that $\theta = [h]_a^k$.*

§6. Cartan's Theory of Systems in Involution

This theory considers the problem of integrating systems of differential equations from the intrinsic point of view, that is, as the problem of finding integral manifolds of some distribution F on a manifold $M (= \mathscr{E})$.

6.1. Polar Systems, Characters and Genres. A *Pfaff system* is a system of the form $\omega_1 = 0, \dots, \omega_s = 0$, where ω_i are differential 1-forms on a smooth manifold M. Let $r_0(a), a \in M$, denote the dimension of the subspace of the cotangent space $T_a^* M$ spanned by the covectors $\omega_{1,a}, \dots, \omega_{s,a}$, and let $s_0 = \max r_0(a)$. A point $a \in M$ is called a *regular point of the Pfaff system* if $r_0(a) = s_0$, and *singular* otherwise. In a neighbourhood of a regular point of the Pfaff system $\omega_1, \dots, \omega_s$ determine a distribution $\mathscr{F}(\omega_1, \dots, \omega_s)$, and in a neighbourhood of a singular point they determine a distribution with singularities.

Examples. 1) Regular points of an ordinary differential equation $\mathscr{E} \subset J^k(1, 1)$ are those points $\theta \in \mathscr{E}$ where the subspaces $T_\theta \mathscr{E}$ and C_θ are transversal. At singular points $T_\theta \mathscr{E} \supset C_\theta$. Thus for the hypergeometric Gauss equation

$$x(x - 1)y'' + [(\alpha + \beta + 1)x - \gamma]y' + \alpha\beta y = 0,$$

where α, β, γ are certain constants and $\alpha\beta \neq 0$, the singular points lie over the points $x = 0$ and $x = 1$. The singular points over the point $x = 0$ form a straight line

$$x = 0, \qquad u = \lambda, \qquad p_1 = \frac{\alpha\beta}{\gamma} \lambda, \qquad p_{11} = \frac{(\alpha\beta)^2}{\gamma(1 + \gamma)} \lambda,$$

where $\lambda \in \mathbb{R}$ is a parameter.

2) This definition of singular points for first-order equations $E \subset J^1(n, 1)$ coincides with the definition of 3.1.

3) We say that a differential equation $\mathscr{E} \subset J^k(n, 1)$ is an *equation of principal type* if the submanifold $\mathscr{E}$ is transversal to fibres of the projection $\pi_{k,k-1} \colon J^k(n, 1) \to J^{k-1}(n, 1)$. All the points of equations of principal type are regular.

Let us fix a regular point $a \in M$ and describe the tangent planes $T_a L$ to integral manifolds $L \subset M$ of the distribution $\mathscr{F} = \mathscr{F}(\omega_1, \ldots, \omega_s)$ that pass through this point. Firstly, we have the inclusion $T_a L \subset \mathscr{F}_a$. Also, since

$$(d\omega_i)|_L = d(\omega_i|_L) = 0,$$

the exterior 2-form Ω_i that is the restriction of the 2-form $d_a \omega_i$ to the subspace $\mathscr{F}_a \subset T_a M$ vanishes on vectors of the tangent space $T_a L$. Bearing this in mind, we say that vectors $\xi, \eta \in \mathscr{F}_a$ are *in involution* if $\Omega_i(\xi, \eta) = 0$ for all i such that $1 \leqslant i \leqslant s$. A subspace $N \subset \mathscr{F}_a$ is called an *integral subspace* if any two vectors $\xi, \eta \in N$ are in involution.

Thus, the construction of integral manifolds suggests the possibility of constructing integral subspaces. The solubility of the latter problem can be investigated by induction on the dimension of integral subspaces.

First of all we observe that every 1-dimensional subspace $N_1 \subset \mathscr{F}_a$ is integral. To construct an integral 2-dimensional subspace N_2 containing N_1 we consider the *polar system* of linear equations

$$\Omega_{1,\xi}(\eta) - \cdots = \Omega_{s,\xi}(\eta) = 0 \qquad (15)$$

with respect to the vector $\eta \in \mathscr{F}_a$, where ξ is a generator of the subspace N_1, and $\Omega_{i,\xi} = \xi \lrcorner \Omega_i$.

Let $r(N_1)$ be the rank of the system of equations (15) and let $s_1 = \max r(N_1)$. We say that the subspace N_1 is *regular* if $r(N_1) = s_1$, and *singular* otherwise. A 2-dimensional integral subspace N_2 is said to be *ordinary* if N_2 contains a regular subspace.

We note that the subspace N_1 occurs in a number of solutions of its polar system. Hence, if $s_1 \geqslant \dim \mathscr{F}_a - 1$, then there are no ordinary 2-dimensonal subspaces.

The number s_1 is called the *first-order character* of the Pfaff system under consideration.

Suppose that $s_1 < \dim \mathscr{F}_a - 1$, and that N_2 is an ordinary 2-dimensional subspace. To construct 3-dimensional integral subspaces N_3 containing N_2 we consider the polar system of N_2:

$$\Omega_{i,\xi_1}(\eta) = \Omega_{i,\xi_2}(\eta) = 0,$$

where ξ_1, ξ_2 is a basis of the subspace N_2, and $1 \leqslant i \leqslant s$.

Let $r(N_2)$ denote the rank of this system, and let r_2 be the greatest value of $r(N_2)$ on condition that N_2 runs through all ordinary 2-dimensional subspaces. As above, we say that an ordinary subspace N_2 is *regular* if $r_2 = r_2(N)$, and *singular* otherwise. Again, since vectors $\eta \in N_2$ are solutions of the polar system of N_2, 3-dimensional integral subspaces containing N_2 exist on condition that $r(N_2) < \dim \mathscr{F}_a - 2$. We say that a 3-dimensional integral subspace is *ordinary* if it contains a 2-dimensional regular integral subspace. The number $s_2 = r_2 - s_1$ is called the *second-order character*.

In the general case the *polar system of an integral p-dimensional subspace* N_p is the system of linear equations

$$\Omega_{i,\xi_1}(\eta) = \cdots = \Omega_{i,\xi_p}(\eta) = 0,$$

where $\xi_1, \ldots, \xi_p$ is a basis of the subspace N_p, and $1 \leqslant i \leqslant s$.

The rank of this system is denoted by $r(N_p)$. As above, the condition for the existence of a $(p + 1)$-dimensional integral subspace containing N_p is the inequality $r(N_p) < \dim \mathscr{F}_a - p$. By induction on p we define the concepts of an ordinary and a regular p-dimensional integral subspace. Namely, an arbitrary p-dimensional integral subspace is said to be *ordinary* if it contains a $(p - 1)$-dimensonal regular subspace. An integral subspace N_p is *regular* if $r(N_p) = r_p$, where

$$r_p = \max_{N_p} r(N_p),$$

on condition that N_p runs through all possible p-dimensional ordinary subspaces.

The number $s_p = r_p - r_{p-1}$ is called the *p-th order character*. An ordinary p-dimensional subspace exists if

$$s_1 + s_2 + \cdots + s_{p-1} < \dim \mathscr{F}_a - (p - 1).$$

The smallest integer p for which

$$s_1 + s_2 + \cdots + s_p = \dim \mathscr{F}_a - p$$

is called the *genre* of the Pfaff system under consideration at the point a. In other words, the genre of the system is equal to p if there are regular integral p-dimensional subspaces and no $(p + 1)$-dimensional integral subspaces containing them.

Example. Let $\mathscr{E} \subset J^2(2, 1)$ be a differential equation of principal type on the plane, and $\mathscr{F} = C(\mathscr{E})$ the restriction of the Cartan distribution. Then all points $\theta \in \mathscr{E}$ are regular. If $\mathscr{E}$ is a differential equation of elliptic type, then all the lines $N_1 \subset C_\theta(\mathscr{E})$ are regular, and all 2-dimensional integral subspaces are ordinary. Here $s_0 = 3$, $s_1 = 2$, and the genre of the distribution $C(\mathscr{E})$ is equal to 2. In the case when $\mathscr{E}$ is a *hyperbolic differential equation*, its characteristics correspond to singular lines N_1. Here also $s_0 = 3$, $s_1 = 2$, and the genre is equal to 2.

6.2. Involutivity and Cartan's Existence Theorems. All the constructions of the previous subsection were carried out at a fixed point $a \in M$. We denote the characters at this point by $s_i(a)$. Let $s_i = \max s_i(a)$. Generalizing the definition of the previous subsection, we say that a point $a \in M$ is *regular* if $s_i = s_i(a)$ for all values of i.

Theorem 1 (Cartan). *Let M be an analytic manifold and $\omega_1 = 0, \ldots, \omega_s = 0$ an analytic Pfaff system (that is, the 1-forms ω_i are analytic). In a neighbourhood of a regular point $a \in M$ of this system, where the genre is equal to p, for each ordinary subspace $N_p \subset T_a M$ containing a regular subspace N_{p-1} and an arbitrary $(p - 1)$-dimensional analytic integral manifold L' touching N_{p-1} there is a unique analytic integral manifold L of dimension p containing L' and touching N_p.*

Theorem 2 (Cartan). *Under the conditions of the previous theorem, for each q-dimensional ordinary subspace $N_q \subset T_a M$, $q \leqslant p$, containing a regular $(q-1)$-dimensional integral subspace N_{q-1}, and an arbitrary $(q-1)$-dimensional analytic integral manifold L' touching N_{q-1}, there is a q-dimensional analytic integral manifold containing L' and touching N_q.*

These theorems are called the first and second existence theorems of Cartan. They enable us to construct by induction analytic integral manifolds up to dimension equal to the genre of the system.

For systems of differential equations $\mathscr{E} \subset J^k(n, m)$ the main interest is in integral manifolds corresponding to smooth functions, that is, that project diffeomorphically onto the base space $\mathbb{R}^n$. Considering only such integral manifolds, we say that a system of differential equations $\mathscr{E}$ is *involutive* (in Cartan's sense) if its genre is at least n and there are n-dimensional ordinary subspaces that project without degeneracy on $\mathbb{R}^n$.

The equivalence of the concepts of involutivity according to Cartan and according to Spencer was proved by Serre.

§ 7. The Geometry of Infinitely Prolonged Equations

7.1. What is a Differential Equation? Can we assume that an adequate geometrical image of a differential equation is a submanifold in the space of jets equipped with the Cartan distribution? Bearing this question in mind, we turn our attention to the differences between the geometry of scalar first-order equations and the geometry of equations of higher order in the sense that we have understood up to now.

1. The question of the consistency of a system of scalar equations $\mathscr{E} \subset J^1(n, 1)$ can be solved without going outside the limits of the space $J^1(n, 1)$ by using the Lagrange bracket (see 3.4). In the general case "intrinsic" methods are not sufficient and we need to consider prolongations of the original system that lie in spaces of jets of higher order.

2. The group of Lie transformations that (according to Klein) controls the geometry of spaces $J^k(n, m)$ does not change as k increases, starting from 1. Thus, the complication of differential equations associated with a growth in their order is not accompanied by an enrichment of the controlling group of automorphisms. The rigidity theorem (see 4.4) shows that the transition to the intrinsic point of view does not change the essence of the matter.

These and other considerations show that a submanifold $\mathscr{E} \subset J^k(n, m)$ (the extrinsic point of view) or a pair $(\mathscr{E}, \tilde{C}(\mathscr{E}))$ are not a valuable geometrical equivalent of the concept of a differential equation. The following considerations from elementary algebraic geometry suggest the way of finding an equivalent.

In the theory of algebraic equations it is useful to make no distinction between equivalent (that is, that reduce to one another by algebraic operations) systems of equations

$$f_i(x_1, \ldots, x_n) = 0, \qquad i = 1, \ldots, p, \quad f_i \in A. \tag{16}$$

Here $A = k[x_1, \ldots, x_n]$, where k is the ground field. For this we need to take the point of view that the expression (16), like any other equivalent to it, is only a visiting card of the object that should be called a "system of equations". As is now well known, such an object is the ideal $\mathscr{I} \in A$, one possible system of generators of which consists of the polynomials $f_1, \ldots, f_p$. The corresponding geometrical object is the algebraic variety $\text{Spec}_k A/\mathscr{I}$ (see Ch. 1, § 2).

In the theory of differential equations expressions of the form

$$F_i\left(x, u, \frac{\partial u}{\partial x}, \ldots\right) = 0 \tag{17}$$

are assumed to be equivalent if they can be obtained from one another by means of both algebraic and differential operations. Successively differentiating the equations (17), we can obtain an infinite system

$$D^\tau F_i = 0, \qquad 0 \leqslant |\tau| < \infty. \tag{18}$$

Algebraic equivalence of systems of the form (18) is obviously tantamount to "differential" equivalence of systems of the form (17). For this reason, from the algebraic point of view the object covered by the words "differential equation" must be the ideal $\mathscr{I}$ generated by the functions $D^\tau F_i$ for all i and τ. The question that arises in this connection is to make precise the algebra of which $\mathscr{I}$ is an ideal!

Consider the space of jets of infinite order $J^\infty = J^\infty(n, m)$ and the natural projection $\pi_{\infty,k}: J^\infty \to J^k$. Any function on J^k can be lifted to J^∞: $\varphi \mapsto \pi^*_{\infty,k}(\varphi)$. We say that functions of the form $\pi^*_{\infty,k}(\varphi)$ on J^∞ are *smooth*, and we consider the algebra $\mathscr{F}$ $(= C^\infty(J^\infty))$ consisting of them. It is the answer to the question posed above.

The ideal $\mathscr{I}$ in question is *differentially* closed, that is, $D^\tau F \in \mathscr{I}$ if $F \in \mathscr{I}$. Thus, algebraically a "differential equation" as an object is a differentially closed ideal $\mathscr{I} \subset \mathscr{F}$, and geometrically it is the maximal spectrum of the algebraic $\mathscr{F}/\mathscr{I}$. Of course, what we have said leaves many details to be refined. The most important of these is that the algebra $\mathscr{F}/\mathscr{I}$ that arises in this way has an additional structure induced by the Cartan distribution on J^∞; see 7.2.

7.2. Infinitely Prolonged Equations. The constructive part of the maximal spectrum of the algebra $\mathscr{F}/\mathscr{I}$ can be obtained as the inverse limit, denoted by $\mathscr{E}^{(\infty)}$, of the chain of smooth maps

$$\mathscr{E} \overset{\alpha_1}{\leftarrow} \mathscr{E}^{(1)} \overset{\alpha_1}{\leftarrow} \cdots \overset{\alpha_l}{\leftarrow} \mathscr{E}^{(l)} \leftarrow \cdots,$$

where $\alpha_l = \pi_{l+k, l+k-1}|_{\mathscr{E}^{(l)}}$ if $\mathscr{E} \subset J^k$. We recall that an element $\theta \in \mathscr{E}^{(\infty)}$ is a sequence $\{\theta_l\}$ of points $\theta_l \in \mathscr{E}^{(l)}$ such that $\alpha_l(\theta_l) = \theta_{l-1}$. Analytically θ is a power series that is a formal solution of the system under consideration.

Let $\beta_l: \mathscr{E}^{(\infty)} \to \mathscr{E}^{(l)}$ be the natural projection, $\theta \mapsto \theta_l$. Functions of the form $\beta^*_l(\varphi)$, $\varphi \in C^\infty(\mathscr{E}^{(l)})$, on $\mathscr{E}^{(\infty)}$ will be called *smooth*. The totality of them forms an algebra $\mathscr{F}(\mathscr{E})$, filtered by subalgebras $\mathscr{F}_l(\mathscr{E}) = \beta^*_l(C^\infty(\mathscr{E}^{(l)}))$. If $\mathscr{I}$ is an ideal of the equation

$\mathscr{E}$ (see the previous subsection), then the algebra $\mathscr{F}/\mathscr{I}$ is identical to the algebra $\mathscr{F}(\mathscr{E})$.

Covariant geometrical objects on $\mathscr{E}^{(\infty)}$, like functions, are lifted from the manifolds $\mathscr{E}^{(l)}, l < \infty$. For example, *differential forms* on $\mathscr{E}^{(\infty)}$ are objects that can be denoted by $\beta_l^*(\omega)$, $\omega \in \Lambda^*(\mathscr{E}^{(l)})$. Contravariant objects, on the other hand, are represented as inverse limits. For example, the *tangent vector* ζ at a point $\theta \in \mathscr{E}^{(\infty)}$ is a sequence of vectors $\zeta_l \in T_{\theta_l}(\mathscr{E}^{(l)})$ such that $(d_{\theta_l}\alpha_l)(\zeta_l) = \zeta_{l-1}$. The totality of vectors thus defined forms a linear space, called the *tangent space* and denoted by $T_\theta\mathscr{E}^{(\infty)}$.

By definition the *Cartan plane* $C_\theta(\mathscr{E}^{(\infty)}) \subset T_\theta\mathscr{E}^\infty$ consists of vectors $\zeta = \{\zeta_l\}$ such that $\zeta_l \in C_{\theta_l}(\mathscr{E}^{(l)})$. This definition is meaningful, since the map α_l takes the distribution $C(\mathscr{E}^{(l)})$ into the distribution $C(\mathscr{E}^{(l-1)})$. The distribution $\theta \mapsto C_\theta(\mathscr{E}^{(\infty)})$ on $\mathscr{E}^{(\infty)}$ is called the *Cartan distribution*. Its dimension does not exceed n, despite the fact that the dimension of the distributions $C(\mathscr{E}^{(l)})$, whose inverse limit it is, increases without limit as $l \to \infty$. It also satisfies the conditions of the theorem of Frobenius (see 3.1), although the theorem itself is false, generally speaking, if $\dim \mathscr{E}^{(\infty)} = \infty$. Finally, integral manifolds of the Cartan distribution on $\mathscr{E}^{(\infty)}$ are identified in a natural way with the solutions of the equation $\mathscr{E}$.

Let us sum up. The "manifold" $\mathscr{E}^{(\infty)}$, endowed with the Cartan distribution, is a geometrical form that is adequate for the concept of a differential equation.

On the manifolds of the form $\mathscr{E}^{(\infty)}$ there is a meaningful differential calculus. To construct it we need to take the algebraic point of view (see Ch. 1, §2) and consider a filtration of the algebra $\mathscr{F}$. For example, a *vector field* on $\mathscr{E}^{(\infty)}$ must be a differentiation $X: \mathscr{F} \to \mathscr{F}$ that is compatible with the filtration in the sense that $X(\mathscr{F}_l) \subset \mathscr{F}_s$ for some $s = s(l)$.

The space $J^\infty(n, m)$ is a special case of the construction described above, corresponding to the trivial equation $0 = 0$, or $\mathscr{E} \equiv J^k(n, m) \subset J^k(n, m)$. It is also obvious that $\mathscr{E}^{(\infty)} \subset J^\infty$.

7.3. C-Maps and Higher Symmetries. A map $F: \mathscr{E}_1^{(\infty)} \to \mathscr{E}_2^{(\infty)}$ is said to be *smooth* if $F^*\mathscr{F}(\mathscr{E}_2) \subset \mathscr{F}(\mathscr{E}_1)$ and a *C-map* if in addition it maps the distribution $C(\mathscr{E}_1)$ into $C(\mathscr{E}_2)$. C-maps take integral manifolds into integral manifolds, thereby generating a map of solutions of $\mathscr{E}_1$ into solutions of $\mathscr{E}_2$. Thus, any C-map generates an operator on the space of local solutions of $\mathscr{E}_1$. We can show that it is a differential map and uniquely determines the map F itself.

An invertible C-map $F: \mathscr{E}^{(\infty)} \to \mathscr{E}^{(\infty)}$ is called a *higher symmetry* of the equation $\mathscr{E}$. Symmetries of $\mathscr{E}$ understood in the classical sense (see 4.3) can be lifted to $\mathscr{E}^{(\infty)}$. This gives the possibility of interpreting them as higher symmetries.

We carry out the procedure of factorizing a differential equation $\mathscr{E}$ by the group G of its higher symmetries as an illustration of the high quality of the geometrical model introduced above. The distribution $C(\mathscr{E}^{(\infty)})$ naturally generates the distribution $C_G(\mathscr{E}^{(\infty)})$ on the quotient space $\mathscr{E}^{(\infty)}/G$ (it is defined, generally speaking, only locally). The pair $(\mathscr{E}^{(\infty)}/G, C_G(\mathscr{E}^{(\infty)}))$ (locally) has the form $(\mathscr{E}_1^{(\infty)}, C(\mathscr{E}_1^{(\infty)}))$. The equation $\mathscr{E}_1$ is called the *quotient equation* of the equation $\mathscr{E}$ by its symmetry group G. We draw attention to the fact that the procedure for factorization cannot be carried out by understanding a differential equation as a submanifold of the space of jets J^k.

Examples. 1) The wave equation $\dfrac{\partial^2 u}{\partial x \partial t} = 0$ is invariant under the group G of translations of the (x, t)-plane. The corresponding quotient equation is again the wave equation.

2) The transformations $u \mapsto cu$, $c \in \mathbb{R} \setminus \{0\}$, form a one-parameter group of transformations of the heat equation $\dfrac{\partial u}{\partial t} = \dfrac{\partial^2 u}{\partial x^2}$. The corresponding quotient-equation is the Burgers equation $\dfrac{\partial v}{\partial t} = \dfrac{\partial^2 v}{\partial x^2} + v \dfrac{\partial v}{\partial x}$.

Vector fields on $\mathscr{E}^{(\infty)}$, that is, those compatible with the filtration of differentiation of the algebra $\mathscr{F}(\mathscr{E})$, which preserve the Cartan distribution, are called *C-fields*. Any vector field on $\mathscr{E}^{(\infty)}$ belonging to the Cartan distribution is automatically a *C*-field. Such a field has the form $\sum f_i \bar{D}_i$, where $f_i \in \mathscr{F}(\mathscr{E}^{(\infty)})$, and the $\bar{D}_i$ are the restrictions of the total derivative operators to $\mathscr{E}^{(\infty)}$, and is said to be *trivial*. On objects of the form $\mathscr{E}^{(\infty)}$ the vector field X, generally speaking, does not generate any current A_t. However, if this were not so, then the corresponding current $\hat{A}_t$ in the space of solutions of the equation $\mathscr{E}$ (see 2.4) would be trivial (that is, $\hat{A}_t = \mathrm{id}$) for a trivial *C*-field. Hence the Lie algebra of all *higher infinitesimal symmetries* of $\mathscr{E}$ is defined as the quotient-algebra of the Lie algebra of all *C*-fields on $\mathscr{E}^{(\infty)}$ by a trivial field.

Any higher symmetry of the equation $\mathscr{E}$ is uniquely determined by its *generating function* $f \in \mathscr{F}(\mathscr{E})$, which has the same sense as the generating function of a contact vector field (see 2.4). However, in contrast to the latter, it may depend on derivatives of arbitrarily higher order.

Example. Let $\mathscr{E} = \{p_{(0,1)} = p_{(2,0)} + p_{(0,0)}p_{(1,0)}\} \subset J^2(2, 1)$ be the Burgers equation (see the previous example). Here $x = x_1$, $t = x_2$ and (k, l) is a multi-index. It is easy to see that the functions x_1, x_2 and $p(k, 0)$, $k \geqslant 0$, can be chosen as coordinates on $\mathscr{E}^{(\infty)}$. The algebra of higher symmetries of $\mathscr{E}$, if we describe it in terms of generating functions, is additively generated by functions of the form $f_k^i = x_2^i p(k, 0) + o(k)$, where $k \geqslant 1$, $0 \leqslant i \leqslant k$ and $o(k)$ is a polynomial in the variables $x_1, x_2, p(j, 0)$, $j < k$. It is isomorphic to the Lie algebra of polynomial zero-divergence vector fields on the plane. The classical part of this algebra, that is, symmetries in the sense of 4.3, is generated by elements f_k^i, $k = 1, 2$.

We emphasize that a higher symmetry, in contrast to a classical symmetry, is not an individual vector field but a whole class of them. This is the simplest concept of the secondary differential calculus which we mentioned in Ch. 1, § 5. Secondary vector fields act on secondary functions. The latter are represented by symbols of the form $\int L(x, u, u_x, \dots) \, dx_1 \dots dx_n$ and are cohomology classes of the so-called horizontal de Rham complex on $\mathscr{E}^{(\infty)}$. Thus, secondary differential calculus begins where the theory of higher symmetries ends.

Chapter 6
Geometric Structures

> "I believe that at present the most important method
> by which a mathematician derives the greatest
> benefit from his work as an investigator of nature is
> the systematic classification of quantities".
>
> J.C. Maxwell

§1. Geometric Quantities and Geometric Structures

1.1. What is a Geometric Quantity? According to Riemann's ideas, a geometry is specified by a field quantity. Above we became acquainted with examples of such quantities – various tensor fields and connections. Now is the time to analyse the very concept of a field quantity in more detail.

We first suppose that a coordinate system is fixed in space. Then we can regard a field quantity as a vector-valued function that assigns a set of numbers to a point of space, for example the components of the velocity vector, the stress tensor, the Christoffel symbols, and so on. A quantity is called a *scalar* if it does not depend on the choice of coordinate system (say the temperature). Most physical quantities are not scalars, but change their values when the coordinates are changed. However, their dependence on the coordinates has one peculiarity, namely the principle of short range: the value of a quantity at a point p depends only on the infinitesimal structure of the coordinate system in a neighbourhood of p. In other words, a quantity takes the same value at a point p in coordinates $x: U \to M$ and $y: V \to M$ that have tangency of some order k at p. This means that the transition functions $y^{-1} \circ x$ have the same k-jet at the point $x^{-1}(p) \in U \subset \mathbb{R}^n$ as the identity map $x^i \to x^i$. The number k depends on the quantity under consideration and is called its (*differential*) *order*. For example, tensor quantities have order 1.

1.2. Bundles of Frames and Coframes. Before giving a strict definition of a geometric quantity we introduce the concepts of frame and coframe, which play an important role in differential geometry.

Let $x: U \to x(U) \subset M$ be a chart with centre at the point $p = x(0)$, and $x^{-1}: x(U) \to U \subset \mathbb{R}^n$ the inverse map. The k-jet $j_0^k x$ of the chart at the point 0 (that is, the class of charts that touch each other with order k at p) is called an *infinitesimal coordinate system at p* or a *frame of order k with origin at p*. Similarly, the k-jet $j_p^k(x^{-1})$ of the vector-valued function x^{-1} is called a *coframe of order* k with origin at p. The set $\text{Rep}^k M$ of frames and the set $\text{Rep}_k M$ of coframes of order k are equipped in a natural way with the structures of smooth manifolds. The projections of these manifolds on M, under which a coframe is associated with its origin, are smooth bundles. They are called the *bundle of frames* and the *bundle of coframes* of order k of the manifold M. These bundles are canonically isomorphic. For many reasons it is more convenient to deal with the bundle of coframes, which we now do.

Examples. 1) A frame of order 1 at a point p on a manifold M is identified with the isomorphism $\mathbb{R}^n \to T_p M$ of vector spaces, which in turn is uniquely determined by a choice of basis of the tangent space and is often identified with such a basis. Similarly, a coframe of order 1 is identified with a basis $e^1, \ldots, e^n$ of the cotangent space $T_p^* M$ or, in other words, with an $\mathbb{R}^n$-valued linear form on $T_p M$ that is non-degenerate (in the sense that it specifies an isomorphism $T_p M \to \mathbb{R}^n$).

2) Let $M = \mathbb{R}^n$. A coframe of order k at the point $0 \in \mathbb{R}^n$ is the k-jet $j_0^k(\varphi)$ of a diffeomorphism $\varphi \colon \mathbb{R}^n \to \mathbb{R}^n$ that preserves the origin 0. It is identified with the polynomial map

$$j_0^k(\varphi) = \varphi_j^i x^j + \varphi_{j_1 j_2}^i x^{j_1} x^{j_2} + \cdots + \varphi_{j_1 \ldots j_k}^i x^{j_1} \ldots x^{j_k}.$$

The coefficients $\varphi_j^i, \ldots, \varphi_{j_1, \ldots, j_k}^i$ can be regarded as the components of the coframe $j_0^k(\varphi)$. They satisfy just the one condition $\det \| \varphi_j^i \| \neq 0$. The set of such coframes forms a Lie group with respect to the multiplication operation

$$j_0^k(\varphi) \cdot j_0^k(\psi) = j_0^k(\varphi \circ \psi),$$

It is denoted by $\mathbf{G}^k(n)$ (or simply $\mathbf{G}^k$) and called the *differential group of order* k. Obviously $\mathbf{G}^1(n) = GL(n)$.

1.3. Geometric Quantities (Structures) as Equivariant Functions on the Manifold of Coframes. Since a coframe of order k describes the infinitesimal structure of a local coordinate system, we can say that a field quantity of order k is a vector-valued function on the manifold of coframes $\text{Rep}_k M$.

Example. We can regard a Riemannian metric as a function on the manifold $\text{Rep}_1 M$ that assigns to a coframe $e^* = (e^1, \ldots, e^n)$ the Gram matrix $g(e) = \| g_{ij} \| = \| g(e_i, e_j) \|$, where $e = (e_1, \ldots, e_n)$ is the frame dual to e^*.

The question arises of how to characterize field geometric quantities among all vector-valued functions on the manifold of coframes. To answer this question we observe that the full linear group $GL(n)$ acts freely (that is, with trivial stabilizers) on the manifold of coframes $\text{Rep}_1 M \colon (A\theta)(\xi) = A(\theta(\xi))$, $A \in GL(n)$, where $\theta \colon T_p M \to \mathbb{R}^n$ is a coframe and $\xi \in T_p M$ or if $\theta = (e^1, \ldots, e^n)$, then $(A\theta)^1 = A_j^i e^j$. The orbits of this group are the fibres of the bundle of coframes $\text{Rep}_1 M \to M$.

As an example, consider a Riemannian metric g. On going over from an infinitesimal coordinate system $\theta = (e^1, \ldots, e^n)$ at the point p to a new coordinate system $\theta' = A\theta$, $A \in GL(n)$, the value of g at p is transformed by a representation of the group $GL(n)$ in the space of matrices $M(n)$, that is,

$$g(A\theta) = T_A g(\theta),$$

where $A \to T_A$ is the representation of $GL(n)$ in the space $M(n)$ given by the formula

$$T_A G = A'GA \qquad \text{or} \qquad T_A \|g_{ij}\| = \|A_i^k g_{kl} A_j^l\|.$$

In other words, the vector-valued function $g\colon \mathrm{Rep}_1 M \to M(n)$ is equivariant under the actions of the group $GL(n)$ described above. The condition of equivariance means that on going over to another infinitesimal coordinate system the value of a field quantity is rewritten in some natural form. It is the only condition that distinguishes field geometric quantities among all vector-valued functions. More precisely, let W be a manifold with fixed action of the group $GL(n)$ or, more briefly, a $GL(n)$-manifold. A *field of geometric quantities of order* 1 *and type* W is a $GL(n)$-equivariant W-valued function on the manifold $\mathrm{Rep}_1 M$ of coframes.

Example. Let $V = \mathbb{R}^n$, and let $W = V_q^p$ be the space of tensors of type (p, q). A geometric structure of type W is simply a tensor field of type (p, q).

To define geometric structures of order k it is sufficient to replace the group $GL(n) = \mathbf{G}^1(n)$ in the definition by the differential group $\mathbf{G}^k(n)$ of order k. This group acts freely in the manifold $\mathrm{Rep}_k M$ according to the formula

$$j_0^k(\varphi) \cdot j_p^k(u) = j_p^k(\varphi \circ u),$$

$$\varphi\colon \mathbb{R}^n \to \mathbb{R}^n, \quad \varphi(0) = 0, \quad u\colon M \to \mathbb{R}^n, \quad u(p) = 0, \quad j_0^k \varphi \in \mathbf{G}^k(n),$$

$$j_p^k u \in \mathrm{Rep}_k M.$$

Its orbits are the fibres of the bundle $\mathrm{Rep}_k M \to M$. We thus arrive at the following definition.

Let W be a $\mathbf{G}^k(n)$-manifold. A *geometric structure* or *field of geometric quantities of type* W *and order at most* k is a $\mathbf{G}^k(n)$-equivariant W-valued function $F\colon \mathrm{Rep}_k M \to W$ on the manifold of coframes.

Since when $l > k$ there is a natural epimorphism of Lie groups $\mathbf{G}^l(N) \to \mathbf{G}^k(n)$ (for details see 4.4), any $\mathbf{G}^k(n)$-manifold can be regarded as a $\mathbf{G}^l(n)$-manifold. In particular, the natural projection $\mathrm{Rep}_l M \to \mathrm{Rep}_k M$ is a $\mathbf{G}^l(n)$-equivariant map. Hence it follows that a geometric structure $F\colon \mathrm{Rep}_k M - W$ of order at most k can be understood as a geometric structure $F\colon \mathrm{Rep}_l M \to W$ of order at most l when $l > k$. A geometric structure F has order k if it is not a structure of order at most $k - 1$.

Let W be a vector (respectively, affine) space on which the group $\mathbf{G}^k(n)$ acts linearly (respectively, affinely). In this case we say that W is a *space of linear* (respectively, *affine*) *geometric quantities*, and geometric structures of type W are said to be *linear* (respectively, *affine*).

1.4. Examples. Infinitesimally Homogeneous Geometric Structures and G-Structures. Let us consider a number of examples of geometric structures.

1. Let $W = J_0^k(\mathbb{R}^n, \mathbb{R})$ be a space of k-jets of functions at the point $0 \in \mathbb{R}^n$ defined in some neighbourhood of it. The group $\mathbf{G}^k(n)$ acts linearly in the space W according to the formula $j_0^k(\varphi) \cdot j_0^k(f) = j_0^k(f \circ \varphi^{-1})$, $j_0^k(\varphi) \in \mathbf{G}^k(n)$, $j_0^k(f) \in W$.

Linear geometric structures of type W on a manifold M are identified with sections of the bundle of k-jets of functions on M (see 4.3 below).

2. The adjoint space $W^* = \text{Hom}(J_0^k(\mathbb{R}^n, \mathbb{R}), \mathbb{R})$ is also a linear $\mathbf{G}^k(n)$-space. The corresponding geometric structures are linear differential operators of order k.

3. Let $W = J_0^k(\mathbb{R}, \mathbb{R}^n)$ be the space of k-jets at the origin of curves $\gamma: [-1, 1] \to \mathbb{R}^n$, $\gamma(0) = 0$. The group $\mathbf{G}^k(n)$ acts linearly in W. The corresponding linear geometric structures are fields of *tangent vectors of order* k or k-*velocities*.

4. The group $\mathbf{G}^2(n)$ splits into the semidirect product $\mathbf{G}^2(n) = GL(n) \cdot \mathbf{G}_1^2(n)$, where $\mathbf{G}_1^2(n)$ is a commutative normal divisor consisting of 2-jets at the origin of diffeomorphisms whose principal linear part is the identity map. The group $\mathbf{G}_1^2(n)$ is identified with the vector group of the space of quadratic maps $\mathbf{G}_1^2(n) = \{\varphi = \varphi_{jk}^i x^j x^k, \varphi_{jk}^i = \varphi_{kj}^i\}$, and the group $\mathbf{G}^2(n)$ acts in this space as a group of affine transformations. The corresponding affine geometric structures of type $\mathbf{G}_1^2(n)$ have order 2 and are identified with symmetric linear connections. For this reason linear connections are often known as affine.

5. Let $G \subset \mathbf{G}^k(n)$ be a subgroup. Geometric structures of type W, where $W = \mathbf{G}^k(n)/G$ is a homogeneous space of geometric quantities, are said to be *infinitesimally homogeneous*. Let $F: \text{Rep}_k M \to W$ be such a structure. As we can easily verify, the inverse image $P = F^{-1}(w_0)$, where $w_0 \in W$ is the residue class of the identity mod G, is a subbundle of the bundle of coframes whose fibres are orbits of the stationary group G. Such subbundles of the bundle of coframes are called G-*structures*. Conversely, a geometric structure F can be uniquely restored from a G-structure P as a $\mathbf{G}^k(n)$-equivariant extension of the constant map $F: P \to w_0$. For this reason infinitesimally homogeneous structures are often identified with the corresponding G-structures.

Examples. 1) A Riemannian metric can be regarded as an infinitesimally homogeneous structure corresponding to the homogeneous space $W = GL(n)/O(n)$, which is identified with the cone of positive definite matrices. The corresponding G-structure consists of all orthonormal coframes.

2) A field of endomorphisms A on a manifold M is an infinitesimally homogeneous structure if and only if the operators $A_x: T_x M \to T_x M$ for all $x \in M$ have the same Jordan form $J \in M(n)$. In this case $W = GL(n)/C(J)$, where $G = C(J)$ is the centralizer of the matrix J in the full linear group, and the G-structure consists of all coframes with respect to which the endomorphism has matrix J. Suppose, for example, that $J^2 = -\text{id}$. Then the field of endomorphisms A is called *an almost complex structure* on M, since it determines in each tangent space $T_x M$ the structure of a complex vector space.

3) Let W be the Grassmann manifold of m-dimensional subspaces of $\mathbb{R}^n$. It is identified with the homogeneous space $W = GL(n)/GL(n, m)$, where the subgroup $GL(n, m) \subset GL(n)$ consists of transformations that preserve a fixed sub-

space $\mathbb{R}^m \subset \mathbb{R}^n$. Geometric structures of type W are m-dimensional distributions on n-dimensional manifolds.

1.5. Natural Geometric Structures and the Principle of Covariance. Above we defined geometric structures as W-valued equivariant functions on the manifold of coframes. The action of a group of diffeomorphisms of a manifold M is naturally lifted to an action in the manifold of coframes $\mathrm{Rep}_k M$ that commutes with the action of the group $\mathbf{G}^k(n)$, and thereby to an action in the space of $\mathbf{G}^k(n)$-equivariant W-valued functions (that is, geometric structures of type W). This action in the space of linear (affine) geometric structures will be linear (affine), as before. From the physical point of view a diffeomorphism of space can be interpreted in two ways: as a "passive" transformation corresponding to transition to another system of coordinates (or observers), and as an "active" transformation corresponding to a transformation of the laboratory together with the phenomenon under investigation, for example a change in the velocity of motion of the laboratory. In any case the requirement that it should be possible to recalculate the measurable value of a physical quantity under a controlled transformation of the system of observers or the object under observation is very natural. It was the basis of the principle of covariance, which was widely discussed at the beginning of this century. The mathematical formulation of this principle consists in postulating the fact that physical field quantities must be *natural*, that is, a group of diffeomorphisms of physical space-time must act in the space of such quantities, and the quantities themselves must be transformed in accordance with this action. Below we shall see that all natural field quantities are exhausted by the geometric structures defined above. Hence the principle of covariance essentially means that physical quantities must be geometric structures.

The development of physics has demonstrated the limitations of this principle and the need to consider more general ("unnatural") field quantities. Independently of this, the logic of the development of differential geometry and differential topology has led to the advisability of studying abstract bundles and, in particular, to the development of the theory of connections in principal bundles and those associated with them.

We mention two facts that point to the need to generalize the principle of covariance:

1. Spinor quantities, discovered by Dirac, which describe elementary particles with a half-integer spin, for example, electrons are not natural geometric structures.

2. An attempt to geometrize an electromagnetic field in the framework of natural geometric structures (special linear connections), undertaken by Weyl, ran into serious physical difficulties. Apparently Kaluza and Klein were the first to go beyond the bounds of natural quantities in physics. Essentially they considered an abstract one-dimensional bundle $P \to M$ over four-dimensional space-time and showed that the Einstein equations $\mathrm{Ric}(g) = 0$ for a suitable pseudo-Riemannian metric g in a 5-dimensional space P are equivalent to the Einstein equations in space-time M that describe (in the framework of relativity theory) the interaction of gravitational and electromagnetic fields.

Later Heisenberg introduced the important concept of an "internal space", whose vectors describe the internal state of an elementary particle. In the simplest case, according to Heisenberg a proton and a neutron form a basis for the internal space of an "isotopic spin" W whose "pure" states they are. From the mathematical point of view we can say that Heisenberg considered a trivial bundle $W \times M \to M$ over space-time with a trivial connection. This means that he assumed that the space W is "absolute" and does not depend on the choice of a point in space-time.

Finally, the decisive step for the geometrization of physical fields and the avoidance of the difficulties that Weyl encountered was taken by Yang and Mills in 1954. They assumed that the space W of "internal degrees of freedom" of a particle depends on the points of space-time. In other words, there is a vector bundle $\pi \colon P \to M$ whose fibres are "spaces of internal degrees of freedom", and there is no way of canonically identifying different fibres (that is, there is no absolute parallelism). They suggested considering a connection in the bundle π, that is, a way of identifying fibres along curves in M, as a physical field (a gauge field or Yang-Mills field), and its holonomy group Γ as the internal group of symmetries of elementary particles. In the simplest case of a one-dimensional complex bundle a connection with an Abelian holonomy group $\Gamma = SO(2)$ is interpreted as an electromagnetic field, and its curvature form is identified with the stress tensor of the electromagnetic field. It turned out that other physical fields that describe the interactions of elementary particles can be interpreted as Yang-Mills fields (that is, connections), but generally speaking with non-Abelian holonomy groups. In this case the field equations suggested by Yang and Mills become non-linear and are a natural generalization of Maxwell's equations (see 3.7 below).

§2. Principal Bundles

Bundles of frames and coframes and more generally G-structures are representatives of a special class of bundles, called *principal bundles*. Their importance is emphasized by the title. G-structures are natural bundles (see 2.1). The abstract concept of a principal bundle was crystallized in the process of realizing the need to go outside the framework of natural bundles (see 2.1). This section is devoted to the elements of the theory of principal bundles.

2.1. Principal Bundles. The action of a group is said to be *free* if all its stabilizers are trivial.

Let $\pi \colon E \to M$ be a smooth bundle. It is called a *principal bundle* if on E we are given the free action of a Lie group G whose orbits coincide with the fibres of this bundle. The group G is called its *structure group*. To emphasize this we need the term *principal G-bundle*. The group G also appears here as a standard fibre.

The specific feature of the description of principal bundles in terms of transition functions consists in the fact that transformations $h_{\alpha\beta}(x) \colon G \to G$ (see Ch. 3, 2.4) are right translations by elements $\varphi_{\alpha\beta}(x)$ of the group G. Consequently, they can be regarded as G-valued functions $\varphi_{\alpha\beta}: U_\alpha \cap U_\beta \to G$, $\varphi_{\alpha\beta}: x \to \varphi_{\alpha\beta}(x)$. The functions $\varphi_{\alpha\beta}$ are called *transition functions* of principal G-bundles.

2.2. Examples of Principal Bundles

1) The trivial bundle $\pi: G \times M \to M$ together with the natural action of the group G: $g(g_1, m) = (gg_1, m)$ is the simplest example of a principal G-bundle.

2) Bundles of frames and coframes of order k are principal $\mathbf{G}^k(n)$-bundles.

3) The action of the multiplicative group $\mathbb{R}^* = \mathbb{R}\backslash\{0\}$ on the manifold $E = \mathbb{R}^{n+1}\backslash\{0\}$, $\lambda \times x \to \lambda x$, $\lambda \in \mathbb{R}^*$, $x \in E$, determines a principal $\mathbb{R}^*$-bundle over projective space $\pi: E \to \mathbb{R}P^n$. Similarly we can construct principal $\mathbb{C}^*$ and $\mathbb{H}^*$ bundles: $\pi: \mathbb{C}^{n+1}\backslash\{0\} \to \mathbb{C}P^n$ and $\pi: \mathbb{H}^{n+1}\backslash\{0\} \to \mathbb{H}P^n$, where $\mathbb{C}^*$ ($\mathbb{H}^*$) is the multiplicative group of non-zero complex numbers (quaternions).

4) Restricting the projection π from the previous example to the unit sphere $S^n = \{x \mid \|x\| = 1\} \subset \mathbb{R}^{n+1}\backslash\{0\}$, we obtain a principal $\mathbb{Z}_2$-bundle over projective space: $\pi: S^n \to \mathbb{R}P^n$, $\mathbb{Z}_2 = \{+1, -1\} \subset \mathbb{R}^*$. This transition to the subgroup $\mathbb{Z}_2 \subset \mathbb{R}^*$ is a special case of the operation of reduction, which is discussed in the next subsection.

Applying a suitable construction to principal bundles over the complex and quaternion projective spaces described in Example 3, we obtain the principal bundles $S^{2n+1} \to \mathbb{C}P^n$ and $S^{4n+1} \to \mathbb{H}P^n$ with structure groups S^1 and S^3 respectively. Here S^1 is the group of complex numbers mod 1, and S^3 is the group of unit quaternions.

5) Let G be a discrete group of transformations of a manifold E. Then, if G acts freely on E, the set of its orbits $M = G\backslash E$ is a smooth manifold, and the natural projection $\pi: E \to M$ that assigns to an element $x \in E$ its orbit Gx is a principal G-bundle. This bundle is called a *covering* with structure group G. The $\mathbb{Z}_2$-bundle $\pi: S^n \to \mathbb{R}P^n$ considered above is an example of such a covering.

6) Multiplication by complex numbers of the form $\exp\dfrac{2\pi i m}{p}$, where p is a fixed prime number and $0 \leqslant m \leqslant p - 1$, determines the free action of the group $\mathbb{Z}_p \subset S^1$ on the sphere $S^{2n+1} = \{z \in \mathbb{C}^{n+1}, |z| = 1\}$. The base of a covering that arises in this way is called a *lens space*.

7) Let G be a Lie group, and H a closed subgroup of it. Then the action of H on G: $(h, g) \to gh^{-1}$, $h \in H$, $g \in G$, turns the homogeneous space G/H into the base of the principal H-bundle $\pi: G \to G/H$.

2.3. Homomorphisms and Reductions.

A morphism of bundles (Ch. 3, 2.4) that commutes with the action of the structure groups is called a *homomorphism* of them. More precisely, we are given a homomorphism $\rho: G \to G'$. Then a ρ-*homomorphism* of a principal G-bundle $\pi: E \to M$ into a principal G'-bundle $\pi': E' \to M'$ is a morphism of bundles $F: E \to E'$ such that $F(gx) = \rho(g)F(x)$ for all $g \in G$, $x \in E$.

The concept of a ρ-*isomorphism* is defined in an obvious way.

In the case when $G = G'$, $\rho = \mathrm{id}$, a ρ-isomorphism of a G-bundle onto itself is called an *automorphism*. Automorphisms of a principal G-bundle that induce identity transformations of the base M are called *gauge transformations*.

If $\rho: G \to G'$ is an embedding of the subgroup G in the ambient group G' and $M = M'$, then a ρ-homomorphism $F: E \to E'$ that induces the identity transformation on M is called a *reduction* of a principal G'-bundle to a G-bundle.

The existence of a reduction is equivalent to the possibility of reducing the transition functions $\varphi_{\alpha\beta}$ of the bundle π' to the subgroup G, that is, to choose them so that $\varphi_{\alpha\beta}(x) \in G \subset G'$.

A reduction $F: E \to E'$ identifies a G-bundle $\pi: E \to M$ with its "G-subbundle" $\pi' | F(E): F(E) \to M$.

Examples. 1) The section $M \ni x \to (e, x) \in G \times M$ of the trivial G-bundle $\pi: G \times M \to M$ is called the *unit section*. Every gauge transformation of this bundle is determined by the image of the unit section. Hence the group of gauge transformations is identified with the group of principal sections of π, or equivalently with the group $C^\infty(M, G)$. The group of all automorphisms of π is the semidirect product of the group of diffeomorphisms of M and the group $C^\infty(M, G)$.

2) The bundle of frames, and also the bundle of coframes, $\pi: \mathrm{Rep}^1 M \to M$ is a natural bundle (see 1.1). Hence every diffeomorphism $F: M \to M$ induces an automorphism $F^{(1)}: \mathrm{Rep}^1 M \to \mathrm{Rep}^1 M$ of this bundle.

3) The embedding of a principal $O(n)$-bundle $\pi: E \to M$ of orthonormal frames on a Riemannian manifold M into the bundle $\mathrm{Rep}^1 M \to M$ determines a reduction of the principal $GL(n, \mathbb{R})$-bundle of frames to the $O(n)$-bundle. An automorphism $F^{(1)}: \mathrm{Rep}^1 M \to \mathrm{Rep}^1 M$ is an automorphism of the $O(n)$-bundle π if and only if F preserves the Riemannian metric (that is, it is an isometry).

4) A reduction of the structure group G of a principal bundle $\pi: E \to M$ to the unit subgroup $\{e\} \subset G$ is equivalent to the choice of e-section $s: M \to E$. Each such section determines a trivialization of the bundle π – an isomorphism of it with trivial bundle $G \times M \to M$:

$$G \times M \ni (g, x) \mapsto gs(x) \in E.$$

2.4. G-Structures as Principal Bundles. Any G-structure on a manifold M can be regarded as the reduction of a principal $\mathbf{G}^k(n)$-bundle of coframes of order k to a principal G-bundle $\pi: E \to M$, $G \subset \mathbf{G}^k(n)$. The number k is called the *order* of the G-structure. Not every principal G-bundle is equivalent to some G-structure. An example is the trivial principal $SO(2)$-bundle on the sphere S^2.

The method described below of internally characterizing those principal G-bundles that are equivalent to G-structures is interesting, apart from anything else, in that it leads to useful generalizations (see 2.5). To this end we observe that on the manifold $\mathrm{Rep}_1 M$ of first-order coframes there is a canonical $\mathbb{R}^n$-valued differential 1-form θ. This form assigns (see Fig. 39) to the tangent vector $\xi \in T_p \mathrm{Rep}_1 M$ at the point $p \in \mathrm{Rep}_1 M$ corresponding to the coframe $p: T_a M \to \mathbb{R}^n$ the vector $\theta(\xi) = p(d_p\pi(\xi)) \in \mathbb{R}^n$. The form $\theta \in \Lambda^1(\mathrm{Rep}_1 M) \otimes \mathbb{R}^n$ is called the *shift form*. It is easy to verify that the form θ has the following properties.

1) θ is strictly horizontal. This means that $\theta(\xi) = 0$ if and only if ξ is a vertical vector, that is, $d_p\pi(\xi) = 0$.

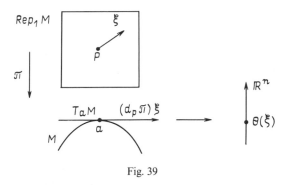

Fig. 39

2) θ is $GL(n, \mathbb{R})$-equivariant, that is, $\theta(g_*(\xi)) = g\theta(\xi)$, $\forall g \in GL(n, \mathbb{R})$.

We now observe that the shift form completely characterizes the bundle of first-order coframes. Namely, if $\pi: E \to M$ is a principal $GL(n, \mathbb{R})$-bundle on which there is given an $\mathbb{R}^n$-valued form θ ($n = \dim M$) satisfying conditions 1) and 2), then the bundle π is isomorphic to the bundle of first-order coframes on M. The required isomorphism is established by the map $F: E \to \mathrm{Rep}_1 M$, $\Gamma. E \ni \lambda \to p_x \in \mathrm{Rep}_1 M$, where $p_x(v) = \theta_x(\xi)$, $v \subset T_{\pi(x)}M$, and $\xi \in T_\lambda E$ is an arbitrary vector projected on v, $d\pi(\xi) = v$.

Let $\pi: E \to M$ be a first-order G-structure and suppose that $E \subset \mathrm{Rep}_1 M$. The restriction of the shift form θ to E determines a differential $\mathbb{R}^n$-valued form on E. This form is called the *shift form* of the G-structure. As before, it will be denoted by θ. This form is strictly horizontal and G-equivariant.

Like the bundle of all coframes, first-order G-structures are characterized (up to an isomorphism) as principal G-bundles equipped with a strictly horizontal G-equivariant $\mathbb{R}^n$-valued differential 1-form.

Similarly, we can describe G-structures of higher orders in an intrinsic way by introducing "higher" shift forms.

2.5. Generalized G-Structures. The abstract definition of a G-structure as a principal bundle with a shift form shows how we can generalize this concept.

Let G be a Lie group and $\rho: G \to GL(n, \mathbb{R})$ a linear representation of it, not necessarily faithful. A *generalized G-structure* on a manifold M is a principal G-bundle $\pi: E \to M$ with a G-equivariant strictly horizontal $\mathbb{R}^n$-valued form $\theta \in \Lambda_1(E) \otimes \mathbb{R}^n$.

If the representation is ρ-faithful, then the group G is identified with the linear group $\rho(G)$, and the generalized G-structure is identified with the "usual" one.

The construction of a map $F: E \to \mathrm{Rep}_1 M$ given in the previous subsection remains in force even when the representation ρ is not faithful. For a generalized G-structure this map is not a diffeomorphism onto its own image, but is a principal H-bundle, where $H = \mathrm{Ker}\, \rho$. The subbundle $F(E) \subset \mathrm{Rep}_1 M$ is obviously a $\rho(G)$-structure. Thus, every element $q \in E$ can be thought of as a frame $F(q)$ having an "intrinsic" structure, whose geometry according to Klein is controlled by the group H. This is how physicists imagine the spin of an electron (see Example 3 below).

Examples. 1) Let $M = G/H$ be the homogeneous space of a Lie group G, and let $\rho: H \to \text{Aut}(\mathfrak{G}/\mathfrak{H})$ be the corresponding isotropy representation. The principal H-bundle $\pi: G \to M$ is a generalized H-structure. The value of the form θ on a vector $\xi \in T_g G$ is equal to the projection of the vector $dl_g(\xi) \in T_e G = \mathfrak{G}$ into the quotient space $\mathbb{R}^n = \mathfrak{G}/\mathfrak{H}$. We note that if the isotropy representation is faithful, then the bundle π can be regarded as an H-structure on M. This observation is the basis of Cartan's moving frame method.

2) Every ρ-homomorphism $F: E \to E'$ of a principal G-bundle $\pi: E \to M$ into a G'-structure $\pi': E' \to M$ that is the identity on the base M enables us to regard π as a generalized G-structure with shift form $\theta = F^*(\theta')$, where θ' is the shift form of the G'-structure π'.

3) The spinor structures needed to describe elementary particles of electron type are a special case of the construction given in Example 2. Here $G' = SO(n)$, $G = \text{Spin}(n)$. The group $\text{Spin}(n)$ is the only non-trivial two-sheeted covering of the group $SO(n)$, and $\rho: \text{Spin}(n) \to SO(n)$ is the covering map. A *spinor structure* on a manifold M is a generalized $\text{Spin}(n)$-structure on it, defined by a ρ-homomorphism into the principal $SO(n)$-bundle of oriented orthonormal frames on M. In contrast to a Riemannian manifold, the given manifold M cannot always be equipped with a spinor structure. This is possible if certain topological conditions are satisfied (more precisely, the second Stiefel-Whitney characteristic class of M must vanish).

2.6. Associated Bundles. With every principal G-bundle $\pi: E \to M$ and G-manifold W there is naturally associated a smooth bundle with fibre W whose transition functions $h_{\alpha\beta}(x): W \to W$ (see Ch. 3, 2.4) are the actions of elements $\varphi_{\alpha\beta}(x) \in G$ on W, where $\varphi_{\alpha\beta}$ are the transition functions of the principal G-bundle π (see 2.1). The resulting bundle $\pi_W: E_G \times W \to M$ is said to be *associated* with the principal bundle π.

Invariantly the space $E_G \times W$ of this bundle is defined as the space of orbits of the natural action of G in the direct product $E \times W$, and the projection π_W as the map

$$E_G \times W \ni (x, w) \to \pi(x) \in M.$$

If an additional G-invariant structure is defined on W (for example, the structure of a vector space), then the fibres of the bundle π_W and the bundle π_W itself inherit this structure.

The bundles associated with the bundle of k-th order coframes and the $G^k(n)$-manifold W are called *bundles of geometric quantities of type W*. The sections of this bundle are called *geometric structures of type W*.

Bundles of geometric quantities are natural bundles.

Examples. 1) All linear representations of the group $GL(n, \mathbb{R})$ are exhausted by representations in spaces of tensors and tensor densities. Hence all bundles of linear first-order geometric quantities are bundles of tensors and tensor densities.

2) Vector bundles associated with principal k-th order bundles, $k \geq 2$, are called *supertensor bundles* according to the terminology of P.K. Rashevskij who, bearing their classification in mind, described the linear representations of the groups $\mathbf{G}^k(n)$.

3) The bundles associated with a generalized Spin(n)-structure and vector Spin(n)-space W are called *spinor bundles*. The sections of these bundles are called *spinor fields*. They are used in field theory to describe elementary particles having a *spin* (the form of a quantum number).

4) The group $GL(n, \mathbb{R})$ acts in a natural way on the Grassmann manifold $W_{n,s}$ of s-dimensional subspaces of $\mathbb{R}^n$. The bundle $\pi_W: G_s(T) \to M$ associated with the bundle of first-order frames and this action of the group $GL(n, \mathbb{R})$ is called a *Grassmann bundle* (see also 4.1).

5) Let $\pi_W: PX_G W \to M$ be the bundle associated with the principal G-bundle $\pi: P \to M$ and the homogeneous G-manifold $W = G/H$. The embedding $P \to P \times_G G/H$, $p \to p \times_G \{H\}$ induces a morphism F of π into π_W that is the identity on the base. Every section s of π_W determines a principal H-subbundle $F^{-1}(s(M)) \subset P$, and conversely Consequently, a principal G-bundle π can be reduced to an H-bundle if and only if π_W has a section. An elementary fact of homotopic topology is that a bundle has a global section if a fibre of it can be contracted over itself to a point. Since the quotient space G/H of a connected Lie group G by a maximally compact subgroup H is diffeomorphic to $\mathbb{R}^n$, and hence contractible, a principal G-bundle can always be reduced to a principal H-bundle. For example, a principal $GL(n)$-bundle can always be reduced to an $O(n)$-bundle.

§3. Connections in Principal Bundles and Vector Bundles

3.1. Connections in a Principal Bundle. A first-order coframe can be represented as a set of n linearly independent covectors. A simultaneous parallel transport of each of them induces a transport of the coframe as a whole. Hence a linear connection on a manifold M induces a connection in the bundle of coframes $\mathrm{Rep}_1 M \to M$.

With each curve $\gamma(t)$ on M and coframe p at the point $\gamma(0)$ we can associate a curve $p(t)$ on $\mathrm{Rep}_1 M$ obtained by parallel transport of the coframe p along the curve γ. It is called the *horizontal lift* of the curve $\gamma(t)$ with origin at p. Let H_p denote the set of tangent vectors at the point p to all possible transports of the coframe p along the curve γ. It is the space $T_p \mathrm{Rep}_1 M$, which is said to be *horizontal*. This subspace is complementary to the tangent space of the fibre $T_p^v \mathrm{Rep}_1 M$, which is said to be *vertical*. The distribution $p \to H_p$ on $\mathrm{Rep}_1 M$ is called the *horizontal distribution* of the connection under consideration. Since the operation of parallel transport of coframes commutes with the action of the

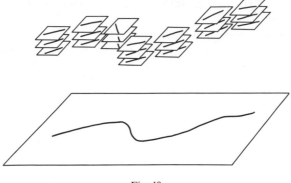

Fig. 40

group $GL(n, \mathbb{R})$, the horizontal distribution is also invariant under this action of the group $GL(n, \mathbb{R})$. Conversely, any $GL(n, \mathbb{R})$-invariant distribution H in $\mathrm{Rep}_1 M$ that is complementary to the vertical distribution $p \to T_p^v \mathrm{Rep}_1 M$ determines a linear connection. Namely, on the inverse image of the curve γ in the bundle of coframes the distribution H cuts out a field of directions (see Fig. 40) whose integral curves define the horizontal lifts of the curve γ.

Guided by what we said above, we define a *connection* in a principal G-bundle $\pi: P \to M$ as a G-invariant distribution $H: P \ni p \mapsto H_p \subset T_pP$ of horizontal subspaces. We say that a subspace is *horizontal* if it is complementary to the tangent space $T_p^v P$ of the fibre. For the same reasons as above a connection H defines a *horizontal lift* of a curve $\gamma(t)$ of the manifold M as a curve $p(t)$ on the manifold P that touches the distribution H and covers the curve γ, that is, $\pi(p(t)) = \gamma(t)$.

All possible lifts of γ that start at points of the fibre $\pi^{-1}(a)$, where $a = \gamma(0)$, determine a diffeomorphism of this fibre onto the fibre $\pi^{-1}(b)$, where $b = \gamma(1)$ (see Fig. 41), which is called a *parallel transport along the curve γ*. The invariance of H under the action of the group G ensures that the operation of parallel transport commutes with this action.

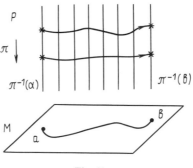

Fig. 41

3.2. Infinitesimal Description of Connections. Let $D(P)^G$ denote the Lie algebra of infinitesimal automorphisms of a G-bundle $\pi\colon P \to M$, that is, the algebra of G-invariant vector fields on P. The projection π defines a natural homomorphism $\pi_*\colon D(P)^G \to D(M)$ of Lie algebras that is simultaneously a homomorphism of $C^\infty(M)$-modules. Its kernel $\mathscr{V}(\pi) = \ker \pi_*$ consists of *vertical* G-invariant vector fields (which are the identity on the base of infinitesimal automorphisms of the G-bundle π) and is called the Lie algebra of *infinitesimal gauge transformations* (or *gauge vector fields*).

Thus, the bundle π determines an extension of the Lie algebra $D(M)$ of vector fields on M by means of the algebra $\mathscr{V}(\pi)$ of gauge fields.

The connection H determines a $C^\infty(M)$-linear map $\hat{H}\colon D(M) \to D(P)^G$, inverse to the projection π_*, that assigns to a vector field $X \in D(M)$ its horizontal lift X^H, the only vector field that belongs to H and covers the field X, that is, $X_p^H \in H_p$, $\forall p \in P$, and $\pi_*(X^H) = X$. Conversely, any such map $\hat{H}$ determines a connection. In other words, a connection can be regarded as a homomorphism $\hat{H}\colon D(M) \to D(P)^G$ of $C^\infty(M)$-modules inverse to the homomorphism $\pi_*\colon D(P)^G \to D(M)$, that is, $\pi_* \circ \hat{H} = \mathrm{id}$.

Generally speaking, a map $\hat{H}\colon D(M) \to D(P)^G$ that specifies a connection is not a homomorphism of Lie algebras. Its deviation from a Lie homomorphism is naturally characterized by a $C^\infty(M)$-bilinear map

$$\Omega\colon D(M) \times D(M) \to \mathscr{V}(\pi), \qquad (X, Y) \mapsto \lfloor X^H, Y^H \rfloor - [X, Y]^H.$$

We shall see below that it is appropriate to interpret this map as the curvature 2-form of the connection under consideration.

One more useful infinitesimal description of a connection is the following. An element A of a Lie algebra $\mathfrak{G}$ of the structure group G determines a one-parameter subgroup $g_t \in G$. The actions of the element g_t in turn determine a current on the manifold P. Let $\bar{A} \in D(P)$ be the velocity field of this current. The field $\bar{A}$ is vertical, that is, it touches the fibres of the projection $\pi\colon P \to M$. It is not difficult to see that the map $A \mapsto \bar{A}_p \in T_p^v P$ is an isomorphism of the Lie algebra $\mathfrak{G}$ on $T_p M$. We denote it by $\hat{p}$.

Any vector $\xi \in T_p P$ can be uniquely represented in the form $\xi = \xi^v + \xi^H$, $\xi^v \in T_p^v P$, $\xi^H \in H_p$. This enables us to define a $\mathfrak{G}$-valued 1-form ω on P whose value on the vector ξ is equal to $\hat{p}^{-1}(\xi^v)$. In other words, $\omega(X) = \hat{p}^{-1}(X^v)$, where X^v is the vertical part of the field $X \in D(P)$. The form ω is called the *connection form*. It is G-equivariant, that is,

$$\omega_{gp}(g(\xi)) = Ad_g \omega_p(\xi), \qquad g \in G, \quad \xi \in T_p P \tag{1}$$

and satisfies the condition

$$\omega(\bar{A}) = A, \qquad A \in \mathfrak{G}. \tag{2}$$

Conversely, any $\mathfrak{G}$-valued 1-form ω on P determines a connection in the bundle $\pi\colon P \to M$:

$$H_p = \{\xi \in T_p P / \omega_p(\xi) = 0\}.$$

Briefly, $H = \ker \omega$.

3.3. Curvature and the Holonomy Group. A connection H is said to be *flat* if the distribution H is completely integrable. This is equivalent to the vanishing of the $\mathfrak{G}$-valued *curvature 2-form of the connection* Ω, which is given by the formula

$$\Omega_p(\xi, \eta) = (d\omega)_p(\xi^H, \eta^H), \qquad \xi, \eta \in T_pP.$$

In fact, by the theorem of Frobenius the complete integrability of the distribution H is equivalent to the commutator $[X, Y]$ of two horizontal (that is, tangent to H) vector fields X and Y being horizontal, that is, $\omega([X, Y]) = 0$ if $\omega(X) = \omega(Y) = 0$. But according to the infinitesimal Stokes formula

$$d\omega(X, Y) = X(\omega(Y)) - Y(\omega(X)) - \omega([X, Y]) = -\omega([X, Y]).$$

Thus, complete integrability of H means that $d\omega(X, Y) = 0$ for any horizontal vector fields X and Y, that is, the curvature form Ω is equal to zero.

We mention the main properties of the curvature form Ω:

1) it is horizontal, that is, $\Omega_p(\xi, \eta) = 0$ if at least one of the arguments ξ, η is vertical;

2) Ω is G-equivariant in the same sense as ω.

3.4. The Holonomy Group. As in the case of a linear connection, for an arbitrary connection H in the principal bundle $\pi: P \to M$ we define the holonomy group Γ_x at a point $x \in M$ as the group of parallel transports τ_γ round loops $\gamma: [0, 1] \to M$, $\gamma(0) = \gamma(1) = x$. Since a parallel transport commutes with the action of the group G, the diffeomorphisms constituting the group Γ_x have this property. Suppose that $p \in \pi^{-1}(x)$. The transformation $\tau_\gamma \in \Gamma_x$ is uniquely determined by the image $\tau_\gamma(p)$ of the point p, since $\tau_\gamma(gp) = g\tau_\gamma(p)$. This determines an embedding

$$\bar{p}: \Gamma_x \hookrightarrow G, \qquad \tau \mapsto g, \quad \text{where } \tau_\gamma(p) = gp,$$

of the holomomy group in the structure group G as a subgroup.

The image $\bar{p}(\Gamma_x)$ is called *the holonomy group at the point* $p \in P$. The holonomy groups of different points are conjugate in the group G. Hence we can talk of the holonomy group Γ of a connection as a subgroup of G defined up to conjugacy.

Let $Q(p)$ denote the set of points of P that can be joined to a point $p \in P$ by horizontal curves. Then $\pi|_{Q(p)}: Q(p) \to M$ is the principal Γ-subbundle of the bundle $\pi: P \to M$, where Γ is the holonomy group. Thus, the connection H determines a reduction of the structure group G to the holonomy group Γ. Since a parallel transport preserves the fibres of the bundle $\pi|_{Q(p)}$, the connection H induces a connection in the reduced bundle $\pi|_{Q(p)}$.

Example. Let H be the Levi-Civita connection of a Riemannian manifold M, regarded as a connection in the bundle of coframes $\pi: \text{Rep}_1 M \to M$ (see §3.1) with holonomy group $\Gamma = SO(n)$. It determines a reduction of the bundle π to a bundle of orthonormal coframes and a connection in this bundle.

The following important theorem of Ambrose and Singer establishes a link between the curvature form and the holonomy algebra (that is, the Lie algebra of the holonomy group Γ) of the connection H.

Theorem. *Let H be a connection in the principal bundle $\pi: P \to M$ with curvature form Ω. The holonomy algebra $\dot{\Gamma} \subset \mathfrak{G}$ at the point $p \in P$ is linearly generated by elements of the form $\Omega_q(\xi, \eta)$, where $q \in Q(p)$, $\xi, \eta \in H_q$.*

3.5. Covariant Differentiation and the Structure Equations. A connection H in the principal bundle $\pi: P \to M$ determines a differential operator D that associates with a k-form ρ on P a horizontal $(k + 1)$-form $D\rho$:

$$D\rho(X_1, \ldots, X_{k+1}) = d\rho(X_1^H, \ldots, X_{k+1}^H)$$

(X^H denotes the horizontal part of the field $X \in D(P)$). The form $D\rho$ is called the *covariant differential* of the form ρ.

Example. The curvature form $\Omega = D\omega$ is the covariant differential of the connection form ω.

The Lie operation in the Lie algebra $\mathfrak{G}$ and the operation of exterior multiplication determine a multiplication of $\mathfrak{G}$-valued differential forms on P which for forms of the type $\rho = \alpha \otimes A, \sigma = \beta \otimes B$, where $A, B \in \mathfrak{G}$ and α, β are real-valued (that is, "ordinary") forms on P, is given by

$$[\rho, \sigma] = \alpha \wedge \beta \otimes [A, B].$$

This turns the space of $\mathfrak{G}$-valued forms into a graded supercommutative algebra:

$$[\rho, \sigma] = (-1)^{pq+1}[\sigma, \rho], \quad \deg \rho = p, \quad \deg \sigma = q.$$

The operator D is a differentiation of this algebra. In terms of this multiplication the covariant differential of a $\mathfrak{G}$-valued $\mathfrak{G}$-equivariant 1-form ρ is written in the form

$$D\rho = d\rho + \tfrac{1}{2}[\omega, \rho].$$

In particular, the connection form ω and the curvature form Ω are linked by the *Cartan structure equation*

$$d\omega + \tfrac{1}{2}[\omega, \omega] = \Omega. \tag{3}$$

The covariant differential taken from both sides of (3) leads to the *Bianchi identity*

$$D\Omega = 0. \tag{4}$$

Example. We can regard a Lie group G as the space of the principal G-bundle over the one-point base $M = \{*\}$. This bundle has a unique connection: $H_p = 0 \in T_pG$. The form ω of this connection is called the Maurer-Cartan form and is a $\mathfrak{G}$-valued right-invariant 1-form on the group G for which $\omega_e(A) = A$, where e is the identity of the group G and we identify $\mathfrak{G}$ and T_eG. The structure equation (3) of this connection has the form

$$d\omega + \tfrac{1}{2}[\omega, \omega] = 0$$

and is called the *Maurer-Cartan equation* of the Lie group G.

3.6. Connections in Associated Bundles. A connection in a principal G-bundle π determines a connection in any bundle π_W associated with it – the law of parallel transport of fibres along curves on the base manifold. To determine it we observe that any point $p \in P$ determines a diffeomorphism $\bar{p}$ of the manifold W onto the fibre $\pi_W^{-1}(x)$, where $x = \pi(p) \in M$:

$$\bar{p}: W \in w \mapsto (p, w) \in P \times_G W.$$

A *parallel transport* $\tau_\gamma^W: \pi_W^{-1}(x) \to \pi_W^{-1}(y)$ of the fibres of the bundle π_W along the curve $\gamma: [0, 1] \to M$, $\gamma(0) = x$, $\gamma(1) = y$, is given by

$$\tau_\gamma^W = \tau_\gamma(p) \circ \bar{p}^{-1},$$

where p is an arbitrary element of the fibre $\pi^{-1}(x)$, and τ_γ is a parallel transport along the curve γ in the principal bundle π. If π_W is the vector bundle associated with the vector G-space W, then the parallel transport τ_γ^W is a linear isomorphism of fibres. The operator ∇_X of infinitesimal parallel transport of sections of the bundle π_W along trajectories of the vector field $X \in D(M)$ (see Ch. 2, 2.6) in this case is a differentiation of the $C^\infty(M)$-module of sections $\Gamma(\pi_W)$. This means that the operator $\nabla_X: \Gamma(\pi_W) \to \Gamma(\pi_W)$ is $\mathbb{R}$-linear and satisfies the "Leibniz rule"

$$\nabla_X(fs) = f\nabla_X(s) + X(f)s, \qquad f \in C^\infty(M), \quad s \in \Gamma(\pi_W).$$

It is called the *covariant derivative* along the field X.

The map $X \mapsto \nabla_X$ is $C^\infty(M)$-linear. This gives the possibility of defining the *covariant differential*

$$D: \Gamma(\pi_W) \to \Lambda^1(M, \pi_W), \tag{5}$$

by putting

$$D(s)(X) = \nabla_X(s); \qquad s \in \Gamma(\pi_W), \quad X \in D(M).$$

Here

$$\Lambda^k(M, \pi_W) = \Lambda^k(M) \otimes_{C^\infty(M)} \Gamma(\pi_W).$$

The k-form $\omega \in \Lambda^k(M, \pi_W)$ is a skew-symmetric $\Gamma(\pi_W)$-valued semilinear form on $D(M)$:

$$\omega(X_1, \ldots, X_k) \in \Gamma(\pi_W), \qquad X_i \in D(M).$$

For example, the value of the element $\omega = \rho \otimes s$ on $X \in D(M)$ is

$$\omega(X) = \rho(X)s, \quad \rho \in \Lambda^1(M), \quad s \in \Gamma(\pi_W).$$

The covariant differential (5) is the $\Gamma(\pi_W)$-valued analogue of the operator $d: C^\infty(M) \to \Lambda^1(M)$. In particular, like d it generates a series of operators

$$D: \Lambda^k(M, \pi_W) \to \Lambda^{k+1}(M, \pi_W), \qquad k \geqslant 0,$$

which act according to the formula

$$D(\rho \otimes s) = d\rho \otimes s + \rho \wedge Ds, \qquad \rho \in \Lambda^1(M), \quad s \in \Gamma(\pi_W).$$

Here we assume that $\rho \wedge (\sigma \otimes s) = (\rho \wedge \sigma)s$ if

$$\rho, \sigma \in \Lambda^1(M), \qquad s \in \Gamma(\pi_W).$$

The operator D can also be defined in the spirit of 3.5 by explicitly using the principal bundle $\pi: P \to M$. To this end we identify the $C^\infty(M)$-module $\Lambda^k(M, \pi_W)$ with the $C^\infty(P)$-module of horizontal G-equivariant W-valued k-forms on P by associating with a form $\sigma \in \Lambda^k(M, \pi_W)$ the form σ on P whose value on the vectors $\xi_1, \ldots, \xi_k \in T_p P$ is defined by

$$\sigma_p(\xi_1, \ldots, \xi_k) = \bar{p}^{-1}\hat{\sigma}_{\pi(p)}(d_p\pi(\xi_1), \ldots, d_p\pi(\xi_k))).$$

Then the operator D, understood as an operator on W-valued differential forms on P, has the form

$$D\sigma(X_1, \ldots, X_{k+1}) = d\sigma(X_1^H, \ldots, X_{k+1}^H),$$

where $X_i \in D(P)$ and $\sigma \in \Lambda^k(P) \otimes_{\mathbb{R}} W$.

Examples. 1) To the curvature form Ω as a horizontal G-equivariant $\mathfrak{G}$-valued 2-form on P there corresponds the 2-form $\hat{\Omega} \in \Lambda^2(M, \pi_\mathfrak{G})$, where $\pi_\mathfrak{G}$ is the associated bundle corresponding to the Lie algebra $\mathfrak{G}$, which is understood as a G-space with respect to the adjoint action of the group G. The space $\Gamma(\pi_\mathfrak{G})$ coincides with the Lie algebra $\mathcal{V}(\pi)$ of gauge vector fields (see 3.1). Gauge fields of $\mathcal{V}(\pi)$ can be regarded as $C^\infty(M)$-linear transformations of the $C^\infty(M)$-module $\Gamma(\pi_W)$. In particular, for any vector fields $X, Y \in D(M)$ the gauge field $\hat{\Omega}(X, Y)$, regarded as such a linear transformation, is identified with the operator

$$\hat{\Omega}(X, Y) = [\nabla_X, \nabla_Y] - \nabla_{[X,Y]},$$

where ∇_X is the covariant derivative acting in the space of sections $\Gamma(\pi_W)$.

2) Let $\Pi(A)$ be an (Ad G)-invariant polynomial of degree k on the Lie algebra $\mathfrak{G}$ and $\Pi(A_1, \ldots, A_k)$ the corresponding symmetric k-linear form. It determines a scalar horizontal equivariant differential $2k$-form $\Pi(\Omega) = \Pi(\Omega, \ldots, \Omega)$ (k times):

$$\Pi(\Omega)(Y_1, \ldots, Y_{2k}) = \Pi(\Omega(Y_1, Y_2), \ldots, \Omega(Y_{2k-1}, Y_{2k})).$$

The corresponding $2k$-form $\widehat{\Pi(\Omega)} \in \Lambda^{2k}(M)$ is closed. This follows from the Bianchi identity (4):

$$d\widehat{\Pi(\Omega)} = D\widehat{\Pi(\Omega)} = \Pi(D\Omega, \Omega, \ldots, \Omega) + \cdots + \Pi(\Omega, \ldots, \Omega, D\Omega) = 0.$$

In contrast to the operator d the square of a covariant differential is not equal to zero, generally speaking. Thus, for any W-valued equivariant function $f: P \to W$

$$(D^2 f)(X, Y) = \Omega(X, Y)f.$$

Hence it follows that the operator

$$D^2: \Gamma(\pi_W) \to \Lambda^2(M, \pi_W)$$

is a homomorphism of $C^\infty(M)$-modules and

$$(D^2 s)(X_1, X_2) = \hat{\Omega}(X_1, X_2)s,$$

where $s \in \Gamma(\pi_W)$, X_1, $X_2 \in D(M)$ and $\hat{\Omega}(X_1, X_2)$ is understood as a gauge transformation acting on $\Gamma(\pi_W)$. More generally, for any k the operator $D^2 \colon \Lambda^k(M, \pi_W) \to \Lambda^{k+2}(M, \pi_W)$ is the operator of multiplication by the curvature form.

3.7. The Yang-Mills Equations. In the case when the base M is a Riemannian manifold, we can define the operator $D^* \colon \Lambda^{k+1}(M, \pi_W) \to \Lambda^k(M, \pi_W)$ conjugate to the operator of covariant differentiation D, just as we have done in Ch. 8, § 11 for the operator of exterior differentiation d. The action of this operator on k-forms is given by $D^* = (-1)^{k+1} * D*$, where $* \colon \Lambda^l(M, \pi_W) \to \Lambda^{n-l}(M, \pi_W)$ is an operator that acts according to the formula

$$*(\sigma \otimes s) = *\sigma \otimes s, \qquad \sigma \in \Lambda^l(M), \quad s \in \Gamma(\pi_W),$$

and the action of the operator $*$ on scalar forms is defined in Ch. 8, § 11.

A connection in the principal bundle over a pseudo-Riemannian manifold M is called a *Yang-Mills field* if its curvature form $\hat{\Omega}$ satisfies the *Yang-Mills equation*

$$D^* \hat{\Omega} = 0.$$

This equation, together with the Bianchi identity $D\hat{\Omega} = 0$, is analogous to the condition for scalar differential forms to be harmonic (see Ch. 8, § 11), so the curvature forms of Yang-Mills fields are said to be *harmonic*.

Examples. 1) The Lie algebra $\mathfrak{G}$ of the group $G = SO(2)$ is isomorphic to the field $\mathbb{R}$ with a trivial Lie operation. Hence the curvature forms Ω of connections on $SO(2)$-bundles can be taken to be scalar, that is, "ordinary". In the identification of $\Lambda^k(M, \pi_\mathfrak{G})$ with $\Lambda^k(M)$ the operator $D \colon \Lambda^k(M, \pi_\mathfrak{G}) \to \Lambda^{k+1}(M, \pi_\mathfrak{G})$ is identified with the operator d and in this case the Yang-Mills equation expresses the "ordinary" harmonicity of Ω. If M is Minkowski space-time, the equations

$$D\hat{\Omega} = 0, \qquad D^* \hat{\Omega} = 0$$

are identical to the system of Maxwell equations for empty space. The Maxwell equations are linear. The same happens with the Yang-Mills equations for an Abelian group G, and then if it is non-Abelian these equations are non-linear.

2) The Levi-Civita connection of a pseudo-Riemannian metric g, regarded as a connection in the bundle of coframes (see 3.1), satisfies the Yang-Mills equations if and only if its Ricci tensor $\text{Ric}(g)$ satisfies the equation

$$(\nabla_X \text{Ric}(g))(Y) = (\nabla_Y \text{Ric}(g))(X), \qquad X, Y \in D(M).$$

For example, the Levi-Civita connection of the Einstein metric is of this type, and so is the Levi-Civita connection of any conformally flat metric with constant scalar curvature.

3) Suppose that the group G is compact. The Cartan-Killing form on the Lie algebra $\mathfrak{G}$ determines a scalar product in the fibres of the bundle $\pi_\mathfrak{G} \colon P_\mathfrak{G} \times G \to M$,

and the Riemannian metric in M determines a scalar product in the bundle $\Lambda^2 T^*M$ of 2-forms. This determines a scalar product $\langle\ ,\ \rangle$ in the bundle of $\pi_{\mathfrak{G}}$-valued 2-forms (with fibre $\Lambda^2 T_x^*M \otimes \pi_{\mathfrak{G}}^{-1}(x)$), whose sections form the space $\Lambda^2(M, \pi_{\mathfrak{G}})$. In particular, there is defined the scalar square $\langle\hat{\Omega}, \hat{\Omega}\rangle \in C^\infty(M)$ of the curvature 2-form $\hat{\Omega}$. The functional $A(H) = \int \langle\hat{\Omega}_x, \hat{\Omega}_x\rangle v_g$, where v_g is the volume form of the Riemannian metric g, is called the *Yang-Mills functional* and characterizes the measure of the deviation of the connection under consideration from a flat connection. The Yang-Mills equations are the Euler-Lagrange equations for this functional. Thus, Yang-Mills connections are connections with an extreme value of the curvature.

In conclusion we note that the Yang-Mills equations occupy a central place in model field theories of elementary particles.

§4. Bundles of Jets

The concept of a jet is fundamental in differential geometry. Namely, jets, organized into manifolds, are a basis of the geometrization of differential calculus, about which we spoke in the introductory chapter.

We have already met various special cases of this concept in this and the preceding chapter. In this section we collect together the basic constructions concerned with bundles of jets.

4.1. Jets of Submanifolds. Let us fix a manifold E and a natural number s, where $0 < s < \dim E$. We say that two submanifolds N_1 and N_2 of E of dimension s are *k-equivalent at a point* $a \in E$ if $a \in N_1 \cap N_2$ and N_1 and N_2 have contact at this point of order at least k, where $0 \leqslant k \leqslant \infty$. A k-equivalence class at a point $a \in E$ of a submanifold $N \subset E$ is denoted by $[N]_a^k$ and called the *k-jet* of N at a (see Fig. 42). The set of all k-jets at a point $a \in E$ of submanifolds of E of dimension s is denoted by $J_a^k(E, s)$, and the set of all k-jets is denoted by $J^k(E, s) = \bigcup_{a \in E} J_a^k(E, s)$.

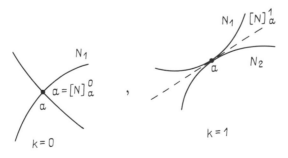

Fig. 42

Just as we did in Ch. 3, 3.3 for the case of the tangent bundle, in the set $J^k(E, s)$ we introduce in a natural way the structure of a smooth manifold with respect to which the projection $\pi_{k,0}: J^k(E, s) \to E$, $\pi_{k0}([N]_a^k) = a$ is a smooth bundle.

For small values of k the manifolds $J^k(E, s)$ admit a simple description. Thus, if $k = 0$ we have $J^0(E, s) = E$. When $k = 1$ the 1-jet of a submanifold N at a point $a \in E$ can be identified with the tangent plane to N at this point. Hence $J_a^1(E, s)$ coincides with the Grassmannian of s-dimensional subspaces of $T_a E$, and $J^1(E, s)$ coincides with the total space of the Grassmann bundle (see §2 above).

To represent the structure of manifolds $J^k(E, s)$ when $k \geqslant 2$ we consider the projections $\pi_{k,l}: J^k(E, s) \to J^l(E, s)$, $k > l$, generated by the operation of reduction of k-jets to l-jets: $[N]_a^k \to [N]_a^l$. These projections are smooth bundles, and the fibres of the projections $\pi_{k,k-1}$ when $k \geqslant '2$ are affine spaces of dimension $m\binom{s+k-1}{k}$, where $m = \dim E - s$ (see 4.5).

Thus, $J^k(E, s)$ is the total space of towers of bundles

$$J^k(E, s) \xrightarrow{\pi_{k,k-1}} J^{k-1}(E, s) \to \cdots \to J^1(E, s) \xrightarrow{\pi_{1,0}} E,$$

all of whose fibres are described above.

Each submanifold $N \subset E$ is lifted to $J^k(E, s)$ by means of a map $j_k(N): N \to J^k(E, s)$, $N \ni x \mapsto [N]_x^k$. The image $N^{(k)}$ of this map is a smooth s-dimensional submanifold of $J^k(E, s)$ diffeomorphic to N. This diffeomorphism is established by means of the projection $\pi_{k,0}$.

As in Chapter 5, a manifold $J^k(E, s)$ carries an additional geometric structure, the *Cartan distribution*, which enables us to describe intrinsically a submanifold of $J^k(E, s)$ of the form $N^{(k)}$. We define the *Cartan plane* $C(\theta) \subset T_\theta(J^k(E, s))$ at the point $\theta = [N]_a^k$ as the linear hull of the union of tangent subspaces at this point to submanifolds of the form $(N^1)^{(k)}$ passing through θ (see Fig. 43); $C(\theta)$ can also be defined as the inverse image of the plane $L(\theta)$ tangent to $N^{(k-1)}$ at the point $\theta' = \pi_{k,k-1}(\theta)$ under the projection $d_\theta \pi_{k,k-1}: T_\theta(J^k(E, s)) \to T_{\theta'}(J^{k-1}(E, s))$. The distribution $C: \theta \to C(\theta)$ on $J^k(E, s)$ is called the *Cartan distribution*.

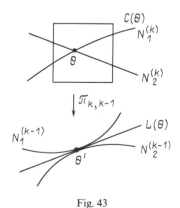

Fig. 43

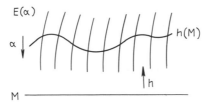

$E(\alpha)$

α

$h(M)$

M

h

Fig. 44

Just as in Chapter 5 we show that submanifolds of the form $N^{(k)}$, and only they, are integral manifolds of the Cartan distribution in $J^k(E, s)$, projected into E without degeneracy.

4.2. Jets of Sections. The image $N = h(M)$ of any (local) section $h\colon M \to E(\alpha)$ of a smooth bundle $\alpha\colon E(\alpha) \to M$ is a submanifold of $E(\alpha)$ that is transversal to fibres of the projection α (see Fig. 44). The k-jet of this submanifold at a point $h(a)$, $a \in M$, is denoted by $[h]_a^k$ and called the *k-jet of the section h* at the point $a \in M$. Let $J_a^k(\alpha)$ denote the set of all k-jets of sections of the bundle α at the point $a \in M$. Obviously, $J_a^k(\alpha) \subset J_{h(a)}^k(E(\alpha), s)$, where $s = \dim M$. The totality of all k-jets of sections of the bundle α

$$J^k(\alpha) = \bigcup_{a \in M} J_a^k(\alpha)$$

is an open everywhere dense subset of $J^k(E(\alpha), s)$. It consists of k-jets of submanifolds transversal to fibres of the bundle α.

Thus, on the spaces $J^k(\alpha)$ there is induced the structure of smooth manifolds with respect to which the projections $\alpha_{k,l}\colon J^k(\alpha) \to J^l(\alpha)$, $k > l$, generated by $\pi_{k,l}$ are smooth bundles. Let us identify $J^0(\alpha)$ and $E(\alpha)$. Then the composition $\alpha_k = \alpha \circ \alpha_{k,0}$ turns $J^k(\alpha)$ into a fibre space over M whose fibres are submanifolds $J_a^k(\alpha)$.

Each section $h\colon M \to E(\alpha)$ generates a section $j_k(h)\colon M \to J^k(\alpha)$ of this bundle: $j_k(h)(a) = [h]_a^k$. As in the case of submanifolds, sections of the form $j_k(h)$ are distinguished among all sections of the bundle α_k by the fact that their images are integral manifolds of the Cartan distribution on $J^k(\alpha)$. The latter is induced by the embedding $J^k(\alpha) \subset J^k(E(\alpha), s)$.

If α is a smooth vector bundle, then so are all the bundles α_k. The structure of a vector space in $J_a^k(\alpha)$ is defined as follows:

$$\lambda_1[h_1]_a^k + \lambda_2[h_2]_a^k \overset{\text{def}}{=} [\lambda_1 h_1 + \lambda_2 h_2]_a^k,$$

where $h_i \in \Gamma(\alpha)$, $\lambda_i \in \mathbb{R}$.

4.3. Jets of Maps. We shall identify smooth maps $f\colon M \to N$ with their graphs S_f, which are sections of the trivial bundle $\alpha\colon M \times N \to M$, $S_f(a) = (a, f(a)) \in M \times N$. We define the *$k$-jet $[f]_a^k$ of the map f* at a point $a \in M$ as the k-jet at the point $a \in M$ of the section S_f. We denote the set of k-jets $[f]_a^k$ of maps such that

$f(a) = b$ by $J_{a,b}^k(M, N)$, and the set of all k-jets by

$$J^k(M, N) = \bigcup_{(a,b) \in M \times N} J_{a,b}^k(M, N).$$

As before, we shall denote the projections generated by reductions of jets by $\pi_{k,l} \colon J^k(M, N) \to J^l(M, N), k > l$.

Examples. 1) $J^0(M, N) = M \times N$. Elements of the fibre of the projection $\pi_{1,0} \colon J^1(M, N) \to M \times N$ at a point $(a, b) \in M \times N$ can be identified with the differentials $d_a f \colon T_a M \to T_b N$ of smooth maps $f \colon M \to N$. Hence

$$\pi_{1,0}^{-1}(a, b) = \operatorname{Hom}(T_a N, T_b M).$$

2) Suppose that $M = \mathbb{R}$. Then the 1-jet of a map $f \colon \mathbb{R} \to N$ at a point $a \in \mathbb{R}$, $b = f(a)$, can be identified with the pair (a, v), where v is the tangent vector to the curve $f(t)$ at the point b. Hence

$$J^1(\mathbb{R}, N) = \mathbb{R} \times TN.$$

Dually, $J^1(M, \mathbb{R}) = T^*M \times \mathbb{R}$.

4.4. The Differential Group. In $J_{a,b}^k(M, M)$ we distinguish the subset $\mathbf{G}_{a,b}^k(M)$ formed by k-jets of local diffeomorphisms. The composition of local diffeomorphisms that preserve the point $a \in M$ determines a group structure in $\mathbf{G}_a^k(M) = \mathbf{G}_{a,a}^k(M)$.

The choice of local chart determines an isomorphism between $\mathbf{G}_a^k(M)$ and $\mathbf{G}^k(n) = \mathbf{G}_0^k(\mathbb{R}^n)$, the group of k-jets of local diffeomorphisms $\mathbb{R}^n \to \mathbb{R}^n$ that preserve the point $0 \in \mathbb{R}^n$.

We call the group $\mathbf{G}_a^k(M)$ or its model $\mathbf{G}^k(n)$ the *total differential group* of order k. Obviously, $\mathbf{G}_a^1(M)$ is the group of linear automorphisms of the tangent space $T_a M$, and $\mathbf{G}^1(n) = GL(n, \mathbb{R})$.

To describe the structure of the groups $\mathbf{G}^k(n), k \geqslant 2$, we consider the projections $\pi_{k,k-1} \colon \mathbf{G}^k(n) \to \mathbf{G}^{k-1}(n)$, which are obviously epimorphisms. The kernel of the homomorphism $\pi_{k,k-1}$ is formed by the k-jets of diffeomorphisms whose $(k-1)$-jets at the point $0 \in \mathbb{R}^n$ coincide with $[\mathrm{id}]_0^{k-1}$. Such diffeomorphisms (up to infinitesimals of order $\geqslant k+1$) in turn can be obtained as follows. Consider a vector field X on $\mathbb{R}^n$ that has a zero of order k at the point $0 \in \mathbb{R}^n$. Let $F_X = A_1$, where A_t is the flow generated by the field X (see Ch. 3, 3.2). Then $[F_X]_0^{k-1} = [\mathrm{id}]_0^{k-1}$ and by dimensional arguments the set of all k-jets of the form $[F_X]_0^k$, $k \geqslant 2$, coincides with the kernel $\pi_{k,k-1}$. It remains to observe that $[F_X]_0^k$ is determined by jets $[X]_0^k$, and $[F_X \circ F_Y]_0^k = [F_{X+Y}]_0^k$ if $k \geqslant 2$.

The space of k-jets of vector fields on $\mathbb{R}^n$ that have a zero of order k at the point $0 \in \mathbb{R}^n$ is identified with the space $S^k(\mathbb{R}^n) \otimes \mathbb{R}^n$ (see Ch. 5, 5.3). This shows that when $k \geqslant 2$ the kernel of the map $\pi_{k,k-1}$ is isomorphic to $S^k(\mathbb{R}^n) \otimes \mathbb{R}^n$, or $S^k(T_a M) \otimes T_a M$ if it is a question of a manifold M (see Ch. 3, 3.4).

Thus, the groups $\mathbf{G}^k(n)$ are obtained by successive extensions of the full linear group $GL(n, \mathbb{R})$ by means of Abelian group $S^k(\mathbb{R}^n) \otimes \mathbb{R}^n, k \geqslant 2$.

4.5. Affine Structures. The groups $\mathbf{G}_a^k(E)$ act in a natural way in spaces of k-jets $J_a^k(E, s)$. If $[F]_a^k \in \mathbf{G}_a^k(E)$, then the image of a jet $[N]_a^k \in J_a^k(E, s)$ under this action is equal to $[F(N)]_a^k$.

Let H_a^k be the kernel of the projection $\pi_{k,k-1}: \mathbf{G}_a^k(E) \to \mathbf{G}_a^{k-1}(E)$. By what we said above, H_a^k is an Abelian group when $k \geqslant 2$. This group acts transitively on the fibre $F(x_{k-1})$ of the projection $\pi_{k,k-1}: J^k(E, s) \to J^{k-1}(E, s)$ over the point $x_{k-1} = [N]_a^{k-1}$. Hence $F(x_{k-1})$ turns into an affine space with which we associate the vector space

$$H_a^k/(\text{the stationary group of the point } x_k),$$

where $x_k = [N]_a^k$.

Direct calculation shows that this space is isomorphic to $S^k(T_a N) \otimes V_a(N)$, where $V_a(N) = T_a E/T_a N$ is the *normal space* to N at the point $a \in E$.

Thus, when $k \geqslant 2$ the bundles $\pi_{k,k-1}: J^k(E, s) \to J^{k-1}(E, s)$ are affine.

Each diffeomorphism $F: E \to E$ generates diffeomorphisms $F^{(k)}: J^k(E, s) \to J^k(E, s)$, where $F^{(k)}([N]_a^k) = [F(N)]_{F(a)}^k$, which are automorphisms of the bundles $\pi_{k,l}$.

From the definition of an affine structure in the bundles $\pi_{k,k-1}$ it follows that when $k \geqslant 2$ the diffeomorphisms $F^{(k)}$ are automorphisms of this structure.

4.6. Differential Equations and Differential Operators. The use of bundles of jets enables us to give the final form of the geometry of differential equations, whose local version was discussed in the previous chapter. Let us dwell once more on the basic concepts.

A *differential equation of order k on submanifolds of the manifold E* is a submanifold $\mathscr{E} \subset J^k(E, s)$. Its solutions are submanifolds $N \subset E$, dim $N = s$, such that $N^{(k)} \subset \mathscr{E}$, or in other words integral manifolds of the Cartan distribution lying in $\mathscr{E}$ and projected without degeneracy into E.

As in Chapter 5, the question of the compatibility of a system of differential equations leads to the necessity of considering extensions $\mathscr{E}^{(l)} \subset J^{k+l}(E, s)$ of the equation $\mathscr{E}$. From the geometrical point of view the *l-th prolongation $\mathscr{E}^{(l)}$* of a differential equation $\mathscr{E}$ is formed by $(k + l)$-jets of submanifolds N at the point $a \in E$ for which the submanifold $N^{(k)}$ touches the equation $\mathscr{E}$ at the point $[N]_a^k \in \mathscr{E}$ with order at least l.

A differential equation $\mathscr{E}$ is said to be *formally integrable* if all its prolongations $\mathscr{E}^{(l)}$, $l = 0, 1, \ldots$, are smooth manifolds, and the natural projections $\pi_{k+l+1,k+l}: \mathscr{E}^{(l+1)} \to \mathscr{E}^{(l)}$, $l = 0, 1, \ldots$, are smooth bundles. As in Chapter 5, 5.4 the conditions of formal integrability are formed in terms of the Spencer δ-cohomology of the symbol of the equation $\mathscr{E}$.

A *k-th order differential operator* acting from sections of the bundle $\alpha: E(\alpha) \to M$ into sections of the bundle $\beta: E(\beta) \to M$ is a morphism of bundles

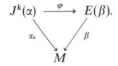

$$J^k(\alpha) \xrightarrow{\;\varphi\;} E(\beta).$$

The link between the ordinary understanding of a differential operator as an operator acting on sections of the bundle α and that mentioned above is given by the formula

$$\Gamma(\alpha) \ni h \mapsto \varphi \circ j_k(h) \in \Gamma(\beta),$$

where $(\varphi \circ j_k(h))(a) = \varphi([h]_a^k)$.

In parallel with the concept of an extension of a differential equation we define a *prolongation of a differential operator* φ as a morphism $\varphi^{(l)}: J^{k+l}(\alpha) \to J^l(\beta)$ of the bundle α_{k+l} into the bundle β_l for which

$$\varphi^{(l)}([h]_a^{k+l}) = [\varphi \circ j_k(h)]_a^l.$$

If a differential equation $\mathscr{E} \subset J^k(\alpha)$ is given by an operator φ, that is, $\mathscr{E} = \varphi^{-1}(h)$, where $h \in \Gamma(\beta)$, then its prolongation $\mathscr{E}^{(l)}$ is given by the operator $\varphi^{(l)}$.

4.7. Spencer Complexes. Spencer complexes are complexes of first-order differential operators associated with linear systems of differential equations. These complexes and the cohomologies determined by them are the most important source of cohomology theories, which have a natural differential-geometric nature. We have already met subcomplexes of them – Spencer δ-complexes – in Chapter 5 in connection with the conditions of formal integrability. In another capacity – in the investigation of the rigidity of geometric structures – Spencer complexes occur in Chapter 8.

Below we give the construction of these complexes.

Let $\alpha: E(\alpha) \to M$ be a smooth vector bundle. Let $\mathscr{J}^k(\alpha)$ denote the module of smooth sections of the bundle α_k. This module is generated by sections of the form $\sum_i f_i j_k(h_i)$, where $f_i \in C^\infty(M)$, $h_i \in \Gamma(\alpha)$.

The *Spencer operator* is the first-order linear differential operator

$$D: \mathscr{J}^k(\alpha) \otimes \Lambda^i(M) \to \mathscr{J}^{k-1}(\alpha) \otimes \Lambda^{i+1}(M),$$

defined by the formula

$$D(j_k(h) \otimes \omega) = j_{k-1}(h) \otimes d\omega,$$

where $h \in \Gamma(\alpha)$, $\omega \in \Lambda^i(M)$, and $d: \Lambda^i(M) \to \Lambda^{i+1}(M)$ is the de Rham operator.

It is not difficult to verify that $D^2 = 0$, that is, the sequence $0 \to \Gamma(\alpha) \overset{j_k}{\to} J^k(\alpha) \overset{D}{\to} \mathscr{J}^{k-1}(\alpha) \otimes \Lambda^1(M) \overset{D}{\to} \mathscr{J}^{k-2}(\alpha) \otimes \Lambda^2(M) \to \cdots$ forms a complex, called the *Spencer complex* of the bundle α.

This complex is exact, that is, $\theta_1 = D(\theta_2)$, $\theta_1 \in \mathscr{J}^{k-1}(\alpha) \otimes \Lambda^{i+1}(M)$, $\theta_2 \in \mathscr{J}^k(\alpha) \otimes \Lambda^i(M)$, if and only if $D\theta_1 = 0$.

If $\mathscr{E} \subset J^k(\alpha)$ is a linear formally integrable system of differential equations, and $\Gamma(\mathscr{E}^{(l)})$ is the module of smooth sections of the bundle $\pi_{k+l}: \mathscr{E}^{(l)} \to M$, then $D(\Gamma(\mathscr{E}^{(l)})) \subset \Gamma(\mathscr{E}^{(l)-1}) \otimes \Lambda^1(M)$, and the Spencer operators D determine a subcomplex

$$\Gamma(\mathscr{E}^{(l)}) \overset{D}{\to} \Gamma(\mathscr{E}^{(l-1)}) \otimes \Lambda^1(M) \overset{D}{\to} \Gamma(\mathscr{E}^{(l-2)}) \otimes \Lambda^2(M) \to \cdots,$$

which is called the *l-th Spencer complex* of the system of differential equations $\mathscr{E}$.

The quotient spaces

$$H^{l,i}(\mathscr{E}) = \frac{\mathrm{Ker}(D: \Gamma(\mathscr{E}^{(l)}) \otimes \Lambda^i(M) \to \Gamma(\mathscr{E}^{(l-1)}) \otimes \Lambda^{i+1}(M))}{\mathrm{Im}(D: \Gamma(\mathscr{E}^{(l+1)}) \otimes \Lambda^{i+1}(M) \to \Gamma(\mathscr{E}^{(l)}) \otimes \Lambda^i(M))}$$

are called the *Spencer cohomologies* of the system $\mathscr{E}$.

For large values of l the spaces $H^{l,i}(M)$ stabilize, that is, they cease to depend on l and are called the *stable Spencer cohomologies*. Their union is $H^{(i)}(\mathscr{E})$.

For definite formally integrable systems of differential equations $\mathscr{E}$ determined by a linear differential operator $\Delta: \Gamma(\alpha) \to \Gamma(\beta)$ the stable Spencer cohomologies are respectively equal to

$$H^0(\mathscr{E}) = \mathrm{Ker}\, \Delta, \qquad H^1(\mathscr{E}) = \mathrm{Coker}\, \Delta, \qquad H^i(\mathscr{E}) = 0, \quad i \geqslant 2.$$

For a system of differential equations determined by an operator $d: C^\infty(M) \to \Lambda^1(M)$, the stable Spencer cohomologies coincide with the de Rham cohomologies (see Ch. 8, §10).

More generally, if a system of differential equations is determined by a natural differential operator, then the stable Spencer cohomologies give global invariants of manifolds with respect to the group of automorphisms. If the system of differential equations is determined by some geometric structure, that is, it is the Lie equation (see Ch. 7) of the given geometric structure, then the stable Spencer cohomologies give important global characteristics of this structure. Thus, for example, for the operator $\bar\partial$ that specifies the complex structure the stable Spencer cohomologies coincide with the Dolbeault cohomologies (see Ch. 8, §16).

Chapter 7
The Equivalence Problem, Differential Invariants and Pseudogroups

§ 1. The Equivalence Problem. A General View

1.1. The Problem of Recognition (Equivalence). As a rule, the concrete definition of mathematical objects is associated with an act of subjectivism, caused by the need to choose some "reference system", for example, a basis in a vector space, a system of generators in a group, local coordinates on a manifold, and so on. Hence the object of interest to us turns out to be represented by many descriptions, which immediately leads to the question: do two concrete descriptions of the same type determine the same object or not?

This problem of recognition is fundamental, since, generally speaking, it is impossible to avoid the use of the "reference systems" already mentioned, just as in physics it is impossible to observe something without using some instruments or other. Moreover, in many cases this problem is the ultimate aim of mathematical research, and called reduction to canonical form, the classification problem, the equivalence problem, and so on. For example, in elementary analytic geometry an interesting question is whether two second-order equations in several unknowns determine the same quadric. It is well known that the answer to this question can be obtained in two ways. Firstly, the two equations, following some algorithm, are reduced to canonical form, and then compared. Secondly, from the coefficients of these equations we can form definite expressions, called invariants, whose coincidence is a necessary and sufficient condition for a positive answer to the question. For differential geometry the second of these approaches has the main significance.

Depending on the context, the problem of recognition may seem fairly simple, but it may lead to difficulties and even to insoluble problems.

Example 1. Speaking informally, a knot is a way of tying a loop of string. Visually, knots are represented by diagrams in the plane, of the type shown in

Fig. 45

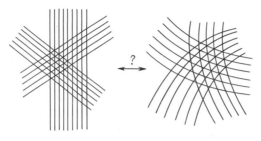

Fig. 46

Fig. 45. It is not known whether there is an algorithm that determines whether two such diagrams represent the same knot. It is not easy to prove that the knots shown in Fig. 45 are different. For this we need to construct a rather subtle theory.

Example 2. A corepresentation of a group is a list of its generators and relations between them. A corepresentation is finite if this list is finite. A widespread method of describing a group consists in giving some corepresentation of it. It has been proved that there is no algorithm that enables us to recognize whether two given corepresentations determine isomorphic groups or not.

Example 3. Three families of curves that stratify the plane are called a 3-*web* if the curves belonging to different families do not touch one another. Two webs are regarded as identical if they can be taken into one another by a certain diffeomorphism of the plane. We can specify a 3-web by three vector fields for which the curves that constitute it are integral curves. Question: do two given triples of vector fields determine identical webs, even locally (see Fig. 46)? It is interesting that research into the dynamics of galaxies is connected with this question, which is characteristic for differential geometry (see the example in 1.5 below).

1.2. The Problem of Triviality. The problem of recognizing the simplest of a class of objects under consideration is a special case of the problem discussed above. It can be called the *problem of triviality*. Let us elucidate it by means of the examples of the previous subsection.

Example 1. The simplest or trivial knot is an "unknotted" curve, for example a circle. The problem of triviality for knots is illustrated in Fig. 47. It has been solved in the sense that an algorithm has been constructed that makes it possible to recognize a trivial knot from its diagram.

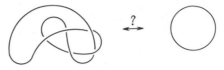

Fig. 47

Example 2. A group consisting of a single element is said to be trivial. It has been proved that recognition of a trivial group from its copresentation is an algorithmically insoluble problem.

Example 3. Three families of parallel lines form a *trivial* 3-web. Fig. 46 can be regarded as an illustration of the problem of triviality in this case. Choosing a local coordinate system in such a way that the coordinate curves coincide with the curves of the first two families of the web under consideration, we can specify the latter by a triple of vector fields $\left(\dfrac{\partial}{\partial x}, \dfrac{\partial}{\partial y}, a(x, y)\dfrac{\partial}{\partial x} + b(x, y)\dfrac{\partial}{\partial y}\right)$. It is not difficult to verify that a web is trivial if and only if the form $k = \dfrac{\partial^2}{\partial x \partial y}\left(\ln \dfrac{a}{b}\right) dx \wedge dy$, which is a differential invariant of it, is identically zero.

From these examples it is clear that the interrelations of problems of triviality and the corresponding equivalence problems can have very different character. The specific form of them in the framework of differential geometry will be discussed below.

1.3. The Equivalence Problem in Differential Geometry. Since differential geometry is concerned with field quantities, the problems of recognition that arise in it admit a localization: we can be interested in equivalence of geometric structures in arbitrarily small domains. This possibility has a fundamental significance not only because it simplifies the general problem, but also because its solution is connected with the "classification of quantities" about which Maxwell spoke in the epigraph to Chapter 6. In fact, if we bear in mind the immense number of coordinate descriptions of a given geometric object, and also the very complicated and intricate character of their transformations, we can hope to solve the recognition problem only by displaying the quantities connected with the object "intrinsically", that is, in a way that does not depend on the choice of coordinate system. In the theory of curves and surfaces (Ch. 2) we have already verified that quantities of this kind are differential invariants. For this reason it is useful to treat the equivalence problem in differential geometry as the problem of finding a complete system of differential invariants (for a more precise definition see below).

The concepts used as a language for formulating theories of specific geometric structures are differential invariants. The curvature and torsion of a curve or the principal curvatures of a surface are examples of those that we can see in a literal sense of the word and which therefore are not perceived as differential invariants. In the unimodular and affine geometry of curves (Ch. 1, 1.4) direct geometrical intuition does not work and a conscious statement of the problem of differential invariants is a necessary prerequisite for constructing the theory.

These considerations establish a link between the first two concepts in the title of this chapter. The last of these – the concept of a pseudogroup – plays a part in this interaction by formalizing the mechanism of sorting out the various coordinate descriptions of a given geometric object with respect to which the question of invariance is being considered.

1.4. Scalar and Non-Scalar Differential Invariants. The Levi-Civita connection and its curvature tensor are differential invariants of the original Riemannian metric. However, the recognition problem for these geometric objects is nothing like as simple as for the metric itself. Hence they cannot immediately be used to this end.

Functions (scalar fields) are the only useful quantities whose components do not change under transformations of the coordinates. For this reason scalar differential invariants (that is, those that take numerical values) do not have this defect and can be used effectively in the recognition problem. Thus, we must make precise the formulation given in the previous subsection and treat the equivalence problem in differential geometry as the problem of constructing a complete system of scalar differential invariants.

Non-scalar differential invariants can be very useful in the problem of triviality. We find an effective example of this in Riemannian geometry, where the vanishing of the curvature tensor is a necessary and sufficient condition for the metric to be locally Euclidean ("trivial") (see Ch. 3, 4.6).

In conclusion we draw attention to the following paradox of mathematical fashion. The theory of differential invariants, despite its fundamental importance, has hardly been developed since the work of the classical 19th century mathematicians Lie, Tresse, Halphen, and others.

1.5. Differential Invariants in Physics. In those cases when a geometric object admits a physical interpretation, its differential invariants also acquire a physical meaning and hence the status of physical quantities. Moreover, many geometric structures are naturally associated with differential equations in general and particularly with differential equations that describe the motions of physical substances. Hence the comparison of these two remarks gives a regular method of extracting physical quantities and concepts from the equations of mathematical physics, that is, the "systematic classification of quantities", whose importance Maxwell spoke of (see the epigraph to Chapter 6). The way in which Riemannian geometry arose from the theory of symbols of differential operators (described in Ch. 3, § 5) is one illustration of what we have said. See also Ch. 3, 1.4.

The insufficient cultivation of the theory of differential invariants of which we spoke above forces us to assume that the set of physical concepts used today is far from complete. Apart from anything else, this restriction of the dictionary substantially limits our possibilities of qualitatively understanding phenomena, particularly non-linear ones.

Example. The stationary motion of the galaxies, if we regard them as thin flat layers, is described by a system of three differential equations in three dependent and two independent variables. With each point of the plane in which the motion occurs we can associate three lines. Two of them form the so-called "Mach cone", and the third touches the flow lines. Mathematically these lines are extracted from the symbol of the system of differential equations. The integral curves of the fields of directions that arise in this way form a 3-web which characterizes the motion. Thus, the differential invariants of 3-webs (for example, the curvature k, see 1.3) become physical characteristics of the dynamics of the galaxies.

§ 2. The General Equivalence Problem in Riemannian Geometry

With the object of preparing and motivating the elements of the general theory, here we consider the local equivalence problem for Riemannian manifolds.

2.1. Preparatory Remarks. The recognition problem in Riemannian geometry consists in determining when two differential forms $g_{ij}(x) \, dx_i \, dx_j$ determine identical metrics. In other words, we need to know when the Riemannian manifolds (M, g) and (M', g') are (locally) *isometric*, that is, there is a diffeomorphism $F: M \to M'$ such that $F^*(g') = g$. If we require in addition that $F(a) = a'$, where $a \in M$, $a' \in M'$ are preassigned points, we come to the "pointwise" equivalence problem, which we have already considered in Ch. 3, 4.4. The complexity of the general problem, by comparison with the "pointwise" problem, is that for a given point $a \in M$ we do not know from the outset the point $a' \in M'$ close to which M' can be isometric to some neighbourhood of the point a.

Any Riemannian manifold (M, g) is infinitesimally homogeneous in the sense that for any two points $a, b \in M$ there is a diffeomorphism $F: M \to M$ that takes a to b and the metric g_a on $T_a M$ to the metric g_b on $T_b M$. For example, a diffeomorphism that maps the normal coordinate systems (see Ch. 3, 4.4) around the points a and b into each other is of this kind. Thus, the field $a \to g_a$ is an ensemble of "identical" (isomorphic) objects, so the peculiarity of their collective behaviour can be studied.

2.2. Two-Dimensional Riemannian Manifolds. The Gaussian (or scalar) curvature $K = K_g$ of a 2-dimensional Riemannian manifold (M, g) is its most important differential invariant. By using it, it is not difficult to construct others. The simplest of these are the geodesic curvature $k = k_g$ of curves $K_g = \text{const}$ and $\kappa = \kappa_g = |\text{grad}_g K|$, where $X = \text{grad}_g \varphi$ (the *gradient of a function* $\varphi \in C^\infty(M)$ with respect to the metric g) is a vector field such that $g(X, Y) = Y(\varphi), \forall Y \in D(M)$. Successively applying the methods by which k and κ were obtained, we can obtain an infinite series of independent differential invariants of the metric g.

We first consider the situaton when two differential invariants of the metric g, say K_g and κ_g, are functionally independent. Then (locally) they can be taken as coordinate functions on M. The components G_{ij}^g of the metric g in these coordinates are also differential invariants of it. The functions $G_{ij}^g(K_g, \kappa_g)$ determine the metric g uniquely up to equivalence and hence give the solution of the recognition problem under consideration. For if $F: M \to M'$ is an isometry, then $F^*(I_{g'}) = I_g$ for any scalar differential invariant I. Hence a map F that claims to be an isometry must satisfy this condition and, in particular, it must match the coordinate system (K_g, κ_g) on M with the coordinate system $(K_{g'}, \kappa_{g'})$ on M'. The condition $F^*(G_{ij}^{g'}) = G_{ij}^g$ guarantees that $F^*(g') = g$. In turn, this is ensured by the fact that $G_{ij}^{g'}$ as a function of $K_{g'}$ and $\kappa_{g'}$ is identical to G_{ij}^g as a function of K_g and κ_g.

However, it may happen that all the scalar differential invariants on M are constants or functions of one of them, say $I = I_g$. In the latter case we choose local coordinates u, v on M in such a way that the coordinate curves are mutually perpendicular and the curves $I_g = \text{const}$ coincide with the curves $u = \text{const}$. In these coordintes the metric g has the form $A(u, v)\, du^2 + B(u, v)\, dv^2$ and $J_g = J(u)$ for any invariant J. Using this as applied to the invariants $J_1 = |\text{grad}_g\, I_g|$ and $J_2 = $ (the geodesic curvature of the curves $I_g = \text{const}$) it is not difficult to verify that $A = A(u)$, $B = \varphi(u)\psi(v)$. Hence in the coordinates U, V, where $dU = \sqrt{A}\, du$, $dV = \sqrt{\psi}\, dv$, the metric g has the form $dU^2 + G(U)\, dV^2$, and we see that the transformations $(U, V) \mapsto (U, V + t)$ form a (local) one-parameter group of isometries, or motions, of it. In other words, the field $\dfrac{\partial}{\partial V}$ that generates this group is the *Killing field of the metric* g, that is, $\mathcal{L}_{\partial/\partial V}(g) = 0$. Up to proportionality this Killing field is unique so long as the metric g is not a metric of constant curvature.

Obviously the differential invariants of a metric must be constants along the orbits of any group of isometries of it. Hence we see that the existence of a unique (up to proportionality) Killing field is equivalent to all the differential invariants of the metric under consideration being functions of one of them, for example K_g. Moreover, direct verification shows that in this case the equivalence problem can be solved in the form of a function $\kappa_g = f(K_g)$.

Finally, if there are no independent differential invariants of the metric, that is, they are all constants, then we are dealing with a manifold of constant curvature. Such manifolds are locally equivalent to a sphere, the Euclidean plane, or the Lobachevskij plane. Hence in this case the recognition problem can be solved by one number – the Gaussian curvature. We also observe that the space of Killing fields of metrics of constant curvature is three-dimensional and they all admit a local transitive group of isometries.

2.3. Multidimensional Riemannian Manifolds. The solution of the equivalence problem for multidimensional Riemannian manifolds does not require the introduction of essentially new ideas in comparison with the two-dimensional case we have already considered, which differs from it in the specific character of the choice of the necessary differential invariants.

We recall (see Ch. 3, 4.7) that n-dimensional Riemannian metrics ($n > 2$) have a non-trivial non-scalar differential invariant – the Ricci tensor R_{ij}. The necessary scalar invariants are extracted from it in the form of the coefficients $I_0, \ldots, I_{n-1}$ of the characteristic polynomial $\det \|R_{ij} - \lambda g_{ij}\|$. For metrics of general form these invariants are independent and they can be taken as local coordinates. The components G_{ij} of the metric tensor g_{ij} in these invariant coordinates are also differential invariants and together with $I_0, \ldots, I_{n-1}$ they form a complete system. This is established by the argument used in the similar situation in the previous subsection.

A vector field $X \in D(M)$ is called a *Killing field* or infinitesimal symmetry of the metric g if $\mathscr{L}_X(g) = 0$. If $I = I_g$ is a differential invariant of the metric g, then obviously $X(I_g) = 0$. Hence a metric that admits n independent differential invariants (for example $I_0, \ldots, I_{n-1}$) does not have non-trivial Killing fields. This can be expressed in another way by saying that its group of isometries is discrete. What we have said is a special case of the following theorem.

Theorem. *Suppose that the maximal number of independent differential invariants of a metric g on M is equal to $k \leqslant n$. Then*

1) *the manifold (M, g) admits a local group of isometries whose orbits have dimension $n - k$, or equivalently M is stratified by $(n - k)$-dimensional submanifolds whose tangent spaces are generated by Killing vectors;*

2) *the metric g is uniquely determined by functions that express its "dependent" differential invariants in terms of the "independent" ones.*

The previous subsection is an illustration of these assertions, the first of which is apparently due to R. Kerr.

§ 3. The General Equivalence Problem for Geometric Structures

3.1. Statement of the Problem. Taking as a model the case of Riemannian geometry considered above, we now pose the equivalence problem for geometric structures in its full generality.

Let ω be a geometric structure on a manifold M, that is, a section of some natural bundle $\pi\colon E \to M$ (see Ch. 6, 1.5). With any diffeomorphism $F\colon M \to M'$ and natural bundle $\pi'\colon E' \to M'$ of the same type as (E, M, π) we canonically associate the bundle map $\hat{F}\colon E \to E'$ that covers the diffeomorphism F (see Ch. 3, 2.3). In particular, $F \circ \pi = \pi' \circ \hat{F}$. The image $F(\omega)$ of the section ω under the diffeomorphism F is a section of the bundle π' that is determined by $F(\omega) = \hat{F} \circ \omega \circ F^{-1}$. Thus, diffeomorphisms enable us to transfer geometric structures from one manifold to another.

The *equivalence problem for geometric structures* (ω and ω') of the same type, defined on manifolds M and M' respectively, consists in determining whether there is a diffeomorphism $F\colon M \to M'$ that takes the first of them into the second, that is, $F(\omega) = \omega'$. If we pose the question of finding such a diffeomorphism only locally, then we talk of the *local equivalence problem*. Obviously the solution of the local problem is a necessary prerequisite for the solution of the problem in the large. This requires, as a rule, the introduction of topological methods. We therefore confine ourselves to the local aspect.

Example 1. A cylinder and a cone with a distant vertex, regarded as surfaces in three-dimensional Euclidean space, are locally isometric and diffeomorphic, but not isometric in the large.

Example 2. Any two non-singular vector fields on manifolds of the same dimension are locally equivalent according to the "rectification theorem". Hence the local equivalence problem for them can be solved trivially. The corresponding problem "in the large" is very non-trivial, and if we discard the theory of singular points it is essentially the subject of the structural theory of dynamical systems (see Volumes 1–5 of the present series). Familiarity with this theory shows that in the given case we cannot hope to obtain any general solution of this problem "in the large".

3.2. Flat Geometric Structures and the Problem of Triviality. Among geometric structures of a given type the simplest ones are the so-called flat structures, defined as follows.

Let $\pi: E \to \mathbb{R}^n$ be a natural bundle and $t_v: \mathbb{R}^n \to \mathbb{R}^n$ the transformation of parallel translation by a vector $v \in \mathbb{R}^n$, that is, $t_v(x) = x + v$, $\forall x \in \mathbb{R}^n$. A section $\omega \in \Gamma(\pi)$ and the geometric structure determined by it are said to be *flat* if $\omega(x) = \hat{t}_x(\omega(0))$, where $\hat{t}_x: E \to E$ is the map induced by the translation t_x (see the previous subsection). Structures on manifolds that are locally equivalent to the one described are also said to be flat.

Examples. 1) A tensor field in $\mathbb{R}^n$ whose components in standard Cartesian coordinates are constants is a flat field, constructed by means of the construction described above.

2) The triple $\left(\dfrac{\partial}{\partial \varphi}, \dfrac{\partial}{\partial \psi}, \dfrac{\partial}{\partial \varphi} + \alpha \dfrac{\partial}{\partial \psi} \right)$, where φ and ψ are the standard angular coordinates on the torus, describes a flat 3-web on it.

The *problem of triviality* for geometric structures is the question of whether a given geometric structure is flat.

3.3. Homogeneous and Non-Homogeneous Equivalence Problems. The equivalence problem for vector fields in its most general formulation obviously contains the problem of classifying its singular points. It is well known that the latter is insoluble if we pose it in full generality. This effect is connected with the fact that vector fields are, generally speaking, fields of non-homogeneous geometric quantities.

Singular points of geometric structures are those in whose neighbourhood there is a change in the type of geometric quantity being considered. The example of vector fields shows that the statement of the equivalence problem close to singular points ceases to be reasonable without further refinements. Finding these refinements is one of the basic problems of singularity theory, which up to now has been used satisfactorily only for the simplest structures of the type of functions and vector fields. For this reason it is advisable to confine ourselves to the consideration of the equivalence problem for fields of homogeneous geometric quantities (see Ch. 6, 1.4), a situation in which we can effectively use differential invariants.

A flat field of homogeneous geometric quantities of a given type is unique up to equivalence. Hence the separation of the homogeneous case is justified from the point of view of the problem of triviality. A vector field that does not vanish anywhere and a vector field that is identically zero are examples of non-equivalent flat structures corresponding to the same non-homogeneous geometric quantity.

We now recall that with any field of homogeneous geometric quantities we can canonically associate a certain G-structure (see Ch. 6, 1.4). The formulations of the problems of equivalence and triviality given above automatically carry over to G-structures, since the latter are geometric structures of a special form. Equivalence (triviality) of the original geometric structures is tantamount to equivalence (triviality) of the corresponding G-structures. Hence both these problems can be investigated by the methods of the theory of G-structures (see § 5 below).

Flat G-structures are usually said to be *integrable*. This terminology originates from the theory of distributions: a distribution is flat if and only if it is integrable (see Ch. 5, 3.1). In this connection the problem of triviality of G-structures is usually called the *integrability problem*.

Thus, there are two general methods of investigating the equivalence problem. One is based on the theory of differential invariants, and the other on the theory of G-structures. They will be considered in the next two sections.

§ 4. Differential Invariants of Geometric Structures and the Equivalence Problem

4.1. Differential Invariants. An accurate coordinate definition of differential invariants is the following. Let ω_α be symbols that denote the components of the geometric quantity ω under consideration, where ω is a specific geometric structure of a suitable type on a manifold M. Fixing the local coordinates $x = (x_1, \ldots, x_n)$ turns the symbols ω_α into specific functions $\omega_\alpha(x_1, \ldots, x_n)$. Consider the function

$$I = I\left(\omega_\alpha, \frac{\partial \omega_\alpha}{\partial x_i}, \frac{\partial^2 \omega_\alpha}{\partial x_i \partial x_j}, \ldots \right),$$

which depends on finitely many formal variables $\omega_\alpha, \dfrac{\partial \omega_\alpha}{\partial x_i}, \ldots$. Substituting the functions $\omega_\alpha(x) = \omega_\alpha(x_1, \ldots, x_n)$ into I instead of the symbols ω_α, we obtain a function $I(\omega, x)$ of the variables x_i. If for any structure ω of the form under consideration and for any other system of coordinates $y = (y_1, \ldots, y_n)$ we have

$$I(\omega, x) = I(\omega, y(x)) \tag{1}$$

then I is called a *differential invariant*. Equality (1) means that $I(\omega, x)$ and $I(\omega, y)$ describe the same function $I(\omega)$ on M in the coordinate systems x and y respectively. $I(\omega)$ is called the *value of the invariant I on the structure ω*.

This definition is inconvenient in that it does not give a method for finding differential invariants. In this connection the following geometrical interpretation of it is useful; this is based on the fact that the function I as a function of the formal symbols $\omega_\alpha, \dfrac{\partial \omega_\alpha}{\partial x_l}, \ldots$ can be understood as a function on the space of jets $J^k(\pi)$ of the corresponding natural bundle $\pi\colon E \to M$ for a suitable choice of the number k.

We recall that any diffeomorphism $F\colon M \to M$ generates a diffeomorphism $\hat{F}\colon E \to E$, and the latter generates a diffeomorphism $\hat{F}^{(k)}\colon J^k(\pi) \to J^k(\pi)$ (see Ch. 6, 4.5). A function $I \in C^\infty(J^k(\pi))$ is called a *differential invariant of order at most k* of geometric structures of the form under consideration if $I \circ \hat{F}^{(k)} = I$ for any diffeomorphism $F\colon M \to M$.

This definition has an equivalent infinitesimal version. Namely, to any vector field X on M there corresponds canonically a field $\hat{X}$ on E, and to the latter there corresponds a field $\hat{X}^{(k)}$ on $J^k(\pi)$ (see Ch. 6, 4.5). $I \in C^\infty(J^k(\pi))$ is a differential invariant if and only if

$$\hat{X}^{(k)}(I) = 0, \qquad \forall X \in D(M). \tag{2}$$

The totality of differential invariants of order at most k obviously forms a subalgebra $\mathscr{A}_k$ in $C^\infty(J^k(\pi))$. Since an invariant of order at most l is an invariant of order at most k if $l \leqslant k$, it follows that $\mathscr{A}_l \subset \mathscr{A}_k$, and we arrive at the algebra $\mathscr{A} = \bigcup_k \mathscr{A}_k$ of all differential invariants.

If $\omega \in \Gamma(\pi)$ is a specific geometric structure on M and $I \in \mathscr{A}_k$, then the function $I(\omega) = I \circ j_k(\omega)$ on M is called the *value of the invariant I on the structure ω*.

Since any two natural bundles $\pi\colon E \to M$ and $\pi'\colon E' \to M$ of the same type are locally equivalent, the corresponding algebras of invariants $\mathscr{A}_k$ and $\mathscr{A}'_k$ defined as above are isomorphic. Hence, if a diffeomorphism $F\colon M \to M'$ establishes an equivalence between the structures ω on M and ω' on M', then

$$I(\omega) = I(\omega') \circ F, \qquad \forall I \in \mathscr{A}. \tag{3}$$

The equivalence problem can be solved by means of differential invariants in those cases when the diffeomorphism F can be restored by starting from (3).

4.2. Calculation of Differential Invariants.

Suppose that a geometric quantity under consideration is homogeneous and has order l. This means that the standard fibre $\mathcal{O}$ of the corresponding natural bundle $\pi\colon E \to M$ is the homogeneous space of the group $\mathbf{G}^l(n)$, where $n = \dim M$ (see Ch. 6, §1.2). Let us fix a point $x_0 \in M$ and identify $\mathcal{O}$ with $\pi^{-1}(x_0)$. If a diffeomorphism $F\colon M \to M$ is such that $F(x_0) = x_0$, then $\hat{F}^{(k)}(\mathcal{O}_k) = \mathcal{O}_k$, $\mathcal{O}_k = \pi_k^{-1}(x_0)$, $\pi_k\colon J^k(\pi) \to M$. Thus there arises the action $F \mapsto \hat{F}^{(k)}|_{\mathcal{O}_k}$ of the group of diffeomorphisms that preserve the point x_0 on $\mathcal{O}_k$. In fact the transformation $\hat{F}^{(k)}|_{\mathcal{O}_k}$ is uniquely determined by the $(k + l)$-jet of the diffeomorphism F. This enables us to define the action of the group $\mathbf{G}^{k+l}(n)$ on $\mathcal{O}_k$. It is not difficult to verify that the algebra of differential invariants $\mathscr{A}_k$ is isomorphic to the algebra of invariants with respect to this action of functions on $\mathcal{O}_k$.

The infinitesimal version of the construction we have described is very useful in finding differential invariants in practice. Namely, the $(k + l)$-jet of the field $X \in D(M)$ that vanishes at the point x_0 uniquely determines a field $\hat{X}^{(k)}|_{\mathcal{O}_k}$. Identifying the Lie algebra $\mathfrak{G}^{k+l}(n)$ of the group $\mathbf{G}^{k+l}(n)$ with the Lie algebra of this kind of jet, we obtain a representation of it $\mathfrak{G}^{k+l}(n) \ni \alpha \mapsto Y_\alpha \in D(\mathcal{O}_k)$. In view of what we have said above and (2), in these terms the algebra $\mathscr{A}_k$ is identified with the algebra of functions I on $\mathcal{O}_k$ that satisfy the system of differential equations

$$Y_{\alpha_i}(I) = 0, \qquad i = 1, \ldots, \dim \mathfrak{G}^{k+l}(n), \tag{4}$$

where $\{\alpha_i\}$ is a basis of the Lie algebra $\mathfrak{G}^{k+l}(n)$.

Manifolds $\mathcal{O}_k$ and vector fields Y_α on them can be described by abstract algebra if we are given a homogeneous space $\mathcal{O}$, which in many cases enables us by using (4) to reduce effectively the calculation of the algebra $\mathscr{A}_k$ to certain problems of linear algebra.

4.3. The Principle of n Invariants. We say that the invariants $I_1, \ldots, I_s \in \mathscr{A}_k$ are independent if they are functionally independent as functions on $J^k(\pi)$ (or $\mathcal{O}_k$). Since any set $I_1, \ldots, I_s \in \mathscr{A}$ actually lies in some subalgebra $\mathscr{A}_k$, we can talk about the independence of functions that occur in it. We say that a structure ω on M has a general form with respect to the invariants $I_1, \ldots, I_s$ if among their values $I_1(\omega), \ldots, I_s(\omega)$ there are $\min(s, n)$ independent ones, where $n = \dim M$.

This is the *principle of n invariants*: if an n-dimensional geometric quantity admits n independent invariants, then for structures of general form with respect to it the equivalence problem is completely soluble by means of differential invariants.

We have already used this principle in § 2. Its truth in the general case follows from the fact that the components $K_1, \ldots, K_m$ of the geometric quantity under consideration in the coordinate system $I_1(\omega), \ldots, I_n(\omega)$ are differential invariants which together with the invariants $I_1, \ldots, I_n$ obviously constitute a complete system.

Example. For 3-webs on the plane $\mathscr{A}_k = \mathbb{R}$ when $k \leqslant 2$, and the algebra $\mathscr{A}_3$ contains exactly two independent invariants.

Vector fields and contact structures are examples of structures that do not have independent invariants.

4.4. Non-General Structures and Symmetries. If $F: M \to M$ is an automorphism of a structure ω on M, then $F \circ I(\omega) = I(\omega)$ for any invariant $I \in \mathscr{A}$. This is a special case of (3). Hence the value of any differential invariant on the structure ω is constant on orbits of the automorphism group of this structure. If the dimensions of these orbits are equal to k, then among the functions $I(\omega)$ on M there are at most $n - k$ independent ones.

Suppose that the algebra $\mathscr{A}$ of invariants of the geometric quantity under consideration contains n independent invariants. The theorem of § 2 and sensible arguments give a basis for the following conjecture.

Suppose that the values $I_1(\omega), \ldots, I_{n-k}(\omega)$ of the differential invariants $I_1, \ldots, I_{n-k}$ on a geometric structure ω are functionally independent, and the value of any other invariant $I \in \mathscr{A}$ can be expressed in terms of them. Then

1) the group of local automorphisms of the structure ω is non-trivial and locally its orbits are the submanifolds given by $I_i(\omega) = \text{const}, i = 1, \ldots, n - k$;

2) the structure ω is uniquely determined up to equivalence by the form of functions that express the values of $I(\omega)$ in terms of $I_1(\omega), \ldots, I_{n-k}(\omega), I \in \mathscr{A}$; it is sufficient to restrict ourselves to finitely many suitably chosen invariants I.

A descriptive justification of this conjecture is given by the geometrical theory of spaces of jets of infinite order (see Ch. 5, § 7). The case $k = 0$ corresponds to the "principle of n invariants" proved in the previous subsection. If, on the contrary, $k = n$, then this conjecture asserts that locally the homogeneous space of a given geometric quantity is uniquely determined by a set of constants – the values of the "fundamental" differential invariants on this structure.

This conjecture can also be represented as a consequence of the following general principle: *situations in differential geometry that are indistinguishable by means of differential invariants are geometrically identical.* This can be taken as a "definition" of geometry if by the latter we understand the theory of geometric structures. We also draw a parallel with the basic methodological statement of modern physics: what is unobservable is non-physical.

§ 5. The Equivalence Problem for G-Structures

5.1. Three Examples. The following three examples, two of which were considered above, enable us to understand informally the point of view on the equivalence problem that follows from the theory of G-structures.

Example 1. A G-structure associated with a Riemannian metric is an $O(n)$-structure (see Ch. 6, 1.4). A necessary and sufficient condition for it to be integrable is that its curvature tensor should be zero.

Example 2. To a distribution of k-dimensional planes in n-dimensional space there corresponds a G-structure with group G consisting of $(n \times n)$-matrices of the form

$$\begin{pmatrix} A & C \\ 0 & B \end{pmatrix},$$

where A and B are non-singular $(k \times k)$- and $(n - k) \times (n - k)$-matrices respectively. The conditions of Frobenius's theorem (see Ch. 5, 3.1) are necessary and sufficient for this G-structure to be integrable.

Example 3. We recall that an almost complex structure on a 2-dimensional manifold M is defined as a field I of linear operators $I_x: T_x M \to T_x M$ satisfying the condition $I_x^2 = -1$. The associated G-structure corresponds to the group $G = GL(n, \mathbb{C})$, regarded as a subgroup of $GL(2n, \mathbb{R})$. The tensor field N of type $(1, 2)$, called the *torsion*, or *Nijenhuis tensor*, of the almost complex structure, is defined as the vector-valued 2-form

$$\tfrac{1}{2} N(X, Y) = [IX, IY] - [X, Y] - I[IX, Y] - I[X, IY].$$

The components of this tensor have the form

$$N_{ij}^k = 2 \left(I_i^s \frac{\partial I_j^k}{\partial x_s} - I_j^s \frac{\partial I_i^s}{\partial x_s} - I_s^k \frac{\partial I_j^s}{\partial x_i} + I_s^k \frac{\partial I_i^s}{\partial x_j} \right).$$

If the almost complex structure is flat (integrable), then in suitable coordinates the operators I_x have constant coefficients, and so $N = 0$. A fundamental theorem of Newlander and Nirenberg asserts that the converse is true.

Integrability of an almost complex structure means that on the manifold M we can introduce local coordinates $x_1, y_1, \ldots, x_n, y_n$ that generate complex coordinates $z_s = x_s + iy_s$, $s = 1, \ldots, n$, in which the operators I_x coincide with the operators of multiplication by i in $T_x M$.

5.2. Structure Functions and Prolongations. The auxiliary geometric objects (curvature tensor and so on) that describe the integrability of the G-structure under consideration in each of the examples given above were obtained from considerations of a special character. One of the main aims of the theory of G-structures consists in revealing the general state of affairs in this kind of problem and, in particular, finding a general mechanism for obtaining integrability conditions. The latter problem is mainly solved by introducing the concepts of the *structure function and prolongations of a G-structure*. In the examples considered above the structure function is the curvature tensor, the conditions of Frobenius's theorem written in a suitable way, and the Nijenhuis tensor, respectively.

If, however, we are precise, then the structure function of the $SO(n)$-structure (Example 1) is always identically zero. This enables us to "prolong" this first-order structure to a second-order structure. The curvature tensor is then the structure function of this prolonged structure. The second-order coframes that form the prolonged structure correspond to 2-jets of local diffeomorphisms $M \to \mathbb{R}^n$ represented by compositions of the form

$$M \xrightarrow{\exp^{-1}} T_x M \xrightarrow{\alpha} \mathbb{R}^n,$$

where α is a linear isometry, and the exponent is constructed from the associated Riemannian metric.

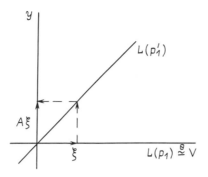

Fig. 48

We now give precise definitions. The bundle of 1-jets $\pi_{1,0}: J^1(\pi) \to P$ of sections of the G-structure $\pi: P \to M$ can be regarded as a G'-structure with vector group $G' = \operatorname{Hom}(V, \mathfrak{G})$, where $V = \mathbb{R}^n$ and $\mathfrak{G}$ is the Lie algebra of the group G. In fact, each element $p_1 \in J^1(\pi)$ determines a plane $L(p_1) \subset T_p P$, $p = \pi_{1,0}(p_1)$, transversal to the fibres of the bundle π (see Ch. 5, 4.2). Let us identify the tangent space to the fibre of π at the point $p \in P$ with the Lie algebra $\mathfrak{G}$ (see Ch. 6, §3.2), and the plane $L(p_1)$ with the vector space V by means of the shift form $\theta: L(p_1) \to V$ (see Ch. 6, 2.4). Then any other plane $L(p_1')$, where $\pi_{1,0}(p_1') = p$, is identified with the graph of the linear map $A: V \to \mathfrak{G}$ (see Fig. 48). This enables us to define the action $A: p_1 \mapsto p_1'$ of the group G' on the manifold $J^1(\pi)$. It turns the bundle $\pi_{1,0}$ into a principal G'-bundle, which can be regarded as a subbundle of the bundle of first-order coframes on P. In fact, each element $p_1 \in J^1(\pi)$ determines an isomorphism of the tangent space $T_p P$ onto the space $V \oplus \mathfrak{G}$ under which a vector $\xi \in T_p P$ is associated with a pair $(\theta(\xi), \xi_{p_1}^v)$, where $\xi_{p_1}^v$ is the projection of the vector ξ along $L(p_1)$ onto the tangent plane to the fibre identified with $\mathfrak{G}$.

The integrability of a G-structure in a neighbourhood of a fixed point $a \in M$ is equivalent to the possibility of choosing a local coordinate system $q_1, \ldots, q_n$ for which the section $s: M \to \operatorname{Rep}_1 M$, $s(x) = (d_x q_1, \ldots, d_x q_n)$, is also a section of the G-structure π. From the definition of the shift form θ it follows (see Ch. 6, 2.4) that $s^*(\theta) = (dq_1, \ldots, dq_n)$, and so $s^*(d\theta) = 0$.

Consequently, the existence of the plane $L(p_1)$ on which the V-valued 2-form $d\theta$ vanishes is a necessary condition for local integrability of the G-structure. Let us dwell on it in more detail. To this end we consider a function t on $J^1(\pi)$ with values in the space $\operatorname{Hom}(\Lambda_2 V, V)$ determined by restrictions of $d\theta$ to the plane $L(p_1)$:

$$t(p_1)(v_1 \wedge v_2) = d\theta(\xi_1, \xi_2);$$

where $\xi_1, \xi_2 \in L(p_1)$, $v_1, v_2 \in V$, and $\theta(\xi_i) = v_i$. It is called the *torsion of the G-structure*.

This function is G'-equivariant, where the action of G' on $\operatorname{Hom}(\Lambda_2 V, V)$ is given by

$$A: \operatorname{Hom}(\Lambda_2 V, V) \ni w \mapsto w + \delta A \in \operatorname{Hom}(\Lambda_2 V, V),$$

where $A \in \operatorname{Hom}(V, \mathfrak{G})$, and $\delta: \operatorname{Hom}(V, \mathfrak{G}) \to \operatorname{Hom}(\Lambda_2 V, V)$ is the Spencer δ-operator (see Ch. 5, 5.4). In the given case $\delta A(v_1 \wedge v_2) = (A(v_1))(v_2) - (A(v_2))(v_1)$, $v_1, v_2 \in V$.

The function

$$c: P = J^1(\pi)/G' \to \operatorname{Hom}(\Lambda_2 V, V)/G',$$

obtained from the torsion function t by factorization with respect to the action of G', is called the *structure function*.

Triviality of the structure function at a point $p \in P$ means that there is a 1-jet $p_1 \in J^1(\pi)$, $\pi_{1,0}(p_1) = p$, for which $d\theta|L(p_1) = 0$. Hence the structure function of locally flat G-structures is equal to zero. Generally speaking the converse is false.

The orbit of $0 \in \operatorname{Hom}(\Lambda_2 V, V)$ under the action of G' is the subspace $\delta(\operatorname{Hom}(V, \mathfrak{G}))$. We choose the subspace $D \subset \operatorname{Hom}(\Lambda^2 V, V)$ complementary to it. The subbundle $P^{(1)} = t^{-1}(D) \subset J^1(\pi)$ is called the *first prolongation* of the G-structure. It is a principal $G^{(1)}$-bundle, where $G^{(1)}$ is an Abelian Lie group that coincides with the vector space $\mathfrak{G}^{(1)} = V^* \otimes \mathfrak{G} \cap S^2 V^* \otimes V$, the *first prolongation of the Lie algebra* $\mathfrak{G}$ as a subsapce in $V^* \otimes V$ (see Ch. 5, 5.3).

Acting in a similar way, with each first-order G-structure $\pi: P \to M$ we associate a tower of $G^{(k)}$-structures

$$\to P^{(k)} \to P^{(k-1)} \to \cdots \to P^{(1)} \to P \to M,$$

where $P^{(k)} \to P^{(k-1)}$ are $G^{(k)}$-structures on the manifold $P^{(k-1)}$, and $G^{(k)} = \mathfrak{G}^{(k)} \subset S^{k+1} V^* \otimes V$ is an Abelian Lie group that coincides with the k-th prolongation of $\mathfrak{G}$ (see Ch. 5, 5.3).

The Lie algebras $\mathfrak{G} \subset V^* \otimes V$ play the same role in the theory of G-structures as symbols in the theory of differential equations. Bearing this parallel in mind, we say that a G-structure is *elliptic* if the Lie algebra $\mathfrak{G} \subset \operatorname{End} V$ does not contain operators of rank 1. This concept is the exact analogue of the concept of an elliptic differential equation. A G-structure is called a *structure of finite type* if $\mathfrak{G}^{(s)} = 0$ for sufficiently large s. For such structures the sequence of prolongations stabilizes at a certain stage; starting from this stage the bundles $P^{(k)} \to P^{(k-1)}$ become $\{e\}$-structures, in other words an absolute parallelism.

5.3. Formal Integrability. We say that a G-structure π is *k-flat* at a point $x \in M$ if there is a local diffeomorphism $F: M \to \mathbb{R}^n$, $F(x) = 0$, for which the G-structure $F(\pi)$ coincides at the point $0 \in \mathbb{R}^n$ with the standard flat G-structure on $\mathbb{R}^n$ up to small quantities of order at least k. When $k = \infty$ a k-flat structure is said to be *formally integrable* at a point $x \in M$.

The condition that a first-order G-structure $\pi: P \to M$ is 1-flat is equivalent to the vanishing of the first structure function (see 5.2), in other words the existence in π of a torsion-free connection.

If this condition is satisfied, then the G-structure is 2-flat if the next structure function is equal to zero.

Obviously, an integrable G-structure is also formally integrable. Hence the conditions for formal integrability are necessary conditions for integrability. For the most important structures these conditions are also sufficient. The first example to show that this is not so in general was constructed by Guillemin [1967] for a group G consisting of 5×5-matrices of the form

$$
\left(
\begin{array}{ccc|c}
\multicolumn{3}{c|}{E_3} & 0 \\
\hline
a & b & c & E_2 \\
-b & a & d &
\end{array}
\right),
$$

where E_k denotes the unit matrix of order k. In this example the problem reduces to the well-known Lewy equation, which is formally integrable but has no solutions.

At first glance the conditions for formal integrability are an infinite set of conditions. In fact this is not so, and we can restrict ourselves to verifying only finitely many of them. This "finiteness theorem" is a consequence of the formal theory of differential equations (see Ch. 5, § 5), since G-structures can be treated as (overdetermined) systems of differential equations of a special form. In particular, the theorems of Cartan-Kähler type (see Ch. 5, 5.5) guarantee the integrability of formally integrable analytic G-structures.

5.4. G-Structures and Differential Invariants. The methods of the theory of G-structures have still not led to any important results on the general equivalence problem. The reason for this is that this theory operates with non-scalar differential invariants like, for example, the structure function (see 1.4). In this connection an approach based on the theory of differential invariants seems preferable. Nevertheless, in the last 50 years the theory of G-structures has been actively developed by many authors, while the theory of differential invariants has practically stood still. This can apparently be explained by the influence of E. Cartan's personality and the attractive mysteriousness of his work, which induced many investigators to reinterpret it. Properly speaking, the very idea of a G-structure was crystallized from the "moving frame method" of Darboux, which was used on a grand scale by Cartan.

§ 6. Pseudogroups, Lie Equations and Their Differential Invariants

The problem of equivalence of geometric structures, considered above, is an important but nevertheless rather special form of recognition problems that arise in a natural way in differential geometry.

For example, the theory of surfaces in Euclidean space, to which Chapter 2 was devoted, can be treated as various aspects of the problem of recognition of descriptions of the form $x_i = x_i(u_1, \ldots, u_k)$, where the outer coordinates x_i are defined up to an orthogonal transformation, and the inner variables u_i are defined up to an arbitrary change of variables.

The concept of a pseudogroup which, as we said above, formalizes the representations of possible mechanisms for sorting out the various coordinate descriptions of objects of interest to us, enables us to encompass all these questions in a single theory.

6.1. Lie Pseudogroups. We say that the totality Γ of local homeomorphisms of a topological space M forms a *pseudogroup* if it contains the identity and the transformation $f \circ g^{-1}$ whenever f, $g \in \Gamma$. It is implied that the domain of definition of the transformation $f \circ g^{-1}$ is $g(U_f) \cap U_g$, where U_h denotes the domain of definition of the local transformation h. We also assume that if $U_f = \bigcup U_i$ and $f|_{U_i} \in \Gamma$, then $f \in \Gamma$.

To introduce this inductively clear topological concept into differential geometry we need to "smooth" it. For this we must first assume that M is a smooth manifold and Γ is the totality of local diffeomorphisms of it. Also, it is natural to replace the latter condition, which defines a pseudogroup, by the infinitesimal version of it. More precisely, if we fix a certain number k, we require that a local diffeomorphism f belongs to Γ whenever the k-jet of this diffeomorphism at any point $x \in U_f$ coincides with the k-jet of some $g \in \Gamma$. Finally, we must require that the totality of local diffeomorphisms that form the pseudogroup Γ should be smooth. This is formalized by the requirement that the totality of k-jets $[f]_x^k$, $f \in \Gamma$, $x \in U_f$, should form a smooth submanifold I of the manifold $\Pi_k(M)$ of k-jets of all local diffeomorphisms of M.

If all these conditions are satisfied, we say that Γ is a *smooth pseudogroup*, or *Lie pseudogroup*, of order k.

The most important and interesting pseudogroups from the viewpoint of differential geometry are the *transitive* ones. This means that for any two points x, $y \in M$ there is a transformation $f \in \Gamma$ such that $y = f(x)$. The property of transitivity is an expression of the homogeneity of the geometrical situation under consideration.

Let G be a Lie group of transformations of a manifold M. The restrictions to a domain $U \subset M$ of the transformations occurring in it determine the pseudogroup $\Gamma(G)$. Pseudogroups of this kind are said to be *globalizable*.

Example. The pseudogroup of conformal transformations of the sphere is globalizable when $n > 2$ but non-globalizable when $n = 1$ or 2.

A *complex pseudogroup* is one that consists of local holomorphic transformations of a complex manifold M. All the "real" concepts discussed above connected with pseudogroups have obvious complex analogues, like all other aspects of the theory discussed in this section.

6.2. Lie Equations. The above definition can be given a more "working" form by observing that $\Pi_k(M)$ is an open domain in $J^k\pi$, where $\pi: M \times M \to M$ is the projection on the first factor. Hence the submanifold $\mathcal{Y} \subset \Pi_k(M) \subset J^k\pi$ considered above can be treated as a system of differential equations of order at most k (see Ch. 5, 4.1) whose solutions are local diffeomorphisms constituting the pseudogroup Γ, as follows from the definition. Thus, instead of a smooth pseudogroup Γ we can consider the system of equations $\mathcal{Y}$ that determine it, which is called a *Lie equation*. This object can be constructively studied by the methods of differential calculus.

(Local) maps from M to M will be identified with their graphs, or equivalently with sections of the bundle $\pi \colon M \times M \to M$. The formal definition of a *Lie equation* consists in requiring that its solutions should form a pseudogroup. More precisely, it is necessary that $[\mathrm{id}]_x^k \in \mathcal{Y}$ (id is the identity transformation $M \to M$) for all $x \in M$, and that $a \cdot b^{-1} \in \mathcal{Y}$ if $a, b \in \mathcal{Y}$ and the composition $a \cdot b^{-1}$ is defined. Here $p \cdot q = [f \circ g]_x^k$ if $p = [f]_{g(x)}^k$, $q = [g]_x^k$, and f, g are local diffeomorphisms.

With any geometric structure ω of order k on a manifold M we naturally associate a smooth pseudogroup Γ_ω consisting of all local automorphisms of this structure. The corresponding Lie equation $\mathcal{Y}_\omega$ has the form $f(\omega) = \omega$.

Example. The equation $\mathcal{Y}_\omega$ for a Riemannian metric g_{ij} has the form

$$g_{ij}(f(x))\frac{\partial f_i}{\partial x_h}\frac{df_j}{\partial x_\partial} = g_{ks}(x),$$

where $f(x) = (f_1(x), \ldots, f_n(x))$ is the expression of the required diffeomorphism f in the coordinates $x = (x_1, \ldots, x_n)$.

6.3. Linear Lie Equations. As the last example shows, Lie equations are complicated non-linear systems of differential equations. A fundamental discovery of Lie is that without loss of information this non-linear system can be replaced by a linear system by taking the infinitesimal viewpoint.

More precisely, a vector field X on M is called an *infinitesimal transformation* belonging to the pseudogroup Γ if the local current generated by it consists of local diffeomorphisms of this pseudogroup. The totality $L(\Gamma)$ of all such fields forms an infinitesimal Lie pseudogroup. It is not difficult to see that $L(\Gamma)$ is a subalgebra of the Lie algebra $D(M)$.

Let $\mathcal{Y}$ be a system of differential equations and $u = u(x)$ a solution of it. We recall that a linearization of this system along the solution $u(x)$ is a linear system of differential equations for the "variation" $h = h(x)$, which expresses the fact that $u + \varepsilon h$ is a solution of the system $\mathcal{Y}$ of the first order in ε. Then $L(\Gamma)$ consists of vector fields that are solutions of the system of differential equations $L(\mathcal{Y})$, which is a linearization of the corresponding pseudogroup Γ of the Lie equation $\mathcal{Y}$ along its trivial solution id: $M \to M$. $L(\mathcal{Y})$ is called the *linear Lie equation* corresponding to the pseudogroup Γ.

Example. Let ω be a certain geometric structure. Then $L(\Gamma_\omega)$ consists of vector fields along which its Lie derivative vanishes. In other words $L_X(\omega) = 0$ is the corresponding linear Lie equation, that is, $L(\mathcal{Y}_\omega)$. In particular, for a Riemannian metric $g_{ij}(x)$ this equation has the form

$$X_k \frac{\partial g_{ij}}{\partial x_k} + g_{ik} \frac{\partial X_k}{\partial x_j} + g_{kj} \frac{\partial X_k}{\partial x_i} = 0,$$

where $X = X_i(x) \dfrac{\partial}{\partial x_i}$.

A remarkable fact established by Lie is that a linear Lie equation $L(\mathcal{Y})$ uniquely determines the original Lie equation $\mathcal{Y}$ just as a Lie algebra determines the corresponding Lie group. In this sense the theory of Lie pseudogroups reduces to the theory of linear Lie equations.

6.4. Differential Invariants of Lie Pseudogroups. Suppose that Γ is a Lie pseudogroup that acts on a manifold M and that we are interested in s-dimensional submanifolds $N \subset M$. Any local diffeomorphism $F: M \to M$ is canonically lifted to a diffeomorphism $F^{(k)}: J^k(M, s) \to J^k(M, s)$ of the manifold of k-jets of s-dimensional submanifolds of M (see Ch. 5, 4.5). Similarly vector fields $X \in D(M)$ are lifted to $J^k(M, s)$. The notation for this is $X^{(k)} \in D(J^k(M, s))$.

A function I on $J^k(M, s)$ is called a differential invariant of Γ of order at most k if

$$I \circ F^{(k)} = I, \qquad \forall F \in \Gamma,$$

or equivalently

$$X^{(k)}(I) = 0, \qquad \forall X \in L(\Gamma).$$

The totality of all differential invariants of order at most k obviously forms an $\mathbb{R}$-algebra $A_k = A_k(\Gamma, s)$; $A_l \subset A_k$ if $l \leqslant k$, and we can talk about the algebra $A = \bigcup_k A_k$ of all differential invariants (cf. 4.1).

The value of a differential invariant $I \in A_k$ on an s-dimensional manifold $N \subset M$ is the function $I(N) = I \circ j_k(N)$ on N. If the submanifolds N and N' are Γ-equivalent, that is, $N' = F(N)$ for some $F \in \Gamma$, then

$$I(L) = I(L') \circ F, \qquad \forall I \in A. \tag{5}$$

The equality (5) is used to solve the problem of Γ-equivalence in exactly the same way as its analogue (3) in the solution of the problem of equivalence of geometric structures. However, this touches on the other aspects discussed above. For example, in the situation under consideration the "principle of n invariants" holds, and so on.

We emphasize that the concept of an invariant of a pseudogroup that we have introduced is defined not only by the pseudogroup itself but also by the number s, the dimension of the submanifolds under consideration. If say Γ is a group of motions of 3-dimensional Euclidean space, then when $s = 1$ we arrive at the differential invariants of curves (curvature, torsion, and so on), and when $s = 2$ we arrive at the differential invariants of surfaces (Gaussian curvature, mean curvature, and so on).

It is not difficult to verify that in order to find differential invariants of order at most k it is sufficient to know only the k-jets of the transformations that form the pseudogroup under consideration. Hence the algebra of invariants A can be described in terms of the corresponding (linear) Lie equation, just like the procedure for finding them.

6.5. On the Structure of the Algebra of Differential Invariants. Differential invariants can be duplicated by means of operators of invariant differentiation. For example, differentiation with respect to the natural parameter in the metric theory of curves is of this kind. In the situation when there are invariants $I_1, \ldots,$ $I_s \in A$ whose values $I_i(N)$ can be taken as coordinates on $N \subset M$ the operators of invariant differentiation have the form

$$\sum I_i \frac{\partial}{\partial I_i}, \qquad I_i \in A.$$

In the case when a Lie pseudogroup Γ is transitive and the corresponding Lie equation Y is formally integrable, we have the following "finiteness theorem".

Theorem. *There are finitely many invariants $I_k \in A$ and operators of invariant differentiation $\Delta_i \colon A \to A$ such that any invariant $I \in A$ is a function of finitely many variables of the form $(\Delta_{i_1} \circ \cdots \circ \Delta_{i_p})(I_k)$.*

This result, usually attributed to Tresse, can be made more precise. Namely, the algebra A, if we neglect some not very important details, coincides with the algebra of smooth functions on some infinitely prolonged differential equation $\mathscr{E}_\infty$ (see Ch. 5, § 7). For the invariants of the theory of surfaces in E^3 such an equation $\mathscr{E}$ is the Gauss-Peterson-Mainardi-Codazzi system, written in invariants (see Ch. 2, 2.9).

§ 7. On the Structure of Lie Pseudogroups

Lie pseudogroups admit a rich structural theory, some characteristic results of which are given below.

7.1. Representation of Isotropy. The set of transformations of a pseudogroup Γ, defined in a neighbourhood of a point $x \in M$ and preserving this point, form a "subpseudogroup" $\Gamma_x \subset \Gamma$, called the *stabilizer* of the point x. The infinitesimal pseudogroup $L(\Gamma_x)$ corresponding to Γ_x consists of all vector fields $X \in L(\Gamma)$ that vanish at x. Moreover, the totality of k-jets at x of transformations that occur in Γ_x forms a Lie group Γ_x^k, which is called the *isotropy group of the pseudogroup* Γ *of order k at the point x*. The group Γ_x^k can be regarded as a linear group of transformations of the vector space $J_x^k(1_M)$ of k-jets of smooth functions on M at the point x. In particular, the group Γ_x^1 is interpreted as a subgroup of the group of all linear transformations of the space T_x^*M.

The natural homomorphism $\Gamma_x \to \Gamma_x^k$ is called the *isotropy representation of order k*. The smallest number k for which this homomorphism does not have a kernel is called the *type* of Γ at x. For transitive pseudogroups this type is the same for all points x and is called the *type* of the pseudogroup itself. If all the isotropy representations have a non-trivial kernel, we say that Γ has *infinite type*.

Examples. 1) The pseudogroup of local isometries of a Riemannian manifold has type 1 at all points.

2) The pseudogroup of all local symplectic transformations of a symplectic manifold has infinite type.

7.2. Examples of Transitive Pseudogroups. The examples given below play an important role in the structural theory of Lie pseudogroups. Bearing in mind what follows (see 7.3), we restrict ourselves here to their complex versions.

1. The pseudogroup $A(\mathbb{C}^n)$ of all local holomorphic transformations of the space $\mathbb{C}^n$.

2. The pseudogroup of transformations $\varphi \in A(\mathbb{C}^n)$ that have constant ("complex") Jacobian.

3. The pseudogroup of transformations $\varphi \in A(\mathbb{C}^n)$ that have unit ("complex") Jacobian.

4. The "Hamiltonian", or symplectic, pseudogroup consisting of transformations $\varphi \in A(\mathbb{C}^{2n})$ that preserve the symplectic complex form $\Omega = \sum_{i=1}^{n} dz_i \wedge dz_{n+i}$.

5. The conformally symplectic pseudogroup consisting of transformations $\varphi \in A(\mathbb{C}^{2n})$ that preserve the form Ω up to multiplication by a function.

6. The contact pseudogroup consisting of transformations $\varphi \in A(\mathbb{C}^{2n+1})$ that preserve the complex distribution on $\mathbb{C}^{2n+1}$ given by the form $dz_0 + \sum_{i=1}^{n} z_{n+i} \, dz_i$, or equivalently that preserve this form up to multiplication by a function.

The pseudogroup in Example 2 is of the second order, and the others are of the first order. Obviously they are all pseudogroups of infinite type.

7.3. Cartan's Classification. A pseudogroup Γ on a manifold M is said to be *primitive* if there is no foliation of dimension other than zero and $n = \dim M$ that is invariant under transformations of Γ, and *irreducible* (at a point $x \in M$) if its isotropy group Γ_x^1, understood as a group of linear transformations of the space $T_x^* M$, does not have non-trivial invariant subspaces.

A primitive pseudogroup is automatically transitive, but it may be reducible. For example, a contact pseudogroup (7.2, Example 6) is of this kind. On the other hand, a pseudogroup that is transitive and irreducible at all points is primitive. All the pseudogroups described in the previous subsection are obviously primitive.

Theorem (Cartan). *Any complex primitive pseudogroup of infinite type is locally isomorphic to one of the pseudogroups described in the previous subsection.*

A similar classification theorem holds for primitive real pseudogroups of infinite type. The corresponding list is longer than in the complex case and consists of the real forms of the complex pseudogroups of Examples 1–6.

The proof of Cartan's theorem is developed along the following lines. Firstly, with each transitive pseudogroup Γ there is associated a *transitive Lie algebra* $\mathfrak{G}(\Gamma)$ consisting of all jets of infinite order of vector fields from $L(\Gamma)$ at some fixed point $x \in M$. The algebra $\mathfrak{G}(\Gamma)$ determines the pseudogroup Γ uniquely up to a local isomorphism in the same way as a Lie algebra determines a Lie group. The primitivity of the pseudogroup Γ is equivalent to the isotropy subalgebra of infinite order $\Gamma_x^\infty \subset \mathfrak{G}(\Gamma)$ being maximal.

The subspace of $\mathfrak{G}(\Gamma)$ consisting of those elements of it whose k-jets are zero at the point under consideration is denoted by $\mathfrak{G}_k(\Gamma)$. Suppose also that

$$\mathfrak{h}_k(\Gamma) = \mathfrak{G}_k(\Gamma)/\mathfrak{G}_{k+1}(\Gamma).$$

Then the Lie operation in $\mathfrak{G}(\Gamma)$ naturally induces a Lie operation in the graded space

$$\mathfrak{h} = \mathfrak{h}(\Gamma) = \sum_{k \geqslant -1} \mathfrak{h}_k(\Gamma),$$

turning it into a graded Lie algebra, that is, $[\mathfrak{h}_l, \mathfrak{h}_k] \subset \mathfrak{h}_{k+l}$. In particular, $\mathfrak{h}_0$ is a subalgebra of this algebra that acts on the subspaces $\mathfrak{h}_k$, since $[\mathfrak{h}_0, \mathfrak{h}_k] \subset \mathfrak{h}_k$. The classification of transitive primitive Lie algebras reduces to the classification of the corresponding graded Lie algebras, which are characterized by the property that the action of $\mathfrak{h}_0$ on $\mathfrak{h}_{-1}$ is irreducible. This is a purely algebraic problem.

7.4. The Jordan-Hölder-Guillemin Decomposition. A central fact of the structural theory of transitive Lie algebras is the existence (established by Guillemin) of a decomposition analogous to the Jordan-Hölder series in group theory. In order to formulate it we need to turn our attention to the following purely topological aspect.

A Lie algebra $\mathfrak{G}(\Gamma)$ can be regarded as a linear topological space if the subspaces $\mathfrak{G}_k(\Gamma)$ are taken to be a fundamental system of neighbourhoods of unity. Since $[\mathfrak{G}_k(\Gamma), \mathfrak{G}_l(\Gamma)] \subset \mathfrak{G}_{k+l}(\Gamma)$, the Lie operation in $\mathfrak{G}(\Gamma)$ is continuous in this topology. Thus, $\mathfrak{G}(\Gamma)$ turns into a topological Lie algebra and it makes sense to talk about its closed subalgebras, ideals, and so on.

Theorem (Guillemin). *Let $\mathfrak{G}$ be a transitive Lie algebra and $\mathfrak{I} \subset \mathfrak{G}$ a closed ideal of it. Then there is a chain $\mathfrak{I} = \mathfrak{I}_0 \supset \mathfrak{I}_1 \supset \cdots \supset \mathfrak{I}_p = 0$ of closed ideals of $\mathfrak{G}$ having the property that either the factors $\mathfrak{I}_s/\mathfrak{I}_{s+1}$ are Abelian Lie algebras, or there are no ideals of $\mathfrak{G}$ between $\mathfrak{I}_s$ and $\mathfrak{I}_{s+1}$. Non-Abelian factors of the form $\mathfrak{I}_s/\mathfrak{I}_{s+1}$ are uniquely determined by the algebra $\mathfrak{G}$, that is, they do not depend on the choice of chain of the given type.*

7.5. Pseudogroups of Finite Type. Pseudogroups of finite type admit an exhaustive description in a certain sense. Namely, with any Lie algebra of vector fields $\mathscr{H} \subset D(M)$ we can associate the pseudogroup $\Gamma(\mathscr{H})$ generated by shift operators along the fields $X \in \mathscr{H}$. Pseudogroups of finite type have the form $\Gamma = \Gamma(\mathscr{H})$, where $\mathscr{H} \subset D(M)$ and $\dim \mathscr{H} < \infty$. This follows from the result of the next paragraph.

The fact that a pseudogroup Γ is of finite type means that for sufficiently large k the symbol of the k-th prolongation of the corresponding linear Lie equation is zero. This in turn means that the Cartan distribution (see Ch. 5, 4.2) induces an n-dimensional completely integrable distribution on any such extension. The integral manfiolds of this distribution, and only they, are solutions of the original Lie equation. It is obvious from what we have said that they form a finite-parameter family. Hence the algebra $L(\Gamma)$ (see 6.3) is finite-dimensional.

The pseudogroups of local automorphisms of many important geometric structures have finite type. For example, linear connections, pseudo-Riemannian metrics, and projective and conformal structures on manifolds of dimension at least 3 are of this kind. Hence the automorphism groups of these structures, like all G-structures of finite type in general, are Lie groups.

The Lie equations of elliptic G-structures (see 5.2) are elliptic systems of differential equations. As we know, the space of smooth solutions of such systems on compact manifolds is finite-dimensional. The next result of Ochiai follows from this.

Theorem. *The group of automorphisms of an elliptic G-structure on a compact manifold is a Lie group.*

An almost complex structure is elliptic. Hence the group of automorphisms of a compact almost complex manifold is a Lie group.

Chapter 8
Global Aspects of Differential Geometry

> "All the various changes in first principles, spaces, times, is enough to drive us mad... When will it all end?"
>
> A. Blok

The problems of global differential geometry or, as we say, differential geometry in the large, traditionally form one of the most exciting branches of mathematics, attractive for its depth and beauty. Here differential geometry interacts closely with topology (set-theoretic, algebraic and differential) and functional analysis. The investigation of some specific problems of geometry in the large has often led to the rise of whole geometrical schools and trends. The remaining articles of this book will be devoted to the results of some of them. The aim of this chapter is to give by means of examples a general and as far as possible elementary presentation of this branch of mathematics.

Most of the results of global differential geometry are connected with some kind of integral formulae. This indicates that in these questions we can hope to construct a rich general theory. In fact, integral formulae are always connected with various natural cohomological systems of differential calculus. Hence to distinguish these systems in pure form, starting from concrete problems and by way of conceptual analysis, is one of the most interesting problems. In this context we mention various versions of Spencer cohomology (see Vinogradov, Krasil'schik and Lychagin [1986], Goldschmidt and Spencer [1967–78], and Spencer [1985]) – a comparatively recently discovered example of this kind.

§ 1. The Four Vertices Theorem

A smooth closed convex curve in a plane will be called an *oval*. An example of an oval is an ellipse. An ellipse has exactly four points, namely its vertices, at which its curvature is stationary, that is, $k'(s) = 0$, where k is the curvature and s is the natural parameter. The four vertices theorem says that on any oval there are at least four such points. The proof of it is a simple integral consequence of the Frenet formulae and we give it here.

In the case under consideration the Frenet formulae have the form $x'' = -ky'$, $y'' = kx'$, where $(x(s), y(s))$ is the parametric representation of our curve Γ, and the primes denote differentiation with respect to the natural parameter s. From these formulae we obtain the integral relation

$$\oint_\Gamma (ax + by + c)k'\, ds = 0, \tag{1}$$

for any a, b, $c \in \mathbb{R}$. The equality $\oint_\Gamma k'\, ds = 0$ is obvious, and the equalities $\oint_\Gamma xk'\, ds = 0$ and $\oint_\Gamma yk'\, ds = 0$ are proved in the same way. Let us prove the first of them. Integrating by parts and using the Frenet formulae, we have

$$\oint_\Gamma xk'\, ds = -\oint_\Gamma x'k\, ds = -\oint_\Gamma y''\, ds = y'|_{s_0}^{s_0} = 0.$$

Also, since the curve Γ is compact, there are two distinct points on it where the function $k(s)$ attains a maximum and a minimum. Let us draw the line l through them, whose equation is $ax + by + c = 0$. If these are the only points where $k'(s) = 0$, then the expression $(ax + by + c)k'$ is non-zero and has the same sign everywhere except at the two points. Hence the integral (1) cannot be equal to zero. This contradiction proves that there are more than two points where $k'(s) = 0$. But a smooth function on a closed curve has an even number (so long as it is not infinite) of changes of sign. Hence there are at least four such points.

Examples show that there are closed curves in space with exactly two points where $k'(s) = 0$.

§2. Carathéodory's Problem About Umbilics

We call that a point on a surface in 3-dimensional Euclidean space is called an *umbilic* if the principal curvatures of the surface are equal there. A surface is said to be *oval* if it bounds a convex bounded body in $\mathbb{R}^3$. Among rotation surfaces it is not difficult to find a "cigar" (see Fig. 49) for which the only umbilics are A and B. Carathéodory conjectured that on any smooth oval surface there are at least two umbilics. Outwardly this problem is very reminiscent of the previous one, but apparently it has not been completely solved up to now. For analytic ovals Hamburger proved this assertion in 1940, and Bol substantially simplified this proof in 1943. Comparatively recently Feldmann, using modern techniques of differential topology, proved that Carathéodory's conjecture is "basically" true. This means, in particular, that by an arbitrarily small deformation of a given surface we can obtain a surface with no fewer than two umbilics.

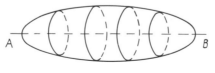

Fig. 49

Feldmann actually proved much more; in particular, he proved that the multi-dimensional analogues of Carathéodory's conjecture are "basically" true.

§ 3. Geodesics on Oval Surfaces

From the Gauss-Bonnet formula it follows that the image of a non-self-intersecting closed geodesic lying on an oval surface P in $\mathbb{R}^3$ under the Gaussian map onto the unit sphere S^2 divides the surface of this sphere in half. Conversely, it is not difficult to show that a closed curve on P that has least length among those whose spherical image divides the area of S^2 in half is a geodesic. From this it follows that on any smooth oval surface there is at least one closed non-self-intersecting geodesic. This result was obtained by Poincaré in connection with his research into celestial mechanics. It can be interpreted, for example, as the existence of a periodic motion in special cases of the three body problem.

A more accurate development of these considerations enables us to show that on an oval surface there are at least three distinct non-self-intersecting closed geodesics. The example of an ellipsoid shows that there can be exactly three of them.

This result was a stimulus for the study of the question of the number of closed geodesics on compact Ricmannian manifolds without boundary. The classic results here are those of Lyusternik and Shnirel'man, and Morse [1934]. Lyusternik and Shnirel'man showed, in particular, that on any Riemannian manifold that is diffeomorphic to a 2-dimensional sphere there are at least three distinct closed non-self-intersecting geodesics. In other words, they removed the condition that the surface is oval: no matter how much a sphere is "creased", there are always three distinct closed geodesics on it.

In contrast to the arguments of Poincaré and his followers, who relied on the classical formulae of differential geometry and the calculus of variations, Lyusternik and Shnirel'man use serious topological methods. Namely, they show first of all that the space Γ of all smooth non-self-intersecting curves on a 2-dimensional sphere can be continuously contracted over itself to a subspace Γ_0 consisting of great circles. Obviously, $\Gamma_0 \approx \mathbb{R}P^2$. Also, any Riemannian metric on S^2 can be used to contract Γ over itself to some smaller part. Analysis of the process of contraction shows that this smaller part is homeomorphic to a point or a 2-dimensional sphere if the metric under consideration has fewer than three closed non-self-intersecting geodesics. In fact in this situation everything reduces to a proof that any smooth function on $\mathbb{R}P^2$ ($\approx \Gamma_0$) has at least three critical points. Lyusternik and Shnirel'man prove the latter result by introducing the important concept of the category of a topological space.

Using modern methods of algebraic topology, mathematicians have recently succeeded in showing that there are infinitely many geometrically distinct closed geodesics on a manifold diffeomorphic to a 2-dimensional sphere (as a rule they are self-intersecting).

From the time of Poincaré the problem of the number of closed geodesics (self-intersecting or not) has been the subject of much research and many interesting results have been obtained here, admittedly not always sufficiently justified. One of these asserts that on any compact Riemannian manifold without boundary there are always three geometrically distinct closed geodesics.

A very important result of this research was the creation of Morse theory, which is at present one of the most powerful tools of global analysis. We shall say a little about it below.

The problem of closed geodesics on a Riemannian manifold has an obvious multidimensional analogue. Namely, geodesics can be regarded as extremals of the functional of 1-dimensional volume (that is, length) or a 1-dimensional action. The extremals of the functional of k-dimensional volume are called *minimal surfaces*, and the extremals of the "action" functional, or Dirichlet functional, are called *harmonic maps*. Hence we can pose the question of the existence of closed minimal submanifolds of a given Riemannian manifold or harmonic maps of one Riemannian manifold into another. Many outstanding results have already been obtained in connection with these questions. Here we recall two results that are 2-dimensional generalizations of the results discussed above about closed geodesics on a sphere. Firstly, Simons and Smith (Simons [1968]) showed that in a 3-dimensional Riemannian manifold diffeomorphic to a sphere there is always a minimal 2-sphere. Secondly, White [1982] established the existence of at least four minimal 2-spheres in such a manifold if its metric is close to the standard metric of a 3-sphere.

§4. Rigidity of Oval Surfaces

In 1838 Minding conjectured that a standard sphere lying in 3-dimensional Euclidean space $\mathbb{R}^3$ is rigid, that is, there is no smooth surface in $\mathbb{R}^3$ different from a sphere but nevertheless isometric to it. This corresponds to our everyday experience that an elastic spherical shell, for example a ping-pong ball, appreciably resists a deformation of its shape. This conjecture of Minding turned out to be unexpectedly difficult, and a proof of it was found only in 1899 by Liebmann.

Shells that have the form of an oval surface also resist deformations that change their original shape. The geometrical equivalent of this is the assertion that an

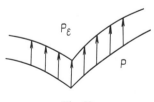

Fig. 50

oval surface is unbendable infinitesimally. More precisely, let $V: x \mapsto V(x)$, $x \in P$, be a smooth vector field on a surface P in $\mathbb{R}^3$ (see Fig. 50). We say that such a field is *trivial* if it is the restriction to P of the velocity field of some motion of the whole space $\mathbb{R}^3$ as a rigid body. Let P_ε (where ε is sufficiently small) be the surface consisting of points of the form $x_\varepsilon = x + \varepsilon V(x)$, where $x \in P$. If the map $x \mapsto x_\varepsilon$ of the surface P onto P_ε is an isometry up to quantities of order ε^2, and the field V is non-trivial, then it is called an *infinitesimal bending* of the surface P. Liebmann also succeeded in proving that oval surfaces do not admit infinitesimal bendings. For a brief proof by Blaschke and Weyl see Blaschke [1930].

Rigidity, that is, the unique determination of the shape of a smooth oval surface, was established in 1934 by Cohn-Vossen. Another short and elegant proof was given by Herglotz [1942].

Pogorelov [1951] showed that the requirement of smoothness here can be dropped, by proving that any closed convex surface, that is, the boundary of a bounded convex body, is uniquely determined up to a motion by its intrinsic metric.

§ 5. Realization of 2-Dimensional Metrics of Positive Curvature (A Problem of H. Weyl)

The problem is the following: can a two-dimensional Riemannian manifold of positive curvature that is diffeomorphic to a sphere be realized as a smooth oval surface in $\mathbb{R}^3$? Weyl [1915] himself suggested an incomplete solution of this problem for analytic manifolds, which was completed by Lewy [1938].

From the general results of A.D. Aleksandrov [1948] it follows that there is a realization as an oval surface whose smoothness is not explicit. This remaining question of smoothness was solved by Pogorelov [1969], who showed that if a realizable Riemannian metric belongs to the class C^r, then its realization has class C^{r-1}.

We shall emphasize that the methods of Aleksandrov by means of which he obtained the result mentioned above are not methods of differential geometry proper, but are a peculiar alloy of topology and synthetic geometry. In particular, one of his main methods is approximation of the original metric by polyhedra.

The question of the realizability of one Riemannian manifold as a submanifold of another is hopeless in its general formulation. However, if one of these manifolds has a simple structure (for example, it is a sphere or a Euclidean space), then we can hope for a reasonable solution. For example, Pogorelov [1961] proved that a two-dimensional metric on a sphere with curvature $K > C$, where $C \leqslant 0$, can be realized in a 3-dimensional Riemannian manifold whose curvature in two-dimensional directions does not exceed C. Of natural interest in this field is the question of the smallest dimension of a Euclidean space such that a given Riemannian manifold can be realized as a submanifold of it, in particular, how does this dimension depend on the curvature of the original manifold? Here very little is known (see §§ 6–7 and Green [1970], for example).

§ 6. Non-Realizability of the Lobachevskij Plane in $\mathbb{R}^3$ and a Theorem of N.V. Efimov

Strictly speaking, Lobachevskij, Bolyai and Gauss did not prove the existence of the non-Euclidean geometry that they discovered. This was actually done when Beltrami realized part of the Lobachevskij plane by a surface of 3-dimensional Euclidean space $\mathbb{R}^3$. In this connection it seems tempting to realize the whole Lobachevskij plane as a surface in $\mathbb{R}^3$, possibly with self-intersections. Hilbert's result, which showed that this is impossible, was unexpected.

Here is an outline of Hilbert's proof. Assuming that the Lobachevskij plane has been realized as a surface $P \subset \mathbb{R}^3$ (possibly self-intersecting), we draw attention to the fact that at each point of this surface there are two distinct asymptotic directions, since the Gaussian curvature at any point of it is negative, say -1. Hence the asymptotic lines, that is, the curves touching the asymptotic directions, form two families of curves on P. We can show that any two curves belonging to different families intersect (this is a very subtle part of the proof). From this it follows that on P there is a global coordinate system, say (u, v), that has the asymptotic lines as the coordinate lines. Using known derivation formulae and the fact that the curvature of P is -1 everywhere, it is not difficult to show that in these coordinates, possibly after some reparametrization, the first quadratic form of the surface P becomes

$$ds^2 = du^2 + 2 \cos \varphi \, du \, dv + dv^2,$$

where $\varphi = \varphi(u, v)$ is the angle between the asymptotic lines.

Gauss's formula, which expresses the total curvature of a surface in terms of the first quadratic form, is in this case

$$k \equiv -1 = -\frac{1}{\sin \varphi} \frac{\partial^2 \varphi}{\partial u \partial v},$$

or

$$\frac{\partial^2 \varphi}{\partial u \partial v} = \sin \varphi \tag{2}$$

(the "sine-Gordon" equation, now famous in the theory of solitons). In view of (2), the area of the quadrilateral D bounded by the coordinate lines $u = u_i$, $v = v_i$, $i = 1, 2$, is equal to

$$S = \iint_D \sin \varphi \, du \, dv = \int_{u_1}^{u_2} du \int_{v_1}^{v_2} \frac{\partial^2 v}{\partial u \partial v} \, dv = \sum_{j=1}^{4} \alpha_j - 2\pi,$$

where α_j are the internal angles of D. Since $0 < \alpha_j < \pi$, it follows that $S < 2\pi$. Choosing D arbitrarily large, we verify that the area of P is at most 2π, which contradicts the fact that the area of the Lobachevskij plane is infinite.

Hilbert's theorem was generalized by Efimov [1948], [1964], [1966], who proved that a complete 2-dimensional Riemannian manifold whose Gaussian curvature does not exceed some fixed negative constant cannot be realized as a surface in 3-dimensional Euclidean space. This is a very subtle result, which is difficult to prove.

§7. Isometric Embeddings in Euclidean Spaces

The problems considered in the last two sections are special cases of the following general question. Let (M, g) be an n-dimensional Riemannian manifold. Can the smooth manifold M be embedded in a Euclidean space of some dimension N in such a way that the metric on M induced by this embedding coincides with g? Such an embedding, if it exists, is said to be *isometric*. A local version of this question is the following: is there an isometric embedding in $\mathbb{R}^N$ of a sufficiently small neighbourhood of a given point $u \in M$?

Clearly, necessary prerequisites for a solution of the problem of isometric embedding are

a) the possibility of a smooth embedding of M in $\mathbb{R}^N$,
b) the possibility of a local isometric embedding for any point $a \in M$.

The combination of these conditions is by no means sufficient. Consider, for example, a flat torus, that is, a torus endowed with a metric $ds^2 = d\varphi^2 + d\psi^2$, where (φ, ψ) are the standard angular coordinates on the torus. Such a torus can be obtained in another way by factorizing the Euclidean plane by means of the group of motions of it consisting of parallel displacements by vectors of the form $pe_1 + qe_2$, where $p, q \in \mathbb{Z}$ and e_1, e_2 are fixed perpendicular unit vectors. The torus can be smoothly embedded in $\mathbb{R}^3$ (like a doughnut). Moreover, a flat torus is locally isometric to the Euclidean plane and hence the local problem of an isometric embedding in $\mathbb{R}^N$, where $N \geqslant 2$, is soluble for any point of it. However, a flat torus does not admit an isometric embedding in $\mathbb{R}^3$. For otherwise such a torus would be a closed surface in $\mathbb{R}^3$ whose Gaussian curvature is zero everywhere. But from the elementary geometry of surfaces it follows that one of the principal curvatures of such a surface is zero, and the corresponding lines of curvature are segments of straight lines in $\mathbb{R}^3$. Hence a complete surface of zero Gaussian curvature in $\mathbb{R}^3$ cannot be compact. We note that a flat torus admits an isometric embedding in $\mathbb{R}^4$, namely $(\varphi, \psi) \mapsto (\sin \varphi, \cos \varphi, \sin \psi, \cos \psi)$.

An example of a non-compact manifold that is not isometrically embeddable in $\mathbb{R}^3$, despite conditions a) and b) being satisfied, is the Lobachevskij plane. Condition a) is satisfied for it in an obvious way, and the Beltrami pseudosphere shows that condition b) is satisfied. Nevertheless, Hilbert's theorem shows that an isometric embedding of the Lobachevskij plane in $\mathbb{R}^3$ is impossible.

The problem of isometric embedding admits an obvious analytic interpretation. Namely, if $g_{ij}(x)$, $x = (x_1, \ldots, x_n)$ are the components of the metric g in local coordinates $x_1, \ldots, x_n$ on M, and $y = (y_1, \ldots, y_N)$ are the standard Cartesian coordinates in $\mathbb{R}^N$, then the condition for an embedding $M \ni x, (y_1(x), \ldots, y_N(x)) \in \mathbb{R}^N$, to be isometric is

$$\sum_{i=1}^{n} \frac{\partial y_k}{\partial x_i} \frac{\partial y_l}{\partial x_i} = g_{kl}(x),$$

that is, we have a system of $\frac{1}{2}n(n + 1)$ non-linear partial differential equations in N unknown functions. If $N = \frac{1}{2}n(n + 1)$, then this system is definite and so we would like to think that it has a solution, whatever the unknown functions $g_{ij}(x)$. This idea was apparently first expressed by Schläfli in 1871 and was later called Schläfli's conjecture. The proof of the local form of this conjecture for analytic Riemannian manifolds was given by Janet [1926] and E. Cartan [1927]. Both these proofs contain certain obscurities. In 1931 Burstin got rid of them in Janet's proof (see Burstin [1931]). Both Janet's proof and Cartan's proof are propositions of the formal theory of differential equations in a specific problem (see Ch. 5); Cartan restated the problem in the language of differential systems and used the Cartan-Kähler theorem. We note that not every Riemannian manifold can be locally isometrically immersed in $\mathbb{R}^N$, where $N \leqslant \frac{1}{2}n(n + 1)$. Thus, the number $\frac{1}{2}n(n + 1)$ is the least possible. The C^∞-version of this result of Janet, Cartan and Burstin has apparently still not been proved. However, Rokhlin and Gromov [1970] and independently Green [1970] established that a local embedding in the class C^∞ is possible in a Euclidean space of dimension $N = n + \frac{1}{2}n(n + 1)$. In this connection we observe that an n-dimensional manifold can automatically be smoothly embedded in $\mathbb{R}^{2n}$. Thus, if we take account of the necessary conditions a) and b), there are grounds for hoping that a smooth isometric embedding of an n-dimensional Riemannian manifold in $\mathbb{R}^N$ is always possible if $N \geqslant n + \frac{1}{2}n(n + 1)$. Nevertheless, the results up to now are quite a long way from this estimate. Let us give them.

A famous theorem of Nash [1956] asserts that any compact Riemannian manifold can be isometrically immersed in a preassigned domain of N-dimensional Euclidean space if $N \geqslant \frac{1}{2}n(3n + 11)$. Nash's method was later improved by Rokhlin and Gromov [1970], who lowered Nash's estimate to $N \geqslant \frac{1}{2}n(n + 1) + 3n + 5$. The proof of Nash's theorem is based on a powerful existence theorem that he proved for non-linear systems of partial differential equations. From the conceptual point of view, this theorem of Nash, strengthened and generalized by Moser and others (see Moser [1961]), is the infinite-dimensional version of Newton's well-known "method of tangents". By means of this many other subtle results of global analysis have been obtained, in particular, some of the results of geometry "in the large" discussed in this chapter.

Nash and his followers also obtained results about isometric immersion of non-compact Riemannian manifolds (see Nash [1956], for example).

We must say that the estimates given above for the dimension of a Euclidean space into which any Riemannian manifold of a given dimension can be isometrically immersed are by no means sharp. For example, it has been proved that a 2-dimensional Riemannian manifold can be isometrically immersed in $\mathbb{R}^{10}$. In addition, it is clear that under various restrictions on the metric or topology of the Riemannian manifold substantial simplifications of the general estimates of Nash type are possible. Here are some examples that illustrate this idea.

1. A two-dimensional Riemannian manifold diffeomorphic to a sphere can be isometrically embedded in $\mathbb{R}^7$.

2. A flat n-dimensional torus can be isometrically embedded in $\mathbb{R}^{2n}$ according to the rule $(\varphi_1, \ldots, \varphi_n) \mapsto (\sin \varphi_1, \cos \varphi_1, \ldots, \sin \varphi_n, \cos \varphi_n)$, but it cannot be embedded in $\mathbb{R}^{2n-1}$.

3. The Lobachevskij space of n dimensions can be isometrically immersed in $\mathbb{R}^{6n-5}$ (Blanuša [1955]). However, this estimate is not sharp, since the Lobachevskij plane can be immersed in $\mathbb{R}^5$ (E.R. Rozendorn), and possibly in $\mathbb{R}^4$.

The theorem on the rigidity of an oval surface should also have multidimensional analogues. We give here a result of Berger, Bryant and Griffiths [1983], which shows that if the dimension of the Euclidean space in which the embedding is carried out is not too high, then local rigidity may be observed, that is, uniqueness of the embedding up to a motion. Namely, suppose that an n-dimensional Riemannian manifold is locally embedded in $\mathbb{R}^{n+r}$, where

$$r \leqslant n \quad \text{if } n > 8; \qquad r \leqslant 6 \quad \text{if } n = 7, 8;$$

$$r \leqslant n \quad \text{if } n = 4 \text{ or } 5; \qquad r \leqslant 3 \quad \text{if } n \leqslant 4.$$

Then this embedding is rigid if the second quadratic form of the embedded submanifold has "general type". We omit a precise definition of generality of type, while emphasizing that this concept in the present context replaces the convexity condition.

Above we have assumed that the Riemannian manifolds and their embeddings are smooth, that is, of class C^∞. In fact, all our results turn out to be true if we consider metrics and embeddings of class C^k, where $k \geqslant 2$ or 3. The position changes abruptly if we consider embeddings of class C^1. Nash [1954] showed that if a manifold M admits a C^1-embedding in $\mathbb{R}^N$, where $N \geqslant n + 2$, then it admits *an isometric C^1-embedding in $\mathbb{R}^N$*. Kuiper [1955] improved this result by showing that it is true when $N \geqslant n + 1$. In proving this theorem, Nash began with a C^1-embedding $J \colon M \hookrightarrow \mathbb{R}^N$. Then, roughly speaking, he perturbed the submanifold $J(M) \subset \mathbb{R}^N$ in a suitable way by waves of small length and small amplitude ("ripples"), trying to get the necessary isometry property.

A compact complex manifold cannot be holomorphically immersed in $\mathbb{C}^N$. For assuming the contrary, we must verify that at least one coordinate function that specifies such an embedding must be different from a constant. However, this contradicts the fact that on a compact complex manifold there are no homomorphic functions other than constants.

Generally speaking, non-compact Kähler manifolds also do not admit isometric holomorphic embeddings in finite-dimensional Euclidean spaces. The first examples of this kind were apparently constructed by Bochner [1947].

§8. Minimal Surfaces. Plateau's Problem

A surface $P \subset \mathbb{R}^3$ is said to be *minimal* if it is an extremal of the "surface area" functional $S = S(P)$. It can be verified directly that the Euler-Lagrange equations for this functional have the form $H = 0$, where H is the mean curvature of P. Thus, minimal surfaces are surfaces of zero mean curvature.

In conformal coordinates on the surface, that is, those in which the first quadratic form is $ds^2 = f(u, v)(du^2 + dv^2)$, the minimality condition is equivalent to the radius vector $r(u, v)$ that describes it being harmonic, that is, $\Delta r = 0$, $\Delta = \dfrac{\partial^2}{\partial u^2} + \dfrac{\partial^2}{\partial v^2}$. This enabled Riemann, Weierstrass and others to establish the exceptionally fruitful connection between the theory of minimal surfaces and the theory of functions of a complex variable.

From the basic principles of mechanics it follows that a film of liquid in equilibrium (we generally talk about a soap film) takes the shape of a minimal surface. Plateau, who investigated this question, posed the following problem in 1886: find a minimal surface whose boundary is a given piecewise-smooth curve in space.

The connection mentioned above with the theory of analytic functions enables us, for example, to reduce Plateau's problem for a polygonal contour to the problem of a conformal map of a spherical polygon onto a planar polygon. This enabled Schwarz in 1867 to find explicitly in terms of elliptic functions the minimal surface spanned by a spatial quadrangle ABCD whose vertices are the vertices of a regular tetrahedron (Fig. 51). However, the general solution of Plateau's problem was obtained much later by the efforts of Rado, Courant, and above all Douglas in 1936–38. Douglas's success owed much to replacing the area functional S by the Dirichlet functonal

$$D(P) = \frac{1}{2} \int \int_P (E + G) du\, dv = \frac{1}{2} \int \int_P (r_u^2 + r_v^2)\, du\, dv,$$

where $r = r(u, v)$ is the radius vector of the surface P. By virtue of the elementary inequality $\frac{1}{2}(E + G) \geqslant \sqrt{EG - F^2}$ we have $D(P) \geqslant S(P)$; equality is attained only when $F = 0$ and $E = G$, that is, in conformal coordinates. This enables us

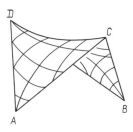

Fig. 51

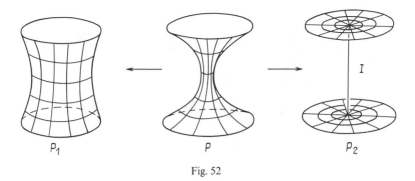

Fig. 52

to replace the original problem by the simpler problem of minimizing the Dirichlet functional.

Plateau's problem can naturally be generalized to higher dimensions. Namely, for a given $(k-1)$-dimensional closed submanifold N of an n-dimensional Riemannian manifold M it is required to find a smooth map $f: W \to M$, where $\dim W = k$ and $f|_{\partial W}: \partial W \to N$ is a diffeomorphism that minimizes the k-dimensional volume functional. We observe that the manifold W is not assumed to be given a priori. As in the classical case, the extremals of this functional are "minimal surfaces" in the sense that their mean curvature vanishes.

Apart from the purely analytical complications connected with the increasing number of variables, in the multidimensional Plateau problem there arises another substantial difficulty, which can be explained visually as follows (see Fig. 52). The films P_1 and P_2 give two solutions of the classical Plateau problem for a contour Γ consisting of two circles. The film P can be deformed either into P_1 or into $P_2 \cup I$. In the latter case $P_2 \cup I = f(W)$, where $W = S^1 \times [0, 1]$ is a cylinder, and f is a map that takes the circle $S^1 \times t$, $t \in [\frac{1}{4}, \frac{3}{4}]$ into one of the points of I and is bijective outside the "subcylinder" $W_0 = S^1 \times [\frac{1}{4}, \frac{3}{4}] \subset W$. Cutting out from W the interior part of the submanifold W_0 and pasting $S \times \frac{1}{4}$ and $S \times \frac{3}{4}$ at a point of the circle, forming its boundary, we obtain a pair of discs, which is diffeomorphically mapped onto P_2 by means of f. This enables us to get rid of the parasitic interval I, which makes no contribution to the surface area. In the multidimensional case a similar operation may be impossible because the space obtained after the cutting out and subsequent pasting is not necessarily a manifold.

This effect justified a modification of Plateau's problem by extending the class of films that paste the given "contour" $N \subset M$. For example, we may require that the film is realized by a singular k-dimensional chain (in the usual topological sense) whose boundary is the submanifold N, also understood as a $(k-1)$-dimensional chain. In this formulation Plateau's problem was solved by Reifenberg [1960] and Almgren [1968] and it was proved that the solution is regular almost everywhere (see Fig. 52). An important contribution to this field was made by Federer and Morrey.

However, it is possible to react to the phenomenon of "flaking" described above in another way. Namely, we can imagine the image $X = f(W)$ (see above) as a stratified manifold, that is, an object of type $P_2 \cup I$ (see Fig. 52) consisting of a suitable image of pasted smooth pieces of dimension $k, k - 1, \ldots, 1, 0$. With this manifold we can associate the vector $V = (v_k, \ldots, v_0)$, called its stratified volume, where v_i is the i-dimensional volume of its i-dimensional part (stratum). We can now pose Plateau's problem as the problem of minimizing the vector V in the lexicographical sense. In this formulation, which corresponds better with the classical formulation, the problem was solved by Fomenko (see his book Fomenko [1982]), where there is an extensive bibliography.[1]

We observe that there are several other versions and formulations of the multidimensional Plateau problem.

§9. Minimal Surfaces. Bernstein's Problem

S.N. Bernstein showed that a minimal surface having the form of the graph of a smooth function $z = f(x, y)$, defined everywhere on the (x, y)-plane, is necessarily a plane. This result is analogous to the well-known elementary fact of complex analysis that a function that is bounded and holomorphic on the whole plane is a constant.

It is natural to ask whether a minimal surface of the form $x_n = f(x_1, \ldots, x_{n-1})$, where f is a smooth function defined for all $(x_1, \ldots, x_{n-1})$, is a hyperplane (Bernstein's problem). An unexpected result discovered by Bombieri, de Giorgi and Giusti [1969] asserts that Bernstein's conjecture is true when $n \leqslant 8$, but not when $n > 8$.

This fact is closely connected with the question of minimal cones. More precisely, let N be a minimal $(n - 2)$-dimensional submanifold of the standard sphere S^{n-1} of Euclidean space $\mathbb{R}^n$. Let us construct a cone $K \subset \mathbb{R}^n$ whose vertex coincides with the centre of the sphere and whose base is N. Can such a cone be an absolutely minimal film that pastes up N? Simons [1968] showed that the answer to this question is negative when $n \leqslant 7$ and positive when $n > 7$.

The given results are among those that connect local properties of the ambient Riemannian manifold and global characteristics of minimal submanifolds contained in it. Let us give an example of this kind: the fundamental group of a compact minimal submanifold of a complete Riemannian manifold of non-positive curvature is infinite (Frankel [1966]).

We also mention that any compact complex submanifold of a Kähler manifold is minimal, and it has minimal volume among all submanifolds homologous to it.

Recently the theory of minimal surfaces has been enriched by subtle new results. For a modern survey of them see the Proceedings of the 1980 Beijing

---

[1] See also the book "Minimal surfaces and Plateau's problem" by A.T. Fomenko and Dao Chong Tkhi (to be published by the American Mathematical Society). Translator's note.

Symposium on Differential Geometry and Differential Equations (Proceedings [1982]).

§10. de Rham Cohomology

The most important global characteristics of geometric structures on manifolds are usually expressed in terms of the de Rham cohomology. Let us consider this concept.

Let $\Lambda^i(M)$ denote the totality of all differential forms of degree i on a manifold M. The natural operations of adding forms and multiplying them by smooth functions $f \in C^\infty(M)$ turn $\Lambda^i(M)$ into a module over the algebra $C^\infty(M)$. In particular, $\Lambda^i(M)$ is a liear space. We shall assume that $\Lambda^i(M) = 0$ if $i < 0$. We also recall that $\Lambda^i(M) = 0$ if $i > n = \dim M$, that is, there are no non-zero forms on M whose degree is greater than the dimension of M. Forms of degree zero are functions on M, that is, $\Lambda^0(M) = C^\infty(M)$.

The differential of a function $f \in C^\infty(M)$ is a form of the first degree, that is, $df \in \Lambda^1(M)$. In other words, the operation $f \mapsto df$ is a linear operator $d: C^\infty(M) = \Lambda^0(M) \to \Lambda^1(M)$. Cartan discovered the existence of the natural analogues of this operator $d = d_i: \Lambda^i(M) \to \Lambda^{i+1}(M)$. The sequence of operators (see Ch. 3)

$$0 \to \Lambda^{-1}(M) \to \Lambda^0(M) \xrightarrow{d = d_0} \Lambda^1(M) \to \cdots \to \Lambda^i(M) \xrightarrow{d = d_i} \Lambda^{i+1}(M) \to \cdots$$

is called the *de Rham complex*, although historically it would be more accurate to call it the Cartan complex. A remarkable property of this sequence is that $d_i \circ d_{i-1} = 0$ for all i. The last property is equivalent to Im $d_{i-1} \subset$ Ker d_i, where Im A and Ker A mean the image and kernel of the operator A, respctively. They are linear spaces if the operator A is linear. In particular, Im d_{i-1} is a linear subspace of Ker d_i, and so there is defined the quotient space

$$H^i(M) = \text{Ker } d_i / \text{Im } d_{i-1},$$

called the *i-dimensional de Rham cohomology group* of the manifold M. The terminology is as follows: if $d\omega = 0$, then the form ω is said to be closed; if $\omega = d\rho$, then the form ω is called exact; the residue class $[\omega]$ of a closed form $\omega \in$ Ker d_i modulo Im d_{i-1}, that is, an element of the space $H^i(M)$, is called its cohomology class. Since Im $d_{-1} = 0$, we have $H^0(M) = \text{Ker } d_0 = \{f \in C^\infty(M) | df = 0\}$. If $df = 0$, then f is locally constant, and so it can be uniquely characterized by a vector $(c_1, \ldots, c_k, \ldots)$, where c_k is its value on the k-th connected component of M. Hence it is obvious that if N is the number of connected components, then

$$H^0(M) = \{(c_1, \ldots, c_i, \ldots)\} = \mathbb{R}^N.$$

Thus, the number $b_0(M) = \dim H^0(M)$ is the simplest global topological characteristic of M – the number of connected components. The *Betti numbers*

$$b_i(M) = \dim_\mathbb{R} H^i(M), \qquad i = 0, 1, \ldots$$

are the most important topological characteristics of M. Roughly speaking, $b_i(M)$ is the number of independent i-dimensional "cycles" on M. For example, for the standard two-dimensional torus the first Betti number $b_1(M)$ is equal to 2. This corresponds to the fact that any one-dimensional cycle, that is, a closed curve on the torus, represents the result of a p-fold circuit round a parallel of it and a q-fold circuit round a meridian of it for some integers p and q. In this sense a parallel and a meridian of the torus generate all its one-dimensional cycles, since they themselves are independent.

We note that $H^i(M) = 0$ if $i > \dim M$, since in this case $\Lambda^i(M) = 0$. It is intuitively obvious that a compact manifold has finitely many independent k-dimensional cycles. For we can prove that $b_i(M) < \infty$ if M is compact or it can be continuously deformed onto a compact subset of itself.

A fundamental fact of the theory of de Rham cohomology is the following.

Let M be a connected compact orientable manifold without boundary and let $n = \dim M$. Then fixing an orientation on M determines a canonical isomorphism $\int \colon H^n(M) \to \mathbb{R}$. A change of orientation changes the sign of this isomorphism.

The significance of this fact and the choice of the symbol $\int$ to denote a canonical isomorphism between $H^n(M)$ and $\mathbb{R}$ are determined by the following: since $\Lambda^{n+1}(M) = 0$, we have $\operatorname{Ker} d_n = \Lambda^n(M)$. Hence

$$H^n(M) = \Lambda^n(M)/\operatorname{Im} d_{n-1} = \Lambda^n(M)/d\Lambda^{n-1}(M),$$

and we can consider the composition of maps

$$\Lambda^n(M) \xrightarrow{\text{factorization}} \Lambda^n(M)/\operatorname{Im} d_{n-1} = H^n(M) \xrightarrow{\int} \mathbb{R}.$$

It turns out that the number that this composition associates with a form $\omega \in \Lambda^n(M)$ is none other than $\int_M \omega$ – the integral of the form ω over M. Thus, the operation of assigning to a form $\omega \in \Lambda^n(M)$ its cohomology class $[\omega] \in H^n(M)$ is the operation of integration up to an isomorphism $\int$. From what we have said it follows, in particular, that

$$\int_M \omega = 0 \Leftrightarrow \omega = d\rho, \qquad \rho \in \Lambda^{n-1}(M).$$

A form $\omega \in \Lambda^n(M)$ is called a *volume form* on M if $\omega_x \neq 0$ for all $x \in M$, that is, it is non-zero everywhere. There are volume forms on M if and only if M is orientable. A fact equivalent to the integral of a positive function being non-zero is that $\int_M \omega \neq 0$ if ω is a volume form.

From the formula

$$d(\omega \wedge \rho) = d\omega \wedge \rho + (-1)^p \omega \wedge d\rho, \qquad p = \deg \omega,$$

it follows that the exterior product of closed forms is closed, and the product of an exact form and a closed form is an exact form. Hence the formula

$$[\omega] \wedge [\rho] = [\omega \wedge \rho], \qquad \omega \in \ker d_i, \quad \rho \in \ker d_j$$

well defines the multiplication

$$H^i(M) \times H^j(M) \to H^{i+j}(M)$$

of cohomology classes.

We now show by a simple typical example how de Rham cohomology can be used in solving problems of differential geometry "in the large".

Problem. Can we intruduce a symplectic structure on an even-dimensional sphere S^{2n}?

Solution. Suppose that a closed form $\omega \in \Lambda^2(S^{2n})$ specifies a symplectic structure on S^{2n}. According to Darboux's lemma (see Ch. 5), in an neighbourhood of an arbitrary point $x \in S^{2n}$ we can construct a canonical coordinate system (p_i, q_i) for ω, that is, $\omega = \sum_{i=1}^{n} dp_i \wedge dq_i$. Hence it is obvious that

$$\omega^n \equiv \omega \wedge \omega \wedge \cdots \wedge \omega = n!dp_1 \wedge dq_1 \wedge \cdots \wedge dp_n \wedge dq_n,$$

and so ω^n is a volume form on S^{2n}. Hence the form ω^n is not exact. On the other hand, if $n > 1$, then $b_2(S^{2n}) = 0$, that is, any closed 2-form on S^{2n} is exact. Hence the form ω^n is also exact. We have thus arrived at a contradiction, and so S^{2n} when $n > 1$ does not admit any symplectic structure. On the other hand, any volume form on S^2 (it exists, since a sphere is orientable) specifies a symplectic structure on it.

This argument shows in fact that if $b_2(M) = 0$, then a compact manifold M without boundary cannot be endowed with a symplectic structure.

It is known that any non-singular complex projective algebraic variety is a Kähler manifold. In turn, a Kähler manifold is symplectic. From this and what we said above it follows that $b_2(M) \geqslant 1$ if M is a non-singular complex projective algebraic variety. Thus, the property of being algebraic predetermines some topological properties of a complex manifold.

This enables us to construct examples of a complex non-algebraic manifold. For example, consider the free action of the group $\mathbb{Z}$ in $\mathbb{C}^n \backslash 0$ generated by the transformation $z \mapsto 2z$. It is not difficult to see that the quotient manifold of this action is diffeomorphic to $S^{2n-1} \times S^1$ (S^k denotes the k-dimensional sphere). Since the action preserves the complex structure $\mathbb{C}^n \backslash 0$, and the projection $\mathbb{C}^n \backslash 0 \to S^{2n-1} \times S^1$ is a local diffeomorphism, we can lower the complex structure from $\mathbb{C}^n \backslash 0$ to $S^{2n-1} \times S^1$. However, $b_2(S^{2n-1} \times S^1) = 0$. Hence the compact complex manifold $S^{2n-1} \times S^1$ is not algebraic.

§11. Harmonic Forms. Hodge Theory

A Riemannian metric on a manifold enables us to introduce a scalar product in the space $\Lambda(M)$ of all differential forms on a compact orientable manifold M. This in turn allows the possibility of constructing an operator $d^*: \Lambda^k(M) \to \Lambda^{k-1}(M)$, conjugate to the operator of exterior differentiation $d: \Lambda^{k-1}(M) \to \Lambda^k(M)$, that is,

$$\langle d\omega, \rho \rangle = \langle \omega, d^*\rho \rangle, \qquad \omega, \rho \in \Lambda(M), \tag{3}$$

where the angle brackets $\langle \ , \ \rangle$ denote the scalar product. From (3) it follows

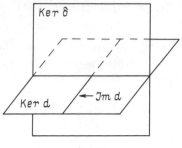

Fig. 53

that the subspaces Im d and Ker d^* are orthogonal complements of each other. If in this infinite-dimensional situation things were the same as in the finite-dimensional case, we could conclude that the orthogonal complement of the subspace Im d in Ker d coincides with Ker $d \cap$ Ker d^* (see Fig. 53). On the other hand, this orthogonal complement is none other than $H(M) = \sum_{i=0}^{n} H^i(M)$. Hence we would have proved that

$$H(M) = \text{Ker } d \cap \text{Ker } d^*, \tag{4}$$

or

$$H^i(M) = \text{Ker } d_i \cap \text{Ker } d_i^*,$$

where $d_i^*: \Lambda^i(M) \to \Lambda^{i-1}(M)$ is the restriction of d^* to $\Lambda^i(M)$. Hodge's theorem (see Spencer [1985], for example) asserts that everything happens as we have described. Hence (4) can be interpreted as follows: in every de Rham cohomology class there is exactly one form ω that satisfies the relation $d^*\omega = 0$.

Such forms, that is, forms ω for which $d\omega = d^*\omega = 0$, are called *harmonic forms*. We have thus established a one-to-one correspondence between harmonic forms and de Rham cohomology classes.

Since the operator $(d^*)^2$ is conjugate to d^2, we have $(d^*)^2 = 0$. The differential operator

$$\Delta = (d + d^*)^2 = dd^* + d^*d: \Lambda^i(M) \to \Lambda^i(M)$$

is called the *Beltrami-Laplace operator* of the Riemannian manifold M. The equality

$$\langle \Delta\omega, \omega \rangle = \langle dd^*\omega + d^*d\omega, \omega \rangle = \langle d\omega, d\omega \rangle + \langle d^*\omega, d^*\omega \rangle$$

shows that the conditions $d\omega = d^*\omega = 0$ and $\Delta\omega = 0$ are equivalent. Hence the last of these can be taken as the definition of a harmonic form, as is usually done.

If $i = 0$ (and then $\Lambda^0(M) = C^\infty(M)$), M is a Euclidean space, and $x_1, \ldots, x_n$ are rectangular coordinates, then

$$\Delta = -\left(\frac{\partial^2}{\partial x_1^2} + \cdots + \frac{\partial^2}{\partial x_n^2} \right),$$

which explains the origin of the terms Laplace operator and harmonic form.

The details of this construction are as follows. Firstly, a Riemannian metric g on M determines a scalar product in the tangent spaces $T_x M$, $x \in M$. This scalar product induces a scalar product in the space of k-multivectors $\Lambda^k T_x M$, $k = 0, \ldots, n$,

$$(\xi_1 \wedge \cdots \wedge \xi_k, \eta_1 \wedge \cdots \wedge \eta_k) = \det \| g_x(\xi_i, \eta_i) \|.$$

The scalar product in the space $\Lambda^k T_x^* M$, dual to $\Lambda^k T_x M$, is defined as the dual. If α, $\beta \in \Lambda^k(M)$, then α_x, $\beta_x \in \Lambda^k T_x^* M$ and so there is defined a function $(\alpha, \beta) \in C^\infty(M)$, whose value at the point $x \in M$ is equal to the scalar product of α_x and β_x defined above. If M is orientable and compact, then (see above)

$$\langle \alpha, \beta \rangle = \int_M (\alpha, \beta) \, dV,$$

where $dV \in \Lambda^n(M)$ is the volume form defined by the Riemannian metric under consideration. More precisely, the value of the n-multivector dV_x on vectors $\alpha_1, \ldots, \alpha_n \in T_x M$ is equal to $\pm \det \| g_x(\alpha_i, \alpha_j) \|^{1/2}$ depending on whether the orientation of the frame $(\alpha_1, \ldots, \alpha_n)$ coincides with the given orientation of M or not.

Next, we define the form $*\rho \in \Lambda^{n-i}(M)$ for all $\rho \in \Lambda^i(M)$ from the condition

$$\alpha \wedge *\rho = (\alpha, \rho) \, dV, \quad \forall \alpha \in \Lambda^i(M).$$

The map

$$* = *_i : \Lambda^i(M) \to \Lambda^{n-i}(M)$$

is an isomorphism of $C^\infty(M)$-modules and

$$*_i^{-1} = (-1)^{i(n-i)} *_{n-1}.$$

Hence it follows in turn that the operator

$$d* = d_i^* = (-1)^{ni+n+1} *d* : \Lambda^i(M) \to \Lambda^{i-1}(M)$$

is conjugate to the operator $d = d_i : \Lambda^{i-1}(M) \to \Lambda^i(M)$, that is, $\langle d_i \alpha, \beta \rangle = \langle \alpha, d_i^* \beta \rangle$.

All these definitions and constructions are from elementary algebra. The difficult part of the theory we have described consists in the following analytic assertion.

The equation $\Delta \rho = \alpha$ for a given form α is soluble if and only if α is orthogonal to the space of harmonic forms.

A subspace of harmonic forms is finite-dimensional, since every cohomology class contains at most one harmonic form. In view of this, there is an operator $H: \Lambda(M) \to \Lambda(M)$ of orthogonal projection onto this subspace. Since for any form ρ the form $\rho - H\rho$ is orthogonal to it, there is a unique solution of the equation $\Delta \alpha = \rho - H\rho$ orthogonal to the space of harmonic forms. Denoting this by $G\rho$, we obtain

$$\rho = H\rho + \Delta G\rho = H\rho + d(d^* G\rho) + d^*(dG\rho).$$

This decomposition of an arbitrary form ρ into harmonic, exact and coexact parts is the main result of Hodge theory.

The Riemannian metric canonically associated with the Kähler metric allows the possibility of developing Hodge theory on any Kähler manifold. The operators $*$, $d*$, H and G interact in a definite way with the complex and symplectic structures of the Kähler manifold, enabling us, in particular, to construct very subtle invariants of it (see § 16).

We also note that the constructions we have presented can be carried over to arbitrary elliptic complexes (Spencer [1985]). Such is the sequence of differential operators

$$0 \longrightarrow \Gamma(\pi_0) \overset{D_0}{\longrightarrow} \cdots \overset{D_{n-1}}{\longrightarrow} \Gamma(\pi_m) \longrightarrow 0$$

of fixed order acting on a section of linear bundles π_i over M, which has the following properties: 1) $D_{i+1} \cdot D_i = 0$ for all i, 2) the corresponding sequence of symbols is exact.

§ 12. Application of the Maximum Principle

The maximum principle asserts that if on a connected compact Riemannian manifold M the functions $\varphi \in C^\infty(M)$ satisfy the condition $\Delta\varphi \leq 0$, then $\varphi = $ const, and so $\Delta\varphi = 0$.

The proof of this follows from the following two remarks. Firstly, $\int_M \Delta(\psi)\,dV = 0$, $\forall \psi \in C^\infty(M)$. In fact, $\Delta\psi = d*d\psi = \pm *d*d\psi$ and $\Delta(\psi)\,dV = \pm d(*d\psi)$, and the integral of an exact form is zero. Since $\Delta\varphi \leq 0$, the integral $\int_M \Delta\phi\,dV$ can vanish only when $\Delta\varphi \equiv 0$.

Secondly, it is not difficult to verify that for $\psi \in C^\infty(M)$

$$\Delta(\psi^2) = 2\psi\Delta\psi + 2(\text{grad }\psi, \text{grad }\psi).$$

Thus, since $\Delta\varphi = 0$, we have $0 = \int_M \Delta(\varphi^2)\,dV = 2\int_M (\text{grad }\varphi, \text{grad }\varphi)\,dV$. Hence it follows that grad $\varphi \equiv 0$, that is, $\varphi = $ const.

The simplest consequence of the maximum principle is that harmonic functions on a connected compact Riemannian manifold are constants. This assertion follows also from Hodge theory, since as we have seen $H^0(M) = \mathbb{R}$.

The maximum principle enables us to obtain results "in the large", starting from various assumptions about the curvature. For example, we can show that in Euclidean space there are no compact minimal submanifolds of positive dimension. Assume for simplicity that $M \subset \mathbb{R}^{n+1}$ is a hypersurface and that $\bar{x}_i = x_i|_M$, where $(x_1, \ldots, x_{n+1})$ are rectangular coordinates in $\mathbb{R}^{n+1}$. Direct calculation shows that $\Delta\bar{x}_i = nk_iH$, where $k = (k_1, \ldots, k_{n+1})$ is the unit normal to the surface and H is its mean curvature. If M is a minimal hypersurface, then $\Delta\bar{x}_i = 0 \Rightarrow \bar{x}_i = $ const.

A simple but very effective method of using the maximum principle was suggested about 40 years ago by Bochner [1946]. Let us illustrate it by proving that $H^1(M) = 0$ if M is a compact Riemannian manifold whose Ricci curvature is non-negative everywhere and is strictly positive at one point at least.

Bearing Hodge theory in mind, it is sufficient to show that there are no non-zero harmonic 1-forms on M. To this end we use the following formula, which holds for harmonic 1-forms ρ:

$$\Delta(|\rho|^2) + 2 \sum_{i=1}^{n} |\nabla V_i(\rho)|^2 + \text{Ric}(X_\rho, X_\rho) = 0, \tag{5}$$

where $V_1, \ldots, V_n$ is an arbitrary orthonormal system of vector fields on M, Ric is the Ricci tensor, and X_ρ is the vector field metrically dual to ρ. This formula shows that if $\text{Ric}(X_\rho, X_\rho) \geqslant 0$, then $\Delta(|\rho|^2) \leqslant 0$. Hence, by the maximum principle, $|\rho|^2 = \text{const}$ and $\Delta(|\rho|^2) = 0$. Taking account of this, it follows from (5) that

$$\sum_{i=1}^{n} |\nabla V_i(\rho)|^2 = 0, \qquad \text{Ric}(X_\rho, X_\rho) = 0. \tag{6}$$

The first equality implies that $\nabla V_i(\rho) = 0$ for all i, and so $\nabla X(\rho) = 0$ for any vector field X. Hence the form ρ is zero everywhere if $\rho_x = 0$ at some point $x \in M$. Such a point x is one where the Ricci tensor is strictly positive by hypothesis. In fact, the second of the equalities (6) at this point shows that $(X_\rho)_x = 0$, and so $\rho_x = 0$. This completes the proof.

If X is a Killing vector field on M, then we have the following analogue of (5):

$$\Delta(|X|^2) + 2 \sum_{i=1}^{n} |\nabla V_i(X)|^2 + 2\text{Ric}(X, X) = 0. \tag{7}$$

If we repeat the argument just given, making the changes $\rho \to X$, (5) → (7) and $(\text{Ric} \geqslant 0) \to (\text{Ric} \leqslant 0)$ in it, we arrive at the following result.

Suppose that the Ricci curvature of a compact Riemannian manifold M is non-positive and strictly negative at one point at least; then there are no non-trivial Killing fields on M. A consequence of this is that the isometry group of a Riemannian manifold satisfying the assumptions we have just made is finite.

Let us turn our attention to the fact that the result we have given is essentially global. In fact, n-dimensional Lobachevskij space has an everywhere strictly negative Ricci tensor and, as is well known, it admits a $\frac{1}{2}n(n + 1)$-dimensional group of motions whose Lie algebra is realized as a $\frac{1}{2}n(n + 1)$-dimensional family of Killing fields on it. Factorizing this space with respect to a suitable discrete group of motions, we obtain a compact manifold with $\text{Ric} < 0$ that admits local Killing fields. Thus, under the assumptions about the Ricci curvature that we have made, non-trivial Killing fields can exist locally, but not in the large, if the manifold is compact.

The way of arguing that we have described enables us to obtain quite subtle results concerning various versions of de Rham cohomology and infinitesimal transformations of certain geometric structures. The most interesting of them concern complex geometry and Kähler geometry.

For a survey of the original achievements in this direction see Yano and Bochner [1953], and for a modern survey see Proceedings of the 1980 Beijing

Symposium on Differential Geometry and Differential Equations (Proceedings [1982]).

§ 13. Curvature and Topology

In this section we assume that all the Riemannian manifolds are complete. Here we consider the question of how far the curvature of a Riemannian manifold predetermines its shape, or in other words can a given manifold be endowed with a Riemannian metric whose curvature (sectional, Ricci or scalar) varies within established limits?

The first important results in this direction can be obtained by starting from the following very simple intuitive arguments. Let $\gamma = \gamma(s)$ be a geodesic referred to the natural parameter, and let $A = \gamma(a)$, $B = \gamma(b)$. We call a smooth family of curves $\gamma_t(s)$ a geodesic variation of it if for any t the curve $\gamma_t(s)$ is a geodesic and $\gamma_0 = \gamma$. If the curves $\alpha(t) = \gamma_t(a)$ and $\beta(t) = \gamma_t(b)$ passing through the points A and B respectively are orthogonal to γ, then the length $d(t)$ of the arc of the geodesic $\gamma_t(s)$ between the points $\alpha(t)$ and $\beta(t)$ in the first approximation does not depend on t as $t \to 0$. This is a consequence of the formula for the first variation. If $\alpha(t) \equiv A$ and $\beta(t) \equiv B$, then $d(t) = d(0) + o(t)$. The same conclusion holds if $\alpha'(t) = 0$, $\beta'(t) = 0$. This is a consequence of the formula for the second variation. If there is a geodesic variation of this kind, then the points A and B are said to be *conjugate along the curve* γ. The North and South Poles of the standard sphere are the simplest example of conjugate points. Consider the situation shown in Fig. 54. The arc AD of the geodesic γ is not a minimal geodesic joining A and D if D is sufficiently close to A. For if the point C is sufficiently close to B and CD is the only minimal geodesic joining C and D, then the quantity $l(t) = d(B, C) + d(B, D) - d(C, D)$ is positive and has second order of smallness with respect to t. Here $d(P, Q)$ denotes the distance between points P and Q, that is, the length of the minimal geodesic joining them. Hence it follows that the quantity $d(0) + d(B, D) - [d(A, C) + d(C, D)]$ is positive and coincides with $l(t)$ up to small quantities of the third order in t if the arc of γ_t is minimal between A and C, and greater than $l(t)$ otherwise for sufficiently small t.

The structure and character of conjugate points are described in the framework of the theory of Jacobi fields. Namely, if $\gamma_t(s)$ is a geodesic variation, then the vector field along γ formed by the tangent vectors of the curves $\alpha_s(t)$: $t \to \gamma_t(s)$ at points $t = 0$ is called a *Jacobi field*. For example, rotating a meridian γ of the

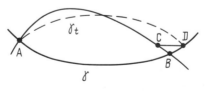

Fig. 54

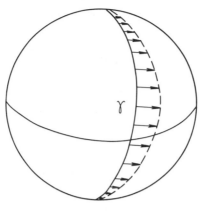

Fig. 55

sphere, fixed at the poles, we obtain a geodesic variation whose Jacobi field is shown in Fig. 55. Any Jacobi field J satisfies the equation

$$\ddot{J} + R(V, J)V = 0, \qquad (8)$$

where V is the unit velocity field along γ and a dot denotes differentiation with respect to the natural parameter. Any Jacobi field can be decomposed into a sum of such fields $J = J_1 + J_2$, where J_1 is perpendicular to γ and J_2 touches γ. The field J_2 corresponds to a trivial geodesic variation when the geodesics $\gamma_t(s)$ represent the reparametrized geodesic γ. Henceforth, when we speak about Jacobi fields, we shall assume that $J = J_1$. With this stipulation, B is conjugate to A along γ if there is a non-zero Jacobi field J such that $J_a = J_b = 0$.

If the curvature is constant and equal to k, the equation (8) takes the form $\ddot{J} + kJ = 0$. Hence the Jacobi field looks like this, assuming that $J_a = 0$ and the point A corresponds to the parameter $s = 0$:

$$J = \alpha \sin \sqrt{ks}P, \quad k > 0; \qquad J = \alpha_s P, \quad k = 0;$$

$$J = \text{sh}(\sqrt{-ks})P, \qquad k < 0,$$

where P is the parallel vector field along γ perpendicular to it.

Equation (8) has the form of the Sturm-Liouville equation and we can prove a comparison theorem of Sturm type for it. In particular, if the curvature $k \geqslant k_0 > 0$, then successive conjugate points along an arbitrary geodesic γ are at a shorter distance from each other than in the case of a manifold with curvature equal to k_0, that is, $\pi/\sqrt{k_0}$. Hence if $k \geqslant k_0 > 0$, then the length of the minimal geodesic between any two points is at most $\pi/\sqrt{k_0}$. This proves that in this case the manifold M is compact and its fundamental group is finite. The latter is obvious from the fact that for the universal covering manifold $\tilde{M}$ of M, which is locally isometric to it, we have the same lower bound on the curvature. Hence M is compact and the covering map is finite-sheeted.

In this context we recall a subtler result of Milnor and Gromov: if the sectional curvature of a manifold M is non-negative, then its fundamental group can be generated by at most $2.5^{n/2}$ elements and has polynomial growth. This means that the number of distinct elements of this group representable by words of length at most s in any fixed set of generators of it does not exceed cs^N, where $c, N \in \mathbb{Z}$.

If the sectional curvature $k \leqslant 0$ on M, then from the comparison theorem it follows that there are no pairs of conjugate points on M. This shows that the universal covering for M is diffeomorphic to $\mathbb{R}^n$. More precisely, the exponential map with centre at any point $x \in M$ is a covering map. This is obvious from the fact that the set of singular points of the map $\exp_x : T_x M \to M$ has the form $\exp_x^{-1}(S)$, where S is the set of points conjugate to x along all possible geodesics that emanate from it. Thus, Riemannian manifolds of non-positive curvature (like smooth manifolds) are quotient spaces of a Euclidean space with respect to discrete groups of diffeomorphisms.

As a contrast to the case $k > 0$, we recall the following result of A.S. Schwarts: if the sectional curvature of a compact manifold M is negative everywhere, then its fundamental group has exponential growth. This means that the number of distinct elements representable by words of length at most s in some finite set of generators is greater than e^m for any $m \in \mathbb{R}$.

The proof of this assertion is based on an estimate of the growth of the volume of a geodesic ball as a function of its radius. Such estimates play an important role in the questions discussed in this section, as well as estimates of the value of the radius for which a geodesic ball is convex. Convexity here means that the minimal geodesic joining any two points of this ball belongs to the ball. An elegant example of the use of this estimate is the following theorem of Weinstein.

Let us fix a number h such that $1 \geqslant h > 0$. Then there are finitely many pairwise non-diffeomorphic smooth manifolds of a given even dimension $n > 4$ admitting a Riemannian metric whose sectional curvature varies between h and 1. If $n = 4$, diffeomorphism above must be replaced by homeomorphism.

The idea of the proof is simple. We first show that a manifold satisfying the conditions of the theorem can be covered by at most c convex open balls, where $c = c(n, h)$ is a universal constant. The intersection of any number of balls of this covering is again convex. For this reason the original manifold is homotopically equivalent to a nerve (that is, roughly speaking, a combinatorial diagram) of this covering. It now remains to observe that there are finitely many combinatorial diagrams with a fixed number of elements, and there are finitely many pairwise non-diffeomorphic (non-homeomorphic if $n = 4$) manifolds homotopically equivalent to the given one. The latter assertion is a subtle result of differential topology.

Many powerful results are obtained by applying the comparison theorem of Toponogov [1959] and some variants of it. In theorems of this kind we compare, for example, the angles of a geodesic triangle on a Riemannian manifold whose sectional curvature is bounded below by a constant h and the angles of a geodesic triangle with the same lengths of sides on a manifold of constant curvature k. Here are two unexpected results obtained by using the theory of comparison.

1 (V.A. Toponogov). Suppose that on a Riemannian manifold M of non-negative sectional curvature there is a complete geodesic that minimizes the

distance between any two of its points. Then M is the Cartesian product of some Riemannian manifold N and $\mathbb{R}^1$.

2 (Praismann). Let M be a compact Riemannian manifold of negative curvature. Then any two elements of its fundamental group commute if and only if they belong to the same cyclic subgroup.

Hence it follows that the direct product of two compact manifolds (for example, a torus) cannot be endowed with a Riemannian metric of negative curvature.

Finally, let us state the famous "theorem on the sphere", whose proof also relies substantially on the comparison theorem.

Let M be a simply-connected Riemannian manifold whose sectional curvature k satisfies the inequalities $1/4 < k \leqslant 1$. Then M is homeomorphic to a sphere.

At present there are a number of results like the theorem on the sphere in Riemannian and Kählerian geometry. The following very difficult theorem of Gromov is one of the examples.

There is a constant $\varepsilon = \varepsilon(n) > 0$ such that any compact Riemannian manifold whose sectional curvature k satisfies the inequalities

$$-\varepsilon/d(M)^2 \leqslant k \leqslant \varepsilon d(M)^2,$$

where $d(M)$ is the maximum distance between points of the manifold, is diffeomorphic to the quotient space of a nilpotent Lie group with respect to a discrete subgroup of it.

The reader can find fairly complete information about the questions discussed above in Cheeger and Ebin [1975].

Until comparatively recently hardly anything was known about the connection between the topology of a Riemannian manifold and various restrictions on its scalar curvature. For example, is there a metric of positive scalar curvature on an n-dimensional torus? A negative answer to this question was obtained by Schoen and Yau in 1979. In their paper (Schoen and Yau [1979]), apart from this they described a wide class of manifolds that do not admit a metric of positive scalar curvature. These authors were also very successful in describing closed three-dimensional manifolds of positive scalar curvature. Namely, they proved (Schoen and Yau [1982]) that any such manifold is the connected sum of a certain number of copies of the manifold $S^2 \times S^1$ and manifolds that have a finite fundamental group. It is very likely that any such manifold is the quotient space of the standard sphere S^3 with respect to a discrete subgroup of its rotation group $SO(4)$.

§ 14. Morse Theory

What is usually called Morse theory today is a method of investigating the structure of a smooth manifold, finite- or infinite-dimensional, based on the study of the singular points of a smooth function defined on it. As well as in differential geometry, it has been successfully used in differential topology and global analysis. In the simplest case it looks like this (see the next paragraph).

Suppose that $f \in C^{\infty}(M)$. A point $a \in M$ is called a *singular point of the function* f if the differential $d_a f$ of the function f at this point vanishes. If $(x_1, \ldots, x_n)$ is a local coordinate system close to a, then what we have said means that $\partial f / \partial x_i(a) = 0$, $i = 1, \ldots, n$. A singular point a is said to be non-degenerate if the Hessian

$$\text{Hess}_a f = \det \| \partial^2 f / \partial x_i \partial x_j(a) \|$$

is non-zero. In this case it is not difficult to find a system of coordinates $(y_1, \ldots, y_n)$ with origin at a in which f is expressed in the form

$$f = y_1^2 + \cdots + y_s^2 - y_{s+1}^2 - \cdots - y_n^2 + \text{const}$$

(the Morse lemma). The number $n - s$ is called the *index of the point a*. We now put

$$M(f, c) = \{x \in M \,|\, f(x) \leqslant c\}.$$

The number c is called a *critical value* of f if there is a singular point $a \in M$ such that $f(a) = c$. If c is non-critical, then by the implicit function theorem $M(f, c)$ is a manifold (possibly empty) with boundary consisting of points $x \in M$ such that $f(x) = c$. The aim of Morse theory is to trace the process of changing the structure of "manifold" $M(f, c)$ as c grows, and on this basis to make deductions about the structure of M itself.

The key argument that enables us to realize this desire constructively is based on the following remark. Suppose that the manifold M can be endowed with a Riemannian metric g in such a way that the vector field grad f with respect to this metric is complete. If, for example, all the sets $M(f, c)$ are compact, then an arbitrary metric is of this kind. Suppose that the interval $[c, d]$, $c < d$, consists of non-critical values of f. Then the manifolds $M(f, c + \varepsilon)$ and $M(f, d - \varepsilon)$, where $\varepsilon > 0$ is sufficiently small, are diffeomorphic. In fact, the one-parameter group of diffeomorphisms $A_t \colon M \to M$ defined in this case by the vector field λ grad f, where the smooth function λ is chosen in such a way that $\lambda(x) = 1$ if $f(x) \notin (c, d)$ and $\lambda(x) = |\text{grad } f|^{-1}$ if $x \in [c + \varepsilon, d - \varepsilon]$, "inflates" $M(f, c + \varepsilon)$ to $M(f, d - \varepsilon)$ in "time" $t = d - c - 2\varepsilon$ (see Fig. 56). Thus, if the critical points of f are situated discretely, it remains to investigate what happens to $M(f, c)$ when c jumps over one critical value.

It is simplest to trace this when for a critical value c_0 there is only one singular point x_0 for which $f(x_0) = c_0$, and this point is non-degenerate.

In this case the manifold $M(f, c_0 + \varepsilon)$ (where ε is sufficiently small) is diffeomorphic to the manifold $M(f, c_0 - \varepsilon)$, to whose boundary there is pasted a "*handle*" *of index* $n - s$. The handle itself is the product $D^{n-s} \times D^s$ of an $(n - s)$-dimensional ball and an s-dimensional ball. It is pasted to $M(f, c_0 - \varepsilon)$ along the part of its boundary having the form $\partial D^{n-s} \times D^s$. For example, if $n = 3$ and $s = 2$, the operation of pasting a handle of index 1 is shown in Fig. 57. Since the boundary of $M(f, c_0 + \varepsilon)$ is smooth (see Fig. 57), after pasting the handle to $M(f, c_0 - \varepsilon)$ we need to smooth the angle thus formed (see Fig. 58).

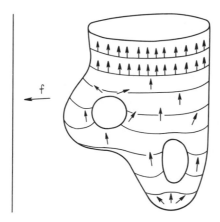

Fig. 56

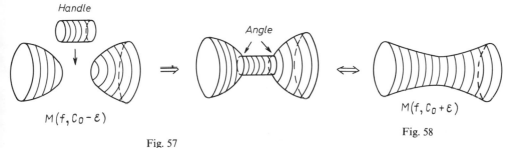

Fig. 57

Fig. 58

$M(f, c_0 - \varepsilon)$

$M(f, c_0 + \varepsilon)$

The passage through a critical value c_0 can also be described in the case when the totality of singular points $S = \{x \in M \mid f(x) = c_0\}$ forms a k-dimensional submanifold, and the quadratic form with matrix

$$\|\partial^2 f/\partial x_i \partial x_j(a)\|$$

on $T_a(M)$ is non-degenerate on the orthogonal complement to the subspace $T_a(S) \subset T_a(M)$. In this case the process of passing through a critical value looks like that of pasting to $M(f, c_0 - \varepsilon)$ a k-parameter family of $(n - k)$-dimensional handles of index $n - s$.

Example. Consider a torus T lying above the plane $z = 0$ and having thickness 1 (see Fig. 59). Suppose that $f = z|_T, c_0 = 1$. Then S is a circle, and what is pasted to $M(f, 1 - \varepsilon)$ (a fragment is shown in Fig. 59) is a 1-parameter family of one-dimensional handles of index 1.

In 1935 Pontryagin used Morse theory to calculate the cohomology of compact Lie groups. For example, for the group $SO(n)$ of orthogonal matrices $\|g_{ij}\|$ with unit determinant he put $f = g_{11}$. In this case the function f has two critical points ± 1 (a maximum and a minimum), and the corresponding manifolds of singular points represent the subgroup $SO(n - 1) \subset SO(n)$ and one of its cosets respectively.

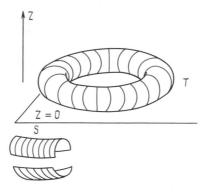

Fig. 59

As an example that illustrates the connection between the structure of a manifold M as a whole and the structure of the singular points of a smooth function on it we mention the so-called Morse inequalities

$$b_i - b_{i-1} + \cdots \pm b_0 \leqslant k_i - k_{i-1} + \cdots \pm k_0,$$

where $b_i = b_i(M)$ is the i-th Betti number of M, and k_i is the number of singular points of index i. We assume that all the singular points are non-degenerate. Moreover, when $i = n$ the Morse inequality becomes an equality.

Example. Let $S^2 = \{x^2 + y^2 + z^2 = 1\}$ be a two-dimensional sphere and let $f = z|_{S^2}$. Then $k_0 = k_2 = 1$, $k_1 = 0$. Since $b_0 = 1$, from the Morse inequality $b_1 - b_0 \leqslant -1$ we find that $b_1 = 0$, and from the equality $b_2 - b_1 + b_0 = 2$ we find that $b_2 = 1$.

A classic example of the application of Morse theory is the study of the loop space $\Omega(M)$ on a Riemannian manifold M. Properly speaking, Morse developed his theory to study this space (see Morse [1934]). We recall that a loop on M is a curve (that is, a map $\gamma: [0, 1] \to M$) starting and ending at a given point $a \in M$. On $\Omega(M)$ there is defined a smooth action functional

$$\Omega(M) \ni \gamma(t) \mapsto \int_0^1 (\dot{\gamma}(t), \dot{\gamma}(t)) \, dt,$$

where $\dot{\gamma}(t)$ is the tangent vector to $\gamma(t)$, and $(\cdot, \cdot)$ is the Riemannian scalar product. The loop γ is a singular point of this functional if and only if it is a geodesic. All that we have said remains true, after some obvious reformulations, for the space of curves joining two given points of M.

In this infinite-dimensional situation we can also define the concept of non-degeneracy of a singular point and its index. It turns out that the index of a geodesic loop γ is equal to the number of points on γ conjugate to the initial point a, counted with their multiplicities. In particular, the index is always finite. As above, we can describe the operation of passing through an isolated non-degenerate singular point of the action functional as pasting a handle $D^\lambda \times D^\infty$

along the part $\partial D^\lambda \times D^\infty$ of its boundary, where λ is the index of the corresponding geodesic loop. This operation can also be investigated in the case when there is a whole manifold of geodesic loops on which the action functional takes a certain constant value. The simplest example of this situation is the family of great circles on a sphere emanating from one point. Here is another famous example.

Example. On the group $SU(n)$ of all unitary matrices with determinant 1 we specify an invariant metric, assuming that for two matrices A and B of its Lie algebra, which we identify with the tangent space to the identity, it has the form

$$(A, B) = \text{Re}(\text{Tr } AB^*),$$

where B^* is the Hermitian conjugate matrix of B, and Re λ means the real part of the number λ. Let $E \in SU(n)$ be the unit matrix. It is not very difficult to show that the manifold of minimal geodesics joining E and $-E$ in $SU(2m)$ is diffeomorphic to the Grassmann manifold of m-dimensional complex subspaces of $\mathbb{C}^{2m}$, and the indices of all the remaining geodesics are at least $2m + 2$. This enabled Bott, using standard arguments of homotopic topology, to establish his famous "periodicity theorem", which asserts that the homotopy groups of the infinite unitary group are trivial for even dimensions and isomorphic to $\mathbb{Z}$ for odd dimensions (that is, they have period 2). Here we understand the infinite unitary group as the direct limit of the sequence

$$U(1) \subset \cdots \subset U(n) \subset U(n + 1) \subset \cdots.$$

The reader can find an informal presentation of the main ideas of Morse theory and new applications of it in Bott [1982].

The study of finite-dimensional manifolds and various spaces of curves on them was the object of Morse theory from the moment of its inception in the 20's and 30's.

Significant progress in the theory of multidimensional minimal surfaces and the theory of harmonic maps made possible the study of spaces of maps of one manifold into another by means of Morse theory. We can expect further interesting results in this direction.

§15. Curvature and Characteristic Classes

In Chapter 2 we mentioned the Gauss-Bonnet formula, the Weyl formula and the Steiner formula, which connect the global properties of manifolds with their local differential-geometric structure. The theory of characteristic classes gives a universal method for obtaining results of this kind.

15.1. Bordisms and Stokes's Formula. The most effective method of calculating integrals is based on Stokes's formula. Bordancy is a dual geometrical concept; it works successfully in a pair having it. Roughly speaking, two manifolds endowed with a geometric structure are *bordant* if they are the boundary of some manifold (a "film") of dimension one higher, clothed with the same geometric structure. The operations of direct product and disconnected sum turn the set of classes of bordant manifolds into a ring. In those cases when the generators of this ring are known, the calculation of integrals, and also the verification of various integral formulae, can be carried out on these generators alone.

Let us illustrate this approach by proving the Gauss-Bonnet formula (see Ch. 2, 2.7).

$$\frac{1}{2\pi} \int_{M_g} \omega = \chi(M_g). \tag{9}$$

Here $\omega\,(= K\,d\sigma)$ is the Gaussian curvature form, and $\chi(M_g) = 2 - 2g$ is the Euler characteristic of an oriented complex surface of genus g.

We recall (see Ch. 2, 2.7) that $\omega = \gamma^*(\sigma)$, where $\gamma: M_g \to S^2$ is the Gaussian or spherical map, and $\sigma \in \Lambda^2(S^2)$ is the standard area form on the sphere.

We observe also that specifying a unit field (not necessarily normal) on a surface is equivalent to specifying a map of it onto the unit sphere.

We say that a compact surface M is *clothed* if a smooth field $v: M \ni x \mapsto N(x) \in S^2$ of unit vectors is fixed on it. With any clothed surface M we can associate the integral $\int_M v^*(\sigma)$, which in the case of a normal clothing represents the left-hand side of (9).

Bearing in mind the application of Stokes's formula to the calculation of these integrals, we say that two oriented compact surfaces M and M', clothed by fields of unit vectors v and v', are *bordant* if they bound a 3-dimensional domain U in $\mathbb{R}^3$ clothed with a field of unit vectors $N: U \to S^2$ in such a way that $N|_M = v$, $N|_{M'} = v'$ (see Fig. 60).

The map $N: U \to S^2$ defines on U a closed differential 2-form $N^*(\sigma)$. Applying Stokes's formula to it, we obtain

$$0 = \int_U dN^*(\sigma) = \int_{\partial U} N^*(\sigma) = \int_M \omega - \int_{M'} \omega'.$$

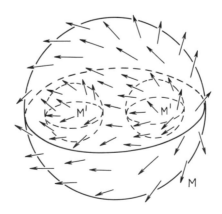

Fig. 60

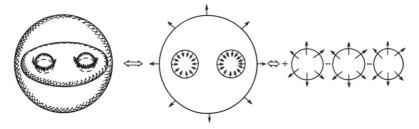

Fig. 61

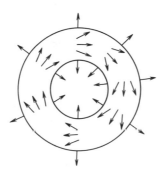

Fig. 62

Consequently, the integral $\int_M \omega$ is a bordism invariant of the clothed surfaces. This remark can form the basis of a proof of the Gauss-Bonnet formula. For this it is sufficient to verify that the normal field on a surface of genus g is bordant to the system of normal vector fields on a system of spheres (see Fig. 61) exterior to the outer sphere and interior to the inner sphere, and that the integral of the field of outward normals over a sphere of arbitrary radius, in which we are interested, is equal to 2.

It is not very easy to describe the field of unit vectors between the surface M_g and the system of spheres. For a torus ($g = 1$) instead of this it is sufficient to show that the field of normal vectors can be extended to a field of unit vectors inside the domain bounded by it.

How to do this is clear from Fig. 62, where the equatorial section of the torus is shown.

In this sketch of the proof of (9), bordancy arguments enable us to reduce the calculation of the integral of interest to us to one special case – the field of unit normals on the standard sphere.

The proof by Hirzebruch of the Riemann-Roch theorem and the original proof of Atiyah and Singer of the theorem about the index of elliptic operators (see § 16) follow this scheme directly.

15.2. The Generalized Gauss-Bonnet Formula. In the multidimensional case the Gauss-Bonnet formula enables us to calculate the Euler characteristic

$$\chi(M) = \sum_{i=0}^{n} (-1)^i b_i(M)$$

of a compact oriented n-dimensional manifold M in terms of the curvature form of the Levi-Civita connection of some Riemannian metric on it.

To generalize (9) to multidimensional manifolds, we analyse the link between the curvature form and the Gaussian curvature form.

The curvature form Ω of the Levi-Civita connection can be regarded as a differential 2-form on the manifold of orthonormal coframes with values in the Lie algebra $so(n)$ of skew-symmetric matrices (see Ch. 7, 3.3). When $n = 2$ this $so(2)$-valued form is

$$\Omega = \begin{bmatrix} 0 & \omega \\ -\omega & 0 \end{bmatrix},$$

where ω is a scalar differential 2-form.

The Gaussian curvature form on the left-hand side of (9) is obtaind by using the construction presented in 3.6 of Ch. 6 to the $SO(2)$-invariant function

$$\mathrm{pf}\colon so(2) \ni \begin{bmatrix} 0 & \lambda \\ -\lambda & 0 \end{bmatrix} \mapsto \lambda \in \mathbb{R}$$

on the Lie algebra $so(2)$, called the *Pfaffian*.

In the multidimensional case $n = 2m$ the *Pfaffian of a matrix $A \in so(2m)$* is the number defined by the formula

$$\omega_A^m = \mathrm{pf}(A) \cdot \mu,$$

where $\omega_A(\xi, \eta) = \langle A\xi, \eta \rangle$ is the exterior 2-form corresponding to the skew-symmetric matrix A, and μ is the oriented volume form, defined by the standard Euclidean metric of the sphere $\mathbb{R}^{2m}$.

We note that $\mathrm{pf}(A)$ is an $SO(2m)$-invariant homogeneous polynomial of degree m on the Lie algebra $so(2m)$. The differential $2m$-form $\mathrm{p\widehat{f}(\Omega)} \in \Lambda^{2m}(M)$ defined by the curvature form Ω of the Levi-Civita connection (see Ch. 6, 3.6) on an oriented $2m$-dimensional Riemannian manifold M is called the *Gaussian curvature form*.

The generalized Gauss-Bonnet formula for even-dmensional compact orientable manifolds has the form

$$\left(\frac{1}{2\pi}\right)^m \int_M \mathrm{p\widehat{f}(\Omega)} = \chi(M),$$

where $2m = \dim M$.

The link between the differential-geometric and topological characteristics of a manifold, established by this formula, can be used in various directions. Here are some examples.

1) On a compact oriented two-dimensional manifold of genus $g \geqslant 1$ there are no metrics of positive curvature. In other words, a compact orientable two-dimensional manifold of positive curvature is diffeomorphic to a sphere.

2) (Chern). If a compact orientable 4-dimensional manifold M has non-positive sectional curvature, then $\chi(M) \geqslant 0$.

15.3. Weil's Homomorphism. The arguments presented above can naturally be generalized to arbitrary G-bundles. For each such bundle $\pi: P \to M$ over a manifold M the curvature form Ω of a connection H is a $\mathfrak{G}$-valued differential 2-form on the bundle space P. According to §3.6 of Ch. 6, with each homogeneous G-invariant polynomial of degree k on the Lie algebra $\mathfrak{G}$ of the group G we can associate a closed $2k$-form $\widehat{f(\Omega)} \in \Lambda^{2k}(M)$. Its de Rham cohomology class will be denoted by $f(\Omega)$. Let $I(\mathfrak{G})$ be the graded algebra of G-invariant polynomials on the algebra $\mathfrak{G}$. The map

$$\omega_H: I(\mathfrak{G}) \ni f \to f(\Omega) \in H^*(M)$$

is called the *Weil homomorphism.*

Theorem. *The Weil homomorphism does not depend on the choice of connection in the principal bundle.*

Algebras of invariant polynomials of the classical Lie groups are constructed as follows.

1) $G = U(n)$ is the Lie group of unitary matrices of order n, and $\mathfrak{G} = U(n)$ is the Lie algebra of skew-Hermitian matrices. The algebra $I(\mathfrak{G})$ is generated by homogeneous polynomials $P_1, \ldots, P_n$, deg $P_j = j$, defined by

$$\det(\lambda + iA) = \lambda^n - P_1(A)\lambda^{n-1} + \cdots + (-1)^{n-1}P_{n-1}(A)\lambda + (-1)^nP_n(A).$$

2) $G = SO(2n)$ is the Lie group of orthogonal matrices of order $2n$ with determinant 1; $so(2n)$ is the Lie algebra of skew-symmetric matrices. The algebra $I(\mathfrak{G})$ is generated by homogeneous polynomials $P_1, \ldots, P_{n-1}$, $Q_n = pf$, deg $P_j = 2j$, $1 \leqslant j \leqslant n - 1$, defined by

$$\det(\lambda - A) = \lambda^{2n} + \lambda^{2(n-1)}P_1(A) + \cdots + \lambda^2 P_{n-1}(A) + (pf(A))^2.$$

3) $G = SO(2n - 1)$, and the algebra $I(\mathfrak{G})$ is generated by polynomials $P_1, \ldots, P_{n-1}$ defined by

$$\det(\lambda - A) = \lambda^{2n-1} + \lambda^{2n-3}P_1(A) + \cdots + \lambda P_{n-1}(A).$$

4) $G = O(n)$ is the Lie group of all orthogonal matrices. The algebra $I(\mathfrak{G})$ is generated by polynomials $P_1, \ldots, P_m$, where $n = 2m$ or $n = 2m + 1$, defined by

$$\det(\lambda - A) = \lambda^n + \lambda^{n-2}P_1(A) + \cdots.$$

5) $G = Sp(2n)$ is the compact symplectic Lie group consisting of unitary matrices $X \in U(2n)$ such that $^tXJX = J$, where $J = \begin{bmatrix} 0 & E \\ -E & 0 \end{bmatrix}$, and E is the unit

$n \times n$-matrix. The algebra $I(\mathfrak{G})$ is generated by polynomials $P_1, \ldots, P_n$, deg $P_j = 2j$, defined by

$$\det(\lambda + iA) = \lambda^{2n} + (-1)P_1(A)\lambda^{2(n-1)} + \cdots + (-1)^n P_n(A).$$

In all the cases listed above the polynomials $\{P_j\}$ are algebraically independent.

15.4. Characteristic Classes. The cohomology classes $f(\Omega)$ on a manifold M corresponding to invariant polynomials $f \in I(\mathfrak{G})$ by virtue of the Weil homomorphism are called *characteristic classes of the given G-bundle*. Applying the Weil homomorphism to the generators of the algebras $I(\mathfrak{G})$ of the classical Lie groups, we obtain the characteristic classes of Chern and Pontryagin.

1) Let $\pi: E(\pi) \to M$ be the complex vector bundle associated with a principal $GL(n, \mathbb{C})$-bundle. The choice of Hermitian structure in the bundle π, that is, a Hermitian metric in fibres of this bundle, determines a reduction of the structure group $GL(n, \mathbb{C})$ to the subgroup $U(n)$. Let Ω be the curvature form of an arbitrary connection of the principal $U(n)$-bundle, and $P_1, \ldots, P_n$ the homogeneous polynomials defined in (1) of 15.3. The cohomology class $c_k(\pi) = P_k\left(\dfrac{1}{2\pi}\Omega\right) \in H_2^k(M)$ is called the *k-th Chern class of the bundle* π.

2) Let $\pi: E(\pi) \to M$ be a real vector bundle. As above, the choice of metric in the bundle π determines a reduction of the corresponding principal $GL(n, \mathbb{R})$-bundle to the principal $O(n)$-bundle with curvature form Ω. The cohomology classes $p_k(\pi) \in H^{4k}(M)$, $p_k(\pi) = P_k\left(\dfrac{1}{2\pi}\Omega\right)$, where $P_1, P_2, \ldots$ are the polynomials defined in (4) of §15.3, are called the *Pontryagin classes of the bundle* π. These classes are connected with the Chern classes of the complexified bundle $\pi^{\mathbb{C}}$ by the formula $p_k(\pi) = (-1)^k c_{2k}(\pi^{\mathbb{C}})$.

If, in addition, the bundle π is oriented and even-dimensional, then the cohomology class $e(\pi) \in H^n(M)$, $n = \dim \pi$, corresponding to the Pfaffian Q_n (see (2) of §15.3) is called the *Euler class of the bundle* π.

The characteristic classes of the tangent bundle are called the *characteristic classes of the manifold*.

15.5. Characteristic Classes and the Gaussian Map. Consider a compact oriented manifold M of dimension n embedded in a Euclidean space of sufficiently high dimension. Carrying over the tangent planes to M in a parallel way to the origin, we obtain a map $\gamma: M \to G_{N,n}^0$ of M into the Grassmannian of n-dimensional oriented subspaces of $\mathbb{R}^N$. This is called the *Gaussian map*; when $N = 3$ and $n = 2$ it coincides with the Gaussian map considered in Ch. 2, 2.7.

The cohomology ring of a manifold $G_{N,n}^0$ for sufficiently large values of N is generated by elements $p_1, p_2, \ldots, p_k \in H^{4k}(G_{N,n}^0)$ – the Pontryagin classes of the tautological bundle over $G_{N,n}^0$, and also by its Euler class if n is even. The images of these classes under the Gaussian map are the Pontryagin classes (and the Euler class) of the manifold M.

For an arbitrary compact Lie group G the role of the Grassmannian $G_{N,n}^0$ is played by the so-called classifying space B_G, whose cohomology ring is isomorphic to $I(\mathfrak{G})$. The Weil homomorphism is an analogue of the cohomology homomorphism defined by the Gaussian map.

§16. The Global Geometry of Elliptic Operators

To investigate not objects themselves, but the maps (morphisms) linking them, is a methodologically useful point of view, and is a consequence of the general category-theoretic approach in mathematics. In the context of interest to us it implies that the centre of gravity must be carried over to the geometry of differential operators. An illustration of the fruitfulness of this step is the modern theory of singularities of smooth maps, which can be treated as the local differential geometry of operators of order zero.

The theory of the index of elliptic operators is a remarkable model of a global theory of this kind.

Let $(E(\alpha), M, \alpha), (E(\beta), M, \beta)$ be a pair of vector bundles on a compact manifold M, and let $D: \Gamma(\alpha) \to \Gamma(\beta)$ be a linear elliptic operator. It is known that its kernel Ker D and cokernel Coker D are finite-dimensional linear spaces. The *index of the operator D* is the number

$$i(D) = \dim \text{Ker } D - \dim \text{Coker } D.$$

An important property of the index is that it does not change under a smooth deformation of the operator D. This led I.M. Gel'fand to the conjecture that the index can be expressed in terms of the topological invariants of the pair (M, D). These invariants were discovered by Atiyah and Singer. Their classic result – the "index theorem" – will be discussed below.

16.1. The Euler Characteristic as an Index. Before stating the Atiyah-Singer index theorem we consider a simple example.

Let M be a compact n-dimensional Riemannian manifold. Let $\Omega^k(M) = \Lambda^k(M) \otimes \mathbb{C}$ denote the space of complex-valued differential k-forms. The operators d, $*$, d^*, $\Delta = dd^* + d^*d$ defined in §11 in the space of real forms $\Lambda^*(M)$ can be extended by linearity to operators in the space $\Omega^*(M)$ of complex-valued forms. Hodge's theorem establishes an isomorphism between the space $H^k(M, \mathbb{C})$ of the complex de Rham cohomology arising in this way and the space $H^k = \ker \Delta | \Omega^k(M)$ of harmonic complex k-forms. Using this isomorphism, we can express the Euler characteristic $\chi(M)$ of the manifold M in terms of the dimension of spaces of harmonic forms:

$$\chi(M) = \sum_{i=0}^{n} (-1)^i \dim H^i = \sum_i \dim H^{2i} - \sum_i \dim H^{2i+1}. \tag{10}$$

The first term in (10) is equal to the dimension of the kernel of the differential operator $D = d + d^*$, which acts from the space $\Omega_e(M)$ of forms of even degree into the space $\Omega_0(M)$ of forms of odd degree, and the second term is equal to the dimension of the kernel Ker $D^* =$ Coker D of the ajoint operator $D^*: \Omega_0(M) \to \Omega_e(M)$. Hence (10) can be written in the form $i(D) = \chi(M)$.

16.2. The Chern Character and the Todd Class. We need these concepts to formulate the general index theorem. Let $\alpha: E(\alpha) \to M$ be a complex vector bundle over M, and $c_1(\alpha), \ldots, c_m(\alpha)$ its Chern classes. We introduce formal variables $x_1, \ldots, x_m$ so that

$$1 + c_1(\alpha) + \cdots + c_m(\alpha) = (1 + x_1) \cdots (1 + x_m).$$

In other words, we shall assume that the Chern classes are elementary symmetric polynomials in the variables $x_1, \ldots, x_m$. Then every symmetric formal series in $x_1, \ldots, x_m$ can be uniquely represented as a formal series in the Chern classes. In particular, this is true for the *Chern character*

$$\mathrm{ch}(\alpha) = 1 + \mathrm{ch}_1(\alpha) + \cdots + \mathrm{ch}_m(\alpha) = e^{x_1} + \cdots + e^{x_m}$$

and the *Todd class*

$$\mathrm{td}(\alpha) = 1 + \mathrm{td}_1(\alpha) + \cdots + \mathrm{td}_m(\alpha) = \frac{x_1}{1 - e^{-x_1}} \cdots \frac{x_m}{1 - e^{-x_m}}.$$

These classes, like the Chern classes, can be expressed in terms of the curvature form Ω of the connection in the bundle α. Thus, for example, the differential form

$$\sum_k \frac{1}{(2\pi i)^k km} \mathrm{tr}\, \Omega^k$$

represents the Chern character $\mathrm{ch}(\alpha)$.

16.3. The Atiyah-Singer Index Theorem. Let us fix a Riemannian metric on an n-dimensional manifold M. Let $\tau: B(M) \to M$ denote the bundle of unit balls in the cotangent spaces $T_x^* M$. Then $B(M)$ is a manifold of dimension $2n$ with boundary. Its boundary is the total space $S(M)$ of the bundle $S(M) \to M$ of unit spheres in $T^* M$.

Let $D: \Gamma(\alpha) \to \Gamma(\beta)$ be a linear elliptic operator acting between spaces of sections of vector bundles α, β of dimension m over M. The fact that D is elliptic implies that for every non-zero covector $p \in T_x^* M$ its symbol $\sigma_p(D): \alpha^{-1}(x) \to \beta^{-1}(x)$ (see Ch. 3, 5.2) is an isomorphism.

Let $C(M)$ denote the manifold obtained by pasting two copies $B_+(M)$ and $B_-(M)$ of the manifold $B(M)$ along their common boundary $S(M)$. The bundles $\tau^*(\alpha)$ and $\tau^*(\beta)$ over $B_+(M)$ and $B_-(M)$, respectively, induced by the bundles α and β by means of the projection τ, are pasted by isomorphisms $\sigma_p(D)$ over $S(M)$ into some vector bundle over the manifold $C(M)$. Let $\mathrm{ch}(D)$ denote the Chern character of this bundle. In this notation the Atiyah-Singer formula for the index of the elliptic operator D has the form

$$i(D) = \int_{C(M)} \mathrm{ch}(D) \wedge \tau_C^*(\mathrm{td}(M)), \tag{11}$$

where in the integrand we have the term of degree $2n$ in the product of the Chern character $ch(D)$ and the inverse image $\tau_C^*(td(M))$ of the Todd class of the complexified tangent bundle under the projection $\tau_C\colon C(M) \to M$.

We note that by using the *Thom isomorphism*, which in the given case reduces to integration along the fibres of the bundle τ_C, we can transform the right-hand side of (11) into an integral over M. Formula (11) was obtained by Atiyah and Singer in this form.

Since the characteristic classes in the integrand are obtained from Chern classes of even degree, $i(D) = 0$ if M is odd-dimensional. This is a far-reaching generalization of the fact that the Euler characteristic of a compact odd-dimensional manifold is always equal to zero.

16.4. The Index Theorem and the Riemann-Roch-Hirzebruch Theorem. The modern version of the Riemann-Roch problem can be stated as the problem of calculating the dimension of the space of holomorphic sections of a holomorphic bundle over a compact complex manifold. The first result of this kind was due to Riemann, who showed that the dimension of the space of holomorphic 1-forms on a Riemann surface (a complex algebraic curve) of genus g is also equal to g. This result of Riemann was generalized many times until it took its definitive form as the Riemann-Roch-Hirzebruch theorem (see below).

This theorem is now a special case of the index theorem. However, historically it preceded the latter and served as a model for its first proof.

The formulation of the Riemann-Roch-Hirzebruch theorem requires some information from differential calculus on complex manifolds, which we give below.

16.5. The Dolbeault Cohomology of Complex Manifolds. A smooth manifold M on which an integrable almost-complex structure is defined is said to be *complex*. In the language of atlases and local coordinates a complex structure on a manifold M can be defined either by means of an atlas $\varphi_j\colon \mathbb{R}^{2n} = \mathbb{C}^n \supset U_j \to M$, whose transition functions $\varphi_j^{-1} \circ \varphi_k\colon U_k \to U_j$ are holomorphic, or in terms of *complex local coordinates* $z_1 = x_1 + iy_1, \ldots, z_n = x_n + iy_n$, connected on the intersections of charts by holomorphic functions. In this case $(x_1, \ldots, x_n, y_1, \ldots, y_n)$ are local coordinates on M.

Let $\Omega^k(M) = \Lambda^k(M) \otimes_{\mathbb{R}} \mathbb{C}$ denote the set of all complex-valued differential k-forms on a manifold M. In local coordinates each k-form $\omega \in \Omega^k(M)$ can be written in the form

$$\omega = \sum_{p+q=k} \omega_{a,b}(x, y)\, dz_{a_1} \wedge \cdots \wedge dz_{a_p} \wedge d\bar{z}_{b_1} \wedge \cdots \wedge d\bar{z}_{b_q}, \tag{12}$$

where we have used the obvious notation $dz_j = dx_j + idy_j$, $d\bar{z}_j = dx_j - idy_j$, $a = (a_1, \ldots, a_p), b = (b_1, \ldots, b_q)$. Hence

$$\Omega^k(M) = \sum_{p+q=k} \Omega^{p,q}(M), \tag{13}$$

where $\Omega^{p,q}(M)$ denotes the set of those k-forms that contain p factors of the form

dz_j and q factors of the form $d\bar{z}_j$ when written in the local form (12). It is not difficult to see that the pair of numbers (p, q) in (12) is an invariant of holomorphic changes of coordinates. The forms in $\Omega^{p,q}(M)$ are called (p, q)-*forms*.

Bearing (13) in mind, we can represent the operator of exterior differentation $d: \Omega^k(M) \to \Omega^{k+1}(M)$ as a sum $d = \partial + \bar{\partial}$, where $\partial: \Omega^{p,q}(M) \to \Omega^{p+1,q}(M)$, $\bar{\partial}: \Omega^{p,q}(M) \to \Omega^{p,q+1}(M)$ are the operators given by

$$\partial\omega = \sum_{a,b,j} \frac{\partial\omega_{a,b}}{\partial z_j} dz_j \wedge dz_{a_1} \wedge \cdots \wedge dz_{a_p} \wedge d\bar{z}_{b_1} \wedge \cdots \wedge d\bar{z}_{b_q},$$

$$\bar{\partial}\omega = \sum_{a,b,j} \frac{\partial\omega_{a,b}}{\partial\bar{z}_j} d\bar{z}_j \wedge dz_{a_1} \wedge \cdots \wedge dz_{a_p} \wedge d\bar{z}_{b_1} \wedge \cdots \wedge d\bar{z}_{b_q},$$

and

$$\frac{\partial}{\partial z_j} = \frac{1}{2}\left(\frac{\partial}{\partial x_j} - i\frac{\partial}{\partial y_j}\right), \quad \frac{\partial}{\partial\bar{z}_j} = \frac{1}{2}\left(\frac{\partial}{\partial x_j} + i\frac{\partial}{\partial y_j}\right).$$

We have $\partial^2 = \bar{\partial}^2 = \partial\bar{\partial} + \bar{\partial}\partial = 0$.

The conditions $\dfrac{\partial f}{\partial\bar{z}_j} = 0, j = 1, \ldots, n$, are the Cauchy-Riemann conditions for the function f to be holomorphic. They can be written more briefly in the form $\bar{\partial}f = 0$. More generally, the differential forms $\omega \in \Omega^{p,0}(M)$ are said to be *holomorphic* if $\bar{\partial}\omega = 0$. Holomorphic forms can be written in local coordinates in the form (12), where $\omega_{a,b}$ are holomorphic functions, and $q = 0$. The operators $\bar{\partial}$ generate a complex

$$0 \to \Omega^{p,0}(M) \xrightarrow{\bar{\partial}} \cdots \to \Omega^{p,q}(M) \xrightarrow{\bar{\partial}} \Omega^{p,q+1}(M) \to \cdots,$$

called the *Dolbeault complex*. Its cohomology at the term $\Omega^{p,q}(M)$ is denoted by

$$H^{p,q}(M) = \frac{\text{Ker}(\bar{\partial}: \Omega^{p,q}(M) \to \Omega^{p,q+1}(M))}{\text{Im}(\bar{\partial}: \Omega^{p,q-1}(M) \to \Omega^{p,q}(M))}$$

and called the *Dolbeault cohomology* of the complex manifold M.

Examples. 1) If $M = \mathbb{C}^n$, then $H^{p,q}(\mathbb{C}^n) = 0$ when $q \geqslant 1$, and the space $H^{p,0}(\mathbb{C}^n)$ is infinite-dimensional and coincides with the space of all holomorphic p-forms on $\mathbb{C}^n$.

2) The space $H^{0,0}(M)$ coincides with the space of all holomorphic functions on the complex manifold M. Hence, by the maximum principle, for a connected compact manifold M we have $H^{0,0}(M) = \mathbb{C}$.

3) If $M = \mathbb{C}P^1$ is the complex projective line, then direct calculation shows that $H^{0,0}(\mathbb{C}P^1) = H^{1,1}(\mathbb{C}P^1) = \mathbb{C}$, $H^{p,q}(\mathbb{C}P^1) = 0$ if $p \neq q$. Similarly for complex projective space: $H^{p,q}(\mathbb{C}P^n) = 0$ if $p \neq q$ and $H^{p,p}(\mathbb{C}P^n) = \mathbb{C}$ if $0 \leqslant p \leqslant n$.

A complex vector bundle $\pi: E \to M$ over a complex manifold M whose transition functions are holomorphic is called a *holomorphic vector bundle*. For such bundles it makes sense to talk about holomorphic sections. Locally they always have a basis consisting of holomorphic sections.

Example. The operators of the complex structure $J_x \colon T_x M \to T_x M$ determine a decomposition of the complexification of the tangent space $T_x M \otimes \mathbb{C}$ into the direct sum of subspaces $T_x^{1,0} M$ and $T_x^{0,1} M$ corresponding to the eigenvalues $+i$ and $-i$ of the operator J_x. The bundle $T^{1,0}(M) = \bigcup_{x \in M} T_x^{1,0} M \to M$ is holomorphic. It is called the *holomorphic tangent bundle*. The operators $\dfrac{\partial}{\partial z_j}, j = 1, \ldots, n$, form a local holomorphic basis of it.

The operator $\bar{\partial} \colon \Omega^{p,q}(M) \to \Omega^{p,q+1}(M)$ admits an extension to the operator $\bar{\partial}_\pi = \bar{\partial} \colon \Omega^{p,q}(M, \pi) \to \Omega^{p,q+1}(M, \pi)$, acting in the space $\Omega^{p,q}(M, \pi) = \Omega^{p,q}(M) \otimes \Gamma(\pi)$ of (p, q)-forms on M with values in sections of the bundle π. In fact, each such form can be written as $\omega = \sum \omega_j \otimes e_j$, where $\{e_j\}$ is a local holomorphic basis. We put $\bar{\partial}\omega = \sum \bar{\partial}\omega_j \otimes e_j$. This is well-defined, since holomorphic bases differ by a matrix with holomorphic coefficients, which are annihilated by the operator $\bar{\partial}$.

The resulting complex

$$0 \to \Omega^{p,0}(M, \pi) \xrightarrow{\bar{\partial}} \cdots \to \Omega^{p,q}(M, \pi) \xrightarrow{\bar{\partial}} \Omega^{p,q+1}(M, \pi) \to \cdots$$

is called the *Dolbeault complex of the holomorphic bundle* π, and its cohomology at the term $\Omega^{p,q}(M, \pi)$ is denoted by $H^{p,q}(M, \pi)$ and called the *Dolbeault cohomology of the manifold with values in the bundle* π.

Let us give the basic facts concerning the Dolbeault cohomology. First of all, if M is compact, then the spaces $H^{p,q}(M, \pi)$ are finite-dimensional (see §15.4). We put

$$h^{p,q}(\pi) = \dim_{\mathbb{C}} H^{p,q}(M, \pi), \qquad h^{p,q} = \dim_{\mathbb{C}} H^{p,q}(M).$$

Serre's duality theorem, which is analogous to Poincaré's duality theorem for the de Rham cohomology, asserts that

$$h^{p,q}(\pi) = h^{n-p,n-q}(\pi^*),$$

where π^* is the complex conjugate bundle of π.

Moreover, generally speaking, $h^{p,q} \neq h^{q,p}$. However, for Kähler manifolds we have $h^{p,q} = h^{q,p}$.

16.6. The Riemann-Roch-Hirzebruch Theorem.

Kodaira carried over Hodge theory (see §11) to the Dolbeault cohomology. Namely, having fixed Hermitian metrics in the tangent bundle to M and the holomorphic bundle π, we define as in §11 the operator $\bar{\partial}^* \colon \Omega^{p,q+1}(M, \pi) \to \Omega^{p,q}(M, \pi)$ conjugate to the differential $\bar{\partial}$. A (p, q)-form $\Omega \in \Omega^{p,q}(M, \pi)$ is said to be *harmonic* if $\bar{\partial}\omega = 0$, $\bar{\partial}^*\omega = 0$. The main result of Hodge theory, as applied to our situation, implies that each Dolbeault cohomology class contains exactly one harmonic representative. Moreover, proceeding as in 16.1, we can regard harmonic forms as elements of the kernel and cokernel of an elliptic differential operator

$$\bar{\partial} + \bar{\partial}^* \colon \Omega_e^p(M, \pi) \to \Omega_0^p(M, \pi),$$

where

$$\Omega_e^p(M, \pi) = \sum_{q \geqslant 0} \Omega^{p, 2q}(M, \pi),$$

$$\Omega_0^p(M, \pi) = \sum_{q \geqslant 0} \Omega^{p, 2q+1}(M, \pi).$$

Hence from the general theory of elliptic operators it follows that on a compact manifold M the Dolbeault cohomology $H^{p,q}(M, \pi)$ is finite-dimensional.

Let $\chi^p(M, \pi)$ denote the index of the operator $\bar{\partial} + \bar{\partial}^*$. This number coincides with the Euler characteristic of the Dolbeault complex:

$$\chi^p(M, \pi) = \sum_{q \geqslant 0} (-1)^q h^{q, p}(\pi).$$

When $p = 0$ the index theorem gives the *Riemann-Roch-Hirzebruch formula*

$$\chi^0(M, \pi) = \int_M \mathrm{ch}(\pi) \wedge \mathrm{td}(M). \tag{14}$$

For arbitrary values of $p \geqslant 1$ the Euler characteristic $\chi^p(M, \pi)$ can also be expressed in terms of the Chern classes of the bundle π and the manifold M. For an explicit expression see Hirzebruch [1966].

The elements of the space $H^{p,0}(M)$ are holomorphic p-forms. They are also called *p-forms of the first kind*. The number $\sum_{p \geqslant 0}(-1)^p h^{p,0}$ is called the *arithmetic genus* of the manifold M. For Kähler manifolds we have $h^{p,0} = h^{0,p}$. Hence the arithmetic genus coincides with $\chi^0(M) = \chi^0(M, 1)$. Thus, (14) shows that the arithmetic genus of a Kähler manifold can be expressed in terms of the topological characteristics of this manifold.

Example. 1) For a connected compact 1-dimensional complex manifold M (a Riemann surface of genus g), which is obviously a Kähler manifold, we obtain

$$h^{0,0} - h^{1,0} = \frac{1}{2} \int_M c_1(M) = \frac{1}{2}(2 - 2g) = 1 - g,$$

since $\mathrm{td}_1(M) = \frac{1}{2} c_1(M)$.

But $h^{0,0} = 1$, so $h^{1,0} = g = \frac{1}{2} b_1(M)$. Consequently, in this case (14) is equivalent to the *Riemann relation* $h^{1,0} = g$.

2) For 2-dimensional compact Kähler manifolds (14) is equivalent to *Noether's formula* $\chi^0(M) = \frac{1}{12} \int_M (c_2(M) + c_1^2(M))$.

The polynomials $\chi_y(M, \pi) = \sum_{p \geqslant 0} \chi^p(M, \pi) y^p$ and $\chi_y(M) = \sum_{p \geqslant 0} \chi^p(M) y^p$ are called the χ_y-*characteristic* of π and the χ_y-*genus* of M.

We can express the topological invariants we have met in terms of these polynomials. Thus, for Kähler manifolds $\chi_0(M)$ is the arithmetic genus and $\chi_1(M)$ is the Euler characteristic.

§ 17. The Space of Geometric Structures and Deformations

17.1. The Moduli Space of Geometric Structures. Physics does not distinguish between solutions of Einstein's equations that are taken into one another by a change of coordinates, since they describe the same universe (or part of it). This is an example of a situation in which we need to identify equivalent geometric structures and it is important to have a description of the space of equivalent classes of geometric structures of a given type on a given manifold. Spaces of this kind are usually called *moduli spaces*. More precisely, a moduli space is the space of orbits of a group of diffeomorphisms on the set of all geometric structures of the type under consideration.

Since differential invariants determine a geometric structure uniquely up to equivalence, the structure of a moduli space is naturally connected with the structure of the corresponding algebra of differential invariants (see Ch. 7, 4.4). More precisely, let $\pi\colon E \to M$ be the natural bundle whose sections are the geometric structures under consideration. The group Diffeo M of diffeomorphisms of the manifold M acts on the space $J^\infty(\pi)$ (see Ch. 7, 4.1). Generally speaking, the quotient space $\mathscr{Y} = J^\infty(\pi)/\text{Diffeo } M$ is a stratified object $\mathscr{Y} = \mathscr{Y}_n \supset \mathscr{Y}_{n-1} \supset \cdots \supset \mathscr{Y}_0$. Its "stratum" $\mathscr{Y}_i - \mathscr{Y}_{i-1}$ (locally) has the form $\mathscr{E}_k^{(\infty)}$, where $\mathscr{E}_k$ is a differential equation in k independent variables. In particular, $\mathscr{Y}_0 = \mathscr{E}_0$ is simply a manifold. The solutions of the equation $\mathscr{E}_k$ correspond to structures that admit a (local) symmetry group with $(n - k)$-dimensional orbits. The algebra of smooth functions on the stratum $\mathscr{Y}_k - \mathscr{Y}_{k-1}$ is identical to the algebra of differential invariants under the assumption that there are exactly k independent ones among them (see Ch. 7, 4.4). Thus the moduli space is the union of spaces of global solutions of the equations $\mathscr{E}_k$, $k = 0, 1, \ldots, n$.

This rough scheme gives an idea of what we can expect if we pose the question of the moduli space in full generality. It is not surprising that existing meaningful results on the structure of the moduli space (which we shall consider below) refer to the structure of the stratum $\mathscr{E}_0$, that is, to the case when obtaining an answer is not connected with a description of the space of global solutions of non-linear systems of differential equations.

17.2. Examples. 1) As a geometric structure on a manifold M we consider the *Euclidean metric* (that is, a complete Riemannian metric of zero curvature). Suppose that M is simply-connected. The moduli space consists of a single point (if M is diffeomorphic to $\mathbb{R}^n$) or is empty. If M is the n-dimensional torus $T^n = \mathbb{R}^n/\mathbb{Z}^n$, then the moduli space is identified with the space of double residue classes $SL(n, \mathbb{Z}) \backslash GL(n \ \mathbb{R})/O(n)$. In fact, the Euclidean metric g in T^n can be specified by the Gram matrix $g_{ij} = (e_i, e_j)$, where e_i is a basis of the lattice $\mathbb{Z}^n$, which is defined up to an integral unimodular transformation. This shows that the moduli space of Eulidean metrics on the torus T^n has dimension $\frac{1}{2}n(n + 1)$. Bieberbach's theorems assert that any compact manifold with a Euclidean metric is covered by a flat torus and there are only finitely many such manifolds that can be covered by the given torus.

2) The moduli space of complex structures on a compact surface M_g of genus $g > 1$ is a smooth manifold, apart from isolated singular points. Its singular points are complex structures that admit a non-trivial group of automorphisms (it is always finite). The natural ramification of the covering of the moduli space that "resolves" the singularities is called the *Teichmüller space*. It is diffeomorphic

to the Euclidean space of dimension $6g - 6$. This dimension was discovered by Riemann. We can regard a complex structure, that is, a $GL(1, \mathbb{C})$-structure, on a surface as a conformal structure, that is, a $CO(2)$-structure, since $GL(1, \mathbb{C}) = CO(2) = R^* \cdot SO(2)$. There is exactly one Riemannian metric of constant negative curvature subordinate to a given conformal structure on M_g. Hence Teichmüller space is simultaneously the moduli space of conformal structures and metrics of constant negative curvature on a surface M_g of genus $g > 1$. We note that the moduli space of complex structures on manifolds of higher dimension have a much more complicated structure. For example, they may have singularities, that is, they may not be smooth manifolds. The dimension of the moduli space of complex structures on two-dimensional complex algebraic surfaces was calculated by Noether in 1888.

3) Generalizing Example 1, as a geometric structure we consider a complete metric of constant curvature k. A simply-connected manifold endowed with such a structure is isometric to the sphere S^n if $k > 0$, to the Euclidean space E^n if $k > 0$, or to the Lobachevskij space L^n if $k < 0$ (see Ch. 3, § 4.8). Hence it follows that on a simply-connected manifold M the moduli space consists of a single point or is empty.

4) The moduli space of metrics of constant negative curvature on a compact manifold M of dimension $n > 2$ consists of a single point or is empty. This follows from a deep theorem of Mostow and Margulis which asserts that any isomorphism of fundamental groups of compact Riemannian manifolds of constant negative curvature and dimension $n > 2$ can be uniquely extended to an isometry of them. When $n = 2$ this is not so (see Example 2).

5) A *group structure* on a manifold M is a complete linear connection of zero curvature with covariantly constant torsion tensor T. The torsion tensor T_x of such a connection at a point $x \in M$ determines the structure of a Lie algebra in the tangent space $V = T_x M$:

$$[\xi, \eta] = T_x(\xi, \eta), \qquad \xi, \eta \in V.$$

If M is simply-connected, then the group structure determines on it the structure of the Lie group (see Ch. 4, 3.2) corresponding to this algebra. Hence it follows that the moduli space of group structures on a simply-connected manifold M is identified with the structure space of Lie algebras on a space V for which the corresponding group is diffeomorphic to M. Lie algebras are considered up to conjugacy. This leads to the problem of describing the moduli space of Lie algebras, that is, the orbit space of the group $GL(n)$ in the structure space of Lie algebras (see 17.5 below).

6) As we remarked in 17.1, two Lorentz metrics that specify the geometry of space-time and differ by a diffeomorphism are physically indistinguishable. Hence the moduli space of Lorentz metrics can be regarded as the space of all possible geometries of space-time M. Einstein's equations, interpreted as equations on this space, distinguish the physically sensible ones.

There is another interpretation of Einstein's equations as a (Hamiltonian) dynamical system with constraints in the moduli space of Riemannian metrics on a three-dimensional manifold S. Roughly speaking, Einstein's equations split into evolutionary equations, which describe the evolution of the metric on a space-like hypersurface S of space-time M as time passes, and constraint equations on the original Cauchy data – the first and second quadratic forms of S. This approach enables us to obtain an existence theorem for solutions of the Einstein equations and to describe the tangent space to the moduli space of Einstein metrics.

17.3. Deformation and Supersymmetries. A one-parameter family $\omega_t, 0 \leqslant t < \varepsilon$, of geometric structures is called a *deformation* of the structure ω_0. Geometrically it determines a curve $\{\omega_t\}$ in the corresponding moduli space $\mathfrak{M}$ (here $\{\omega\} \in \mathfrak{M}$ is a class of equivalent geometric structures). It is appropriate to call the tangent vector to this curve an *infinitesimal deformation* of ω_0.

If the moduli space is smooth close to a point $\{\omega_0\}$, the totality of infinitesimal deformations of ω_0 forms a linear space $T_{\{\omega_0\}}\mathfrak{M}$, called the *space of essential deformations* of this structure. The description of these spaces is the natural first step that must be taken when embarking on a study of the moduli space. Since, as a rule, the description of the moduli space is a very complicated problem, here the problem is this: to describe the space of essential deformations without knowing about the structure of the moduli space.

The solution of this problem led Nijenhuis to discover the device of Lie superalgebras (see 17.4), which together with the device of Spencer cohomology constitutes a basis for the theory of deformations.

It is remarkable that the theory of Lie superalgebras and Lie supergroups, like the "super" terminology, arose independently in modern physics under the influence of the ideas of F.A. Berezin, who succeeded in combining into a unified theory the symmetries of bosons and fermions – particles (or fields) that had been regarded up to then as immiscible in principle. In this way there arose the theory of supersymmetry, which plays such an important role in modern field theories of elementary particles.

17.4. Lie Superalgebras. A $\mathbb{Z}$-graded algebra $A = \sum_{i \in \mathbb{Z}} A_i$ with multiplication $A_i \times A_j \to A_{i+j}$, $a \times b \mapsto [a, b]$ is said to be *supercommutative* if

$$[a, b] = -(-1)^{p(a)p(b)}[b, a],$$

where $p(a)$ and $p(b)$ are the degrees of the homogeneous elements a and b. If the *supercommutative Jacobi identity*

$$(-1)^{p(a)p(b)}[a, [b, c]] + (-1)^{p(b)p(c)}[b, [c, a]] + (-1)^{p(c)p(a)}[c, [a, b]] = 0$$

is satisfied, we say that A is a *(graded) Lie superalgebra*. We note that its even part $A^0 = \sum A_{2i}$ is an ordinary graded Lie algebra.

Example. 1) Let $A = \sum A_i$ be an associative commutative or supercommutative graded algebra. The operation $a \times b \mapsto ab - (-1)^{p(a)p(b)}ba$, $a \in A_{p(a)}$, $b \in A_{p(b)}$, determines the structure of a supercommutative Lie algebra in A.

2) The space Der $A = \sum_j \mathrm{Der}_j\, A$ of all derivations of a graded algebra $A = \sum A_i$ is a Lie superalgebra with respect to the bracket

$$[D, D'] = D \circ D' - (-1)^{p(D)p(D')}D' \circ D.$$

Here by definition a derivation $D \in \mathrm{Der}_j\, A$ of degree $p(D) = j$ is defined as a linear map that raises the degree by j and satisfies the *supercommutative Leibniz rule*

$$D(ab) = D(a) \cdot (b) + (-1)^{jp(a)}a \cdot D(b).$$

For the graded algebra of exterior forms $A = \Lambda^*(V)$ the superalgebra Der A is identified with the tensor product $A \otimes V$. An element $a \otimes v \in A \otimes V$ determines a derivation

$$a \otimes v \colon b \to a \cdot (v\,\lrcorner\, b), \qquad b \in A.$$

The operation $[\ ,\ ]$ in the superalgebra $A \otimes V$ is called the *Nijenhuis algebraic bracket*, and the superalgebra Der $\Lambda^*(V) = \Lambda^*(V) \otimes V$ is called the *superalgebra of vector-valued Nijenhuis forms*.

3) The superalgebra Der $\Lambda^*(M)$ of derivations of the algebra $\Lambda^*(M)$ is described as follows. It has a distinguished element – the operator d of exterior derivation. The *Nijenhuis algebra* $N(M) = \Lambda^*(M) \otimes D(M)$ of vector-valued differential forms can be regarded as the algebra of $C^\infty(M)$-linear derivations of the algebra $\Lambda^*(M)$. The vector-valued form $\omega \otimes X$ acts on the form σ according to the rule

$$(\omega \otimes X)(\sigma) = \omega \wedge (X\,\lrcorner\,\sigma).$$

It is not difficult to verify that

$$\mathrm{Der}\ \Lambda^*(M) = N(M) \oplus [d, N(M)],$$

where the right-hand side is the direct sum of non-commuting subalgebras.

For $A \in N(M)$ we put $d_A = [d, A]$. If A is a 0-form (that is, a vector field X), then $d_A = d \circ (X\,\lrcorner) + (X\,\lrcorner) \circ d = \mathscr{L}_x$ is the Lie derivative operator in the direction of the field X. If $A, B \in N(M)$, then $[d_A, d_B] \in [d, N(M)]$, that is,

$$[d_A, d_B] = d_{[A, B]}.$$

The form $[A, B]$ defined by this equality is called the *Nijenhuis differential bracket* of the forms A and B.

4) Let A be a supercommutative algebra, and $\mathfrak{G}$ a Lie algebra. Then the bracket

$$[a \otimes g, b \otimes h] = ab \otimes [g, h], \qquad a, b \in A; \quad g, h \in \mathfrak{G},$$

turns the tensor product $A \otimes \mathfrak{G}$ into a Lie superalgebra.

Special cases:

a) Let $\mathfrak{G}$ be a Lie algebra, and V a vector space. Then $\Lambda^*(V) \otimes \mathfrak{G}$ is a Lie superalgebra.

b) The space $\Lambda^*(P, \mathfrak{G})$ of $\mathfrak{G}$-valued differential forms on a manifold P with the bracket defined in Ch. 6, 3.5, is a Lie superalgebra. If $\pi\colon P \to M$ is a principal G-bundle, and $\mathfrak{G}$ is the Lie algebra of the Lie group G, then the subspace of G-equivariant horizontal forms is a subalgebra of the Lie superalgebra $\Lambda^*(P, \mathfrak{G})$. Hence it follows that the space $\Lambda^*(M, \pi\mathfrak{G})$ of differential forms on M with values in $\pi\mathfrak{G}$, which is isomorphic to it, is also a Lie superalgebra.

17.5. The Space of Infinitesimal Deformations of a Lie Algebra. Rigidity Conditions. Following Nijenhuis and Richardson, we show how to describe the space of real infinitesimal deformations of the structure of a Lie algebra. It can be interpreted as the tangent space to the moduli space of Lie algebras.

The structure of a Lie algebra in a vector space V is specified by a skew-symmetric bilinear map, that is, by an element $c \in \Lambda^2(V) \otimes V = N^2(V)$ of the superalgebra $N(V) = \Lambda^*(V) \otimes V$. The Jacobi identity implies that $[c, c] = 0$. Thus, the space of Lie algebras $L(V)$ is identified with the space $L(V) = \{c \in N^2(V), [c, c] = 0\}$, and the moduli space is identified with the orbit space of the natural action of the group $GL(V)$ in $L(V)$. An element $c \in L(V)$ determines a derivation ad_c of degree 1 of the superalgebra $N(V)$ for which $\mathrm{ad}_c^2 = 0$, and hence the complex

$$0 \to V \overset{\mathrm{ad}_c}{\to} V^* \otimes V \overset{\mathrm{ad}_c}{\to} \Lambda^2(V) \otimes V \to \cdots .$$

If $\omega \in \Lambda^k(V) \otimes V$, then

$$(\mathrm{ad}_c)(\omega)(\xi_1, \ldots, \xi_{k+1}) = \sum_i (-1)^{i+1}[\xi_i, \omega(\xi_1, \ldots, \hat{\xi}_i, \ldots, \xi_{k+1})]$$

$$+ \sum_{i<j} (-1)^{i+j}\omega([\xi_i, \xi_j], \xi_1, \ldots, \hat{\xi}_i, \ldots, \hat{\xi}_j, \ldots, \xi_{k+1}),$$

where $[\xi, \eta] = c(\xi, \eta)$ is the commutator of vectors in the Lie algebra (V, c). Let

$$H^i(N(V), c) = \frac{\mathrm{Ker}(\mathrm{ad}_c\colon N^i(V) \to N^{i+1}(V))}{\mathrm{Im}(\mathrm{ad}_c\colon N^{i-1}(V) \to N^i(V))}$$

denote the i-th cohomology space of this complex. This cohomology is called the *cohomology of the Lie algebra (V, c) with values in V.*

Since the tangent space to $L(V)$ at the point $c \in L(V)$ has the form

$$T_c L(V) = \{x \in N^2(V), \mathrm{ad}_c x = 0\},$$

and the tangent space to the $GL(V)$-orbit O_c of the point $c \in L(V)$ is

$$T_c O_c = \{y = \mathrm{ad}_c x, x \in N^1(V) = V^* \otimes V\},$$

the space of real infinitesimal deformations of the Lie algebra (V, c) is identified with the space

$$T_c L(V)/T_c O_c = H^1(N(V), c).$$

In particular, if $H^1(N(V), c) = 0$, then O_c is open in the space of Lie algebras $L(V)$ and any small deformation of the Lie algebra (V, c) is inessential, that is, it consists of isomorphic Lie algebras. For example, this is so if (V, c) is semisimple.

This construction admits the following formalization. Let $\mathfrak{A} = \mathfrak{A}_0 + \mathfrak{A}_1 + \cdots$ be a Lie superalgebra with dim $\mathfrak{A}_i < \infty$, and let G be a simply-connected Lie group with Lie algebra $\mathfrak{A}_0$. The Lie group G acts in the space $L(\mathfrak{A}) = \{a \in \mathfrak{A}_1, [a, a] = 0\}$ of "self-commuting" elements. Each such element determines a complex

$$\mathfrak{A}_0 \overset{\mathrm{ad}_a}{\to} \mathfrak{A}_1 \overset{\mathrm{ad}_a}{\to} \mathfrak{A}_2 \to \cdots.$$

It is said to be *deformed*, and its cohomology $H^i(\mathfrak{A}, a)$ is called the *deformed cohomology*. In particular, the space $H^1(\mathfrak{A}, a) = T_a L(\mathfrak{A})/T_a G(a)$ is interpreted as the tangent space at the point $\{G(a)\}$ to the orbit space of the action of the group G in $L(\mathfrak{A})$. If $H^1(\mathfrak{A}, a) = 0$, then the element is "rigid", that is, its orbit $G(a)$ contains a neighbourhood of the point a in $L(\mathfrak{A})$.

17.6. Deformations and Rigidity of Complex Structures. Our construction can be applied, if we take the necessary care, to infinite-dimensional Lie superalgebras. Suppose, for example, that J is an almost complex structure on a manifold M, regarded as a vector-valued 1-form $J \in N(M)$. Its integrability condition has the form $[J, J] = 0$, where $[\ ,\ ]$ is the Nijenhuis differential bracket. The corresponding deformed complex coincides with the Dolbeault complex (see 16.5) of the holomorphic tangent bundle τ of the manifold M. Let $H^{p,q}(M, \tau)$ denote the corresponding Dolbeault cohomology. In an appropriate way the refined arguments of 17.5 lead to the following result of Frölicher and Nijenhuis.

Theorem. *Let J_t, $0 \leqslant t \leqslant \varepsilon$, be a smooth family of complex structures on a manifold M, and suppose that $H^{0,1}(M, \tau) = 0$. Then for sufficiently small ε the structures J_t, $0 \leqslant t \leqslant \varepsilon$, are equivalent to the structure $J = J_0$.*

Example. Let us calculate the dimension of the moduli space of complex structures on a compact surface M of genus g. Its tangent space at the point $[J]$ is identified with $H^{0,1}(M, J)$. By the Riemann-Roch theorem, for the tangent bundle (see 16.5) we have

$$\dim_{\mathbb{C}} H^{0,0}(M, \tau) - \dim_{\mathbb{C}} H^{1,0}(M, \tau) = \int_M \mathrm{ch}(\tau) \wedge \mathrm{td}(M) = \frac{3}{2} \int_M c_1(M)$$

$$= \frac{3}{2}(2 - 2g) = 3 - 3g.$$

Hence, using *Serre duality* (see 16.5) we obtain

$$\dim_{\mathbb{C}} H^{0,1}(M, \tau) = \dim_{\mathbb{C}} H^{1,0}(M, \tau) = \dim_{\mathbb{C}} H^{0,0}(M, \tau) + 3g - 3.$$

The space $H^{0,0}(M, \tau)$ is identified with the space of holomorphic vector fields on (M, J). When $g \geqslant 2$ we have $H^{0,0}(M, \tau) = 0$. This follows from the fact that the sum of the indices of the singular points of a vector field on M is equal to $\chi(M) = 2 - 2g$, and the indices of a holomorphic vector field are non-negative.

This proves Riemann's theorem (see 17.2, Example 4):

$$\dim_{\mathbb{C}} H^{0,1}(M, \tau) = 3g - 3 \qquad \text{if } g \geqslant 2.$$

§18. Minkowski's Problem, Calabi's Conjecture and the Monge-Ampère Equations

In the majority of cases the problems of differential geometry reduce to the solution of systems of partial differential equations that are non-linear as a rule. For example, the classic work of Darboux on the theory of surfaces (Darboux [1894–1925]) is mainly devoted to differential equations rather than surfaces as such. The investigation of the differential equations that arise when considering geometrical problems is lightened to some extent by the accompanying geometrical arguments. The latter, being clothed in analytic form, can usually be generalized with no difficulty, enabling us to obtain the corresponding results for a wider class of equations than the original ones. This in turn can lead to the solution of new more complicated geometrical problems, and so on. We illustrate this process by some geometrical problems that lead to the Monge-Ampère equations.

Let us begin with the well-known problem of Minkowski, which consists in the following. We shall call a closed compact strictly convex hypersurface of Euclidean space an *oval hypersurface*. Strict convexity means that at all points of the surface the Gaussian curvature is positive. Then the *Gaussian* or *spherical map* $f: V \to S^n$ of an oval hypersurface $V \subset \mathbb{R}^{n+1}$ that assigns to a point $x \in V$ the end of the unit normal to V starting from it, laid off from the origin, is a diffeomorphism. In view of this, any function φ on V can be carried over to S^n by means of this diffeomorphism. Suppose, in particular, that $\theta \to K_V(\theta)$, $\theta \in S^n$, is the Gaussian curvature function of V, carried over to S^n in this way. We ask, to what extent does that function $K(\theta)$ determine the original hypersurface V, and can we construct from an a priori given function K on S^n an oval hypersurface V such that $K = K_V$? Since $K_V > 0$, and as we can easily verify $\int_{S^n} \varphi K_V^{-1} \Omega = 0$ for any linear function φ on $\mathbb{R}^{n+1}$ (Ω is the unit volume form on S^n), we need to assume that K satisfies these conditions.

Minkowski himself gave a positive solution to the question of uniqueness, relying on the theory of mixed volumes that the constructed and the remarkable inequalities for them, known nowadays as the Minkowski inequalities (see Aleksandrov [1937–38], Minkowski [1903]). He also solved the question of existence to a certain extent. Namely, Minkowski constructed the required hypersurface as the limit of a sequence of convex polyhedra that give in a certain sense a piecewise-constant approximation of the function K. Minkowski's solution was later extended to arbitrary compact convex hypersurfaces, that is, not necessarily smooth, by Aleksandrov [1984] and Fenchel and Jessen. A deficiency of these solutions is that they do not guarantee that the hypersurface is smooth if the original function K is smooth or even analytic.

A regular solution of Minkowski's problem, that is, a solution free from this deficiency, is due to Pogorelov [1969]. It is based on the reduction of the question to the solution of a non-linear differential equation

$$\det\left\|\frac{\partial^2 u}{\partial x_i \partial x_j}\right\| = F(x_1, \ldots, x_n), \tag{15}$$

where the known function F is connected in a definite way with the curvature function K. This reduction is carried out in the following way.

Let $V \subset \mathbb{R}^{n+1}$ be a smooth oriented hypersurface and $(x_1, \ldots, x_{n+1})$ the standard coordinates in $\mathbb{R}^{n+1}$. We shall identify the standard unit n-dimensional sphere $S^n \subset \mathbb{R}^{n+1}$ with the set of unit vectors in $\mathbb{R}^{n+1}$. With any unit vector e we associate the number $p(e)$, the distance from the origin to the tangent plane of V for which the vector e is the oriented normal. Then $p(e)$ is a smooth and possibly multivalued function on some domain in S^n. Any hypersurface, being the envelope of its family of tangent planes, is uniquely determined by this family. In view of this, V is completely determined by the function $p(e)$. In fact, the tangent plane having e as its oriented normal can be uniquely restored if we know the number $p(e)$. If V is an oval surface, then $p(e)$ is a smooth single-valued function defined on S^n.

From the function $p(e)$ it is convenient to go over to the function $P(\xi)$, $\xi \in \mathbb{R}^{n+1} \setminus 0$, by putting $P(\xi) = |\xi| p(\xi/|\xi|)$. Thus, $P(\xi)$ is a homogeneous function of the first degree, called the Minkowski support function. Like p it uniquely determines the hypersurface V. It is not difficult to show that the principal curvatures $\lambda = \lambda_i$ of V are the roots of the "characteristic equation"

$$\det\left\|\frac{\partial^2 p}{\partial x_i \partial x_j} - \lambda^{-1}\delta_{ij}\right\| = 0,$$

which has apart from them one more "parasitic" root $\lambda^{-1} = 0$. Let us introduce the function

$$u(x_1, \ldots, x_n) = P(1, x_1, \ldots, x_n).$$

Calculating the determinant in terms of the function u, we arrive at the equation

$$\left\| \begin{matrix} -\lambda^{-1}\rho^2 & \lambda^{-1}x_1 & \cdots & \lambda^{-1}x_n \\ \lambda^{-1}x_1 & \rho u_{11} - \lambda^{-1} & \cdots & \rho u_{1n} \\ \vdots & \vdots & & \\ \lambda^{-1}x_n & \rho u_{n1} & \cdots & \rho u_{nn} - \lambda^{-1} \end{matrix} \right\| = 0,$$

where $u_{ij} = \partial^2 u/\partial x_i \partial x_j$ and $\rho = (1 + x_1^2 + \cdots + x_n^2)^{1/2}$. From this it follows that the product of the principal curvatures, that is, the Gaussian curvature K, is equal to

$$K = \det\|u_{ij}\|^{-1}\rho^{-(n+2)}.$$

Hence it is obvious that the function $u(x)$ satisfies (15), where $F = K^{-1}\rho^{-(n+2)}$.

The key step in the solution of (15) is that of obtaining a priori estimates for the unknown function and its derivatives up to the third order inclusive. In this step we use the "accompanying" geometrical arguments. Also, the resulting estimates are used in the proof of convergence of the standard method of successive approximations, by means of which the solution is constructed.

In the last 10–15 years the technique of obtaining a priori estimates has been greatly improved, which has enabled us to solve a whole series of problems of geometry " in the large" that had not yielded to solution for a long time. As an example we mention *Calabi's conjecture*, stated in 1954 and proved by Yau in 1978 (Yau [1978]). Calabi's conjecture is as follows.

Let M be a compact Kähler manifold with metric $g_{k\bar{l}}dz^k d\bar{z}^l$. If $R_{k\bar{l}}dz^k d\bar{z}^l$ is its Ricci tensor, then as we know the differential (1, 1)-form

$$\sigma = \frac{i}{2\pi} R^l_{k\bar{l}} dz^k \wedge d\bar{z}^l$$

is closed and represents the first Chern class of the complex manifold M. We now specify a closed differential form

$$\tilde{\sigma} = \frac{i}{2\pi} \tilde{R}_{k\bar{l}} dz^k \wedge d\bar{z}^l$$

on M having type (1, 1) and representing its first Chern class. We ask, is there a Kähler metric $\tilde{g}_{k\bar{l}}dz^k d\bar{z}^l$ on M such that $\tilde{R}_{k\bar{l}}dz^k d\bar{z}^l$ is its Ricci tensor and the symplectic 2-form $\frac{i}{2}\tilde{g}_{kl}dz^k \wedge d\bar{z}^l$ is homologous to the symplectic form $\frac{i}{2}\tilde{g}_{k\bar{l}}dz^k \wedge d\bar{z}^l$ of the original metric? Calabi [1954] proposed an affirmative answer to this question. Calabi himself was able to prove that this conjecture is true under certain essential additional restrictions. He also proved the uniqueness of the solution.

Calabi's conjecture reduces to the solution of the following equation of Monge-Ampère type for the function φ:

$$\det \left\| g_{k\bar{l}} + \frac{\partial^2 \varphi}{\partial x^k \partial \bar{z}^l} \right\| = ce^F \det \| g_{k\bar{l}} \|, \tag{16}$$

where the known function F is constructed from the Ricci tensor $\tilde{R}_{k\bar{l}}dz^k d\bar{z}^l$, and the constant c is found from the equality

$$c \int_M e^F = (\text{volume of } M).$$

In fact, Yau [1978] was able to establish a priori estimates for a much wider class of equations than in Lychagin [1975], which enabled him to find the solution of a number of other problems of geometry "in the large". For example, he proved the existence of a *Kähler-Einstein metric* on a compact Kähler manifold M with an *ample canonical bundle*. A *canonical bundle* L on an n-dimensional complex manifold M is a one-dimensional holomorphic complex bundle whose holomorphic sections are holomorphic differential n-forms on M. If for any holomorphic vector bundle E on M the bundle $E \otimes L^k$ (where L^k is the k-th tensor power of

L) for sufficiently large k is generated by its global holomorphic sections, then the bundle L is said to be *ample*.

§19. Spectral Geometry

The main problem of spectral geometry is to restore the differential-geometric characteristics of a Riemannian manifold from the spectrum of its Laplace operator.

The source of this kind of problem was the theory of radiation of an absolutely black body, specifically, the problem of justifying Kirchhoff's law. Lorentz was the first to mention the connection between Kirchhoff's law and the spectrum of the Laplace operator in Dirichlet's problem in a series of lectures given at Göttingen in April 1910.

The eigenvalues $\{\lambda_n\}$ of the Laplace operator, regarded as functions of a natural parameter n, can behave in a very irregular way. Kirchhoff's law asserts that the asymptotic behaviour of the sequence λ_n as $n \to \infty$ does not depend on the form of the domain under consideration, but is determined only by its volume. Hence we can expect that its asymptotic behaviour of λ_n as $n \to \infty$ is quite regular. According to the words of Weyl "Lorentz convinced mathematicians that questions of this kind are very important in physics". Recently one could add – in differential geometry also. We note that one of the proofs of the index theorem, given by Atiyah, Bott and Patodi, is based on an analysis of the asymptotic behaviour of the Θ-function of the manifold, which is closely connected with the spectrum of the Laplace operator.

Let us proceed to the exact formulation.

Let (M, g) be a compact Riemannian manifold of dimension n and $\Delta = d^*d: C^\infty(M) \to C^\infty(M)$ its Laplace operator. This is an elliptic self-adjoint positive operator, which can be extended to an (unbounded) operator in $L^2(M)$. Hence Δ has a purely discrete non-negative spectrum $\lambda_0 = 0 < \lambda_1 \leqslant \lambda_2 \leqslant \cdots \leqslant \lambda_n \leqslant \cdots$, called the *spectrum of the Riemannian manifold M*. In this series, each eigenvalue occurs as many times as its multiplicity (it is always finite).

The operator Δ has a complete orthonormal system of eigenfunctions $\{l_j, l_j(x) \in C^\infty(M), j = 0, 1, \ldots, \}, \int_M l_j(x)l_i(x)\omega_g = \delta_{ij}$, where ω_g is the Riemannian volume form.

Examples. 1) $M = T^n = \mathbb{R}^n/\mathbb{Z}^n$ is the n-dimensional torus with the standard flat Riemannian structure. Then

$$\Delta = -\frac{\partial^2}{\partial x_i^2} - \cdots - \frac{\partial^2}{\partial x_n^2}$$

and the eigenfunctions of Δ have the form $\exp 2\pi c \langle k, x \rangle$, where $k = (k_1, \ldots, k_n) \in \mathbb{Z}^n$ and $\lambda_k = 4\pi^2 \langle k, k \rangle$.

2) $M = (S^n, g)$ is the standard Riemannian sphere. Then $\lambda_k = k(n + k - 1)$, $k = 0, 1 \ldots$, and the eigenfunctions are spherical harmonics.

We can also consider the spectrum of a compact Riemannian manifold with boundary, understanding the latter as the spectrum of the Laplace operator in the corresponding Dirichlet problem.

The *spectral function* $N(\lambda)$, $\lambda > 0$, of a manifold M is defined as the number of eigenvalues λ_i less than λ. Weyl's theorem describes the asymptotic behaviour of the functions $N(\lambda)$ for the Dirichlet problem in a bounded domain $M \subset \mathbb{R}^n$:

$$\lim_{\lambda \to \infty} N(\lambda)/\lambda^{n/2} = V_n \cdot \frac{\text{Vol } M}{(2\pi)^n}, \tag{17}$$

where V_n is the volume of the unit ball in $\mathbb{R}^n$ and $\text{Vol } M$ is the volume of M.

The relation (17) is equivalent to the following asymptotic distribution law for the numbers λ_k as $k \to \infty$:

$$\lambda_k \sim V_n(k/\text{Vol } M)^{2/n}. \tag{18}$$

Weyl's formula (17) is also true for an arbitrary compact manifold M if we assume that $\text{Vol } M = \int_M \omega_g$.

The main deficiency of the spectral function in deriving asymptotic distribution laws is that it contains global characteristics of the manifold M. More convenient is the apparatus of Θ-functions, in which local characteristics of a manifold are used explicitly. The transition to Θ-functions from the viewpoint of physics implies transition to the *statistical sum* of the "system of harmonic oscillators", and so the appearance of local characteristics of the manifold M is not unexpected.

We define the Θ-*function of the manifold M* by the relation

$$\Theta(t) = \sum_{k=0}^{\infty} \exp(-\lambda_k t).$$

Using an orthonormal basis of eigenfunctions $\{l_j(x)\}$, we can represent the Θ-function in the form

$$\Theta(t) = \int_M K(t, x, x)\omega,$$

where

$$K(t, x, y) = \sum_{j=0}^{\infty} \exp(-\lambda_j t)l_j(x)l_j(y), \qquad x, y \in M.$$

The character of the behaviour of the kernel $K(t, x, x)$ for small t describes the asymptotic behaviour of $N(\lambda)$ for large λ.

Kac and Pleijel considered the asymptotic behaviour of the function $\Theta(t)$ as $t \to 0$ for the Dirichlet problem in a finite domain $M \subset \mathbb{R}^2$ with a finite number of "handles" g. For a domain with a smooth boundary it has the form

$$\Theta(t) \sim \frac{\text{area of } M}{4\pi t} - \frac{\text{perimeter of } M}{4\sqrt{4\pi t}} + \frac{1}{6}(1 - g) + O(1).$$

If we represent a drum as a domain M, then $\lambda_1, \lambda_2, \ldots$ are its fundamental tones. This comparison enabled Kac to reformulate the problem of spectral geometry in the descriptive form "... can a person having perfect pitch hear the shape of a drum?"

The given formula shows that one can "hear" the area, perimeter and the number of "holes".

For an arbitrary compact Riemannian manifold (without boundary) McKean and Singer showed that the function $K(t, x, x)$ as $t \to 0$ admits the Minakshi-sundaram-Pleijel expansion

$$K(t, x, x) \sim (4\pi t)^{-n/2}(c_0(x) + tc_1(x) + \cdots),$$

whose coefficients $c_j(x)$ are invariants of the metric. They can be expressed in terms of the components of the curvature tensor and its covariant derivatives.

Integrating this relation over M, we can obtain the asymptotic expansion of the Θ-function:

$$\Theta(t) \sim (4\pi t)^{-n/2}(c_0(M) + tc_1(M) + \cdots),$$

whose coefficients $c_i(M)$ are global characteristics of M.

Weyl's formula is equivalent to $C_0(M) = \text{Vol}(M)$. With respect to the remaining coefficients of the asymptotic expansion, Weyl wrote: "I feel that this information about the principal oscillations of membrane, however useful it may be, is very incomplete. I have some propositions on how one should view the complete analysis of the asymptotic behaviour of a membrane. However, since in the course of more than 35 years I have made no serious attempts to prove my propositions, it is better than I keep my conjectures to myself".

The next two terms of the expansion were found by McKean and Singer [1967]. It turned out that

$$C_1(M) = \frac{1}{3} \int_M k(x)\omega g,$$

where $k(x)$ is the scalar curvature, and

$$C_2(M) = \frac{1}{180} \int_M (10A - B + 2C)\omega g,$$

where A, B, C are a special basis in the space of second-order polynomials in the components of the curvature tensor, invariant under the action of the group $O(n)$.

In particular, when $n = 2$ we have $10A - B + 2C = 12K^2$. Hence for two-dimensional compact surfaces the asymptotic behaviour of the Θ-function has the form

$$\Theta(t) \sim \frac{\text{Vol}(M)}{4\pi t} + \frac{1}{6}\chi(M) + \frac{\pi t}{60} \int_M K^2\omega g + O(t^2).$$

Thus, for compact surfaces we can "hear" the volume and the Euler characteristic. Nevertheless, the answer to the fundamental question of spectral geometry

– whether the Riemannian structure can be restored from the spectrum of the Laplace operator – is negative. Milnor constructed two non-isometric 16-dimensional tori on which the spectra of the Laplace operators coincide. However, for metrics in general position Blecker showed that the first variation of the spectrum under a change of the metric determines it on manifolds of dimension at least 5.

It is natural to generalize the preceding constructions to Laplace operators in spaces of p-forms. Let $^p\Theta(t)$ be the Θ-function of this operator. Gaffney showed that as $t \to 0$ it has the Minakshisundaram-Pleijel expansion

$$^p\Theta(t) \sim (4\pi t)^{-1/2}(^pC_0(M) + t{}^pC_1(M) + \cdots),$$

where $^pC_0(M) = C_n^p \, \mathrm{Vol}(M)$, $^pC_1(M) = \dfrac{n-6}{6} \int_M k(x)\omega g$.

We also note that from Hodge theory we can derive the following relations for the dimensions of the eigenspaces F_λ^p of these operators:

$$\sum_{p=0}^{n} (-1)^p \dim F_0^p = \chi(M),$$

$$\sum_{p=0}^{n} (-1)^p \dim F_\lambda^p = 0.$$

Commentary on the References

Differential geometry as a subject was actually founded in the 19th century. The first treatise on differential geometry is usually taken to be the work of Monge [1850]. An excellent survey of the history of the development and the rise of new geometrical ideas was given in the lectures of Klein [1926 a, b], and also in Kolmogorov and Yushkevich [1981].

The classical work on the theory of surfaces is the monograph of Darboux [1894–1896], see also Kagan [1947–48], Norden [1956], Bianchi [1922–24], Blaschke [1930]. The reader will find a history of the rise and development of the ideas of Riemannian geometry in the survey of Kagan [1933]. The books of Rashevskij [1967], Cartan [1928], Eisenhart [1949], Levi-Civita and Ricci [1901], Schouten and Struik [1935–38], Veblen and Whitehead [1932] contain presentations of Riemannian geometry and tensor analysis of the classical period. For a modern presentation see Dubrovin, Novikov and Fomenko [1979], Bishop and Crittenden [1964], Favard [1957], Kobayashi and Nomizu [1969], Sternberg [1964].

At the foundation of the group approach to geometry lie the ideas of Klein's Erlangen programme (Klein [1872]). A further development of them, based on the concept of the holonomy group, is contained in a lecture of E. Cartan [1926]. The three-volume work of Lie and Engel [1888–93] is the first fundamental work on transformation groups. The reader will find a presentation of the current position in the books by Shirokov [1966], Campbell [1903], Cartan [1927], Eisenhart [1933], Helgason [1978], Kobayashi [1972], Lichnerowicz [1958].

The work of Monge [1850] is also the first treatise on the geometry of differential equations. A systematic study of the geometry of differential equations was begun by Lie [1922–37] and E. Cartan [1945]. For the current position see Vinogradov, Krasil'shchik and Lychagin [1986], Ovsyannikov [1978], Rashevskij [1947], Kuranishi [1962], Pommaret [1978], Spencer [1985].

An approach to the study of differential geometry from the point of view of pseudogroups and their differential invariants also had its beginning in the works of Lie [1922–37] and E. Cartan [1945]; see also Amaldi [1942], Kuranishi [1959], Medolaghi [1897].

Important progress in this field became possible after the introduction by Ehresmann of the concept of a jet (Ehresmann [1953]) and the construction by Spencer [1985] of his cohomology theories. A current view of the subject can be found in Goldschmidt [1972–76], Goldschmidt and Spencer [1976–78], Kumpera [1975], Kumpera and Spencer [1972], Malgrange [1972], Singer and Sternberg [1965]. There are fewer papers on differential invariants of pseudogroups Kumpera [1975], Thomas [1934], Veblen [1927], Weitzenbock [1923].

The reader can find surveys on geometrical structures and G-structures in Bernard [1960], Chern [1966], Sternberg [1964].

References*

Aleksandrov, A.D. [1937–38]. On the theory of mixed volumes of convex bodies, Mat. Sb. *44*, 947–972 [Russian], Zbl.17,426; Mat. Sb., Nov. Ser. 2, 1205–1238 [Russian], *3*, 27–46, 227–251 [Russian], Zbl.18,276; Zbl.18,424; Zbl.19,328

Aleksandrov, A.D. [1948] The intrinsic geometry of convex surfaces, Gostekhizdat, Moscow-Leningrad [Russian], Zbl.38,352

Allendorfer, C.B. [1939] Rigidity for spaces of class greater than one, Am. J. Math. *61*, 633–644, Zbl.21,158

Almgren, F.J. [1968] Existence and regularity almost everywhere of solutions to elliptic variational problems among surfaces of varying topological type and singularity structure, Ann. Math., II. Ser. *87*, 321–391, Zbl.162,247

Amaldi, U. [1942] Introduzione alla teoria dei gruppi infiniti di transformazioni, Libreria dell'Univ., Rome

Ambrose, W., Singer, I.M. [1953] A theorem on holonomy, Trans. Am. Math. Soc. *75*, 428–443, Zbl.52,180

Atiyah, M.F., Singer, I.M. [1968–71] The index of elliptic operators. I–V, Ann. Math., II. Ser. *87*, 484–530, 531–545, 546–604; *93*, 119–149, Zbl.164,240; Zbl.164,242; Zbl.164,243; Zbl.212,286

Beez, R. [1986] Zur Theorie der Krümmungsmassen von Mannigfaltigkeiten höherer Ordnung, Z. Math. Phys. *21*, 373–401

Berger, E., Bryant, R.L., Griffiths, P.A. [1983] The Gauss equations and rigidity of isometric embeddings, Duke Math. J. *50*, 803–892, Zbl.526.53018

Berger, M., Gauduchon, P., Mazet, E. [1971] Le spectre d'une variété riemannienne, Lect. Notes Math. *194*, Zbl.223.53034

Bernard, D. [1960] Sur la géométrie différentielle des G-structures, Ann. Inst. Fourier *10*, 151–270, Zbl.95,364

Besse, A.L. [1978] Manifolds all of whose geodesics are closed, Springer-Verlag, Berlin-Heidelberg-New York, Zbl.387.53010

Besse, A.L. [1981] Géométrie Riemannienne en dimension 4, Seminaire Arthur Besse 1978/79, CEDIC, Paris, Zbl.472.00010

Bianchi, L. [1922–24] Lezioni di geometria differenziale, Pisa, Jbuch 49,498

Bishop, R.L., Crittenden, R.J. [1964] Geometry of manifolds, Acad. Press, New York, Zbl.132,160

Blanuša, D. [1955] Über die Einbettung hyperbolischer Räume in euklidische Räume, Monatsh. Math. *59*, 217–229, Zbl.67,144

Blaschke, W. [1930] Differentialgeometrie, Springer-Verlag, Berlin, 5th ed. 1973, Zbl.264.53001

Blaschke, W. [1955] Einführung in die Geometrie der Waben, Birkhäuser-Verlag, Basel-Stuttgart, Zbl.68,365

Bleecker, D. [1985] Determination of a Riemannian metric from the first variation of its spectrum, Am. J. Math. *107*, 815–831, Zbl.577.58032

Bochner, S. [1946] Vector fields and Ricci curvature, Bull. Am. Math. Soc. *52*, 776–797, Zbl.60,383

Bochner, S. [1947] Curvature in Hermitian metric, Bull. Am. Math. Soc. *53*, 179–195, Zbl.35,104

---

*For the convenience of the reader, references to reviews in Zentralblatt für Mathematik (Zbl.), compiled using the MATH database, and Jahrbuch über die Fortschritte der Mathematik (Jbuch) have, as far as possible, been included in this bibliography.

Bombieri, E., De Giorgi, E., Giusti, E. [1969] Minimal cones and the Bernstein problem, Invent. Math. 7, 243–268, Zbl.183,259

Bonnesen, T., Fenchel, W. [1934] Theorie der konvexen Körper, Springer-Verlag, Berlin, Zbl.8,77

Bonnet, O. [1855] Sur quelques propriétés des lignes géodésiques, C.R. Acad. Sci. Paris 40, 1311–1313

Bott, R. [1982] Lectures on Morse theory, old and new, Proc. 1980 Beijing Vol. 1, 168–218 (also appeared in Bull. Am. Math. Soc., New. Ser. 7, 331–358 (1982)), Zbl.505.58001

Brieskorn, E. [1966] Beispiele zur Differentialtopologie von Singularitäten, Invent. Math. 2, 1–14, Zbl.145,178

Burstin, C. [1931] Ein Betrag zum Problem der Einbettung der Riemannschen Räume in euklidische Räume, Mat. Sb. 38, No. 3/4, 74–85, Zbl.6,80

Buser, P., Karcher, H. [1981]. Gromov's almost flat manifolds, Astérisque 81, 148p., Zbl.459.53031

Calabi, E. [1954] The space of Kähler metrices, Proc. Int. Congr. Amsterdam, Vol. 2, 206–207

Campbell, J.E. [1903] Introductory treatise on Lie's theory of finite continuous transformation groups, Chelsea, New York, Jbuch 34,390

Carathéodory, C. [1956] Variationsrechnung und partielle Differentialgleichungen erster Ordnung I, Teubner, Leipzig, Zbl.70,316

Cartan, E. [1926] Les groupes d'holonomie des espaces généralisés, Acta Math. 48, 1–42, Jbuch 52,723

Cartan, E. [1927] Sur la possibilité de plonger un espace riemannien donné dans un espace euclidien, Ann. Soc. Math. Pol. 6, 1–7, Jbuch 54,763

Cartan, E. [1928] Leçons sur la géométrie des espaces de Riemann, Gauthier-Villars, Paris (2nd ed. 1946), Jbuch 54,755; Zbl.60,381

Cartan, E. [1945] Les systèmes différentiels éxtérieurs et leurs applications géométriques, Hermann, Paris

Cartan, E. [1927] La géométrie des groupes de transformations, J. Math. Pures Appl. 6, 1–119

Cartan, E. [1953] Oeuvres complètes, Gauthier-Villars, Paris, Zbl.58,83

Cheeger, J., Ebin, D.G. [1975] Comparison theorems in Riemannian geometry, North-Holland Publ. Co., Amsterdam, Zbl.309.53035

Chern, S.S. [1946] Characteristic classes of Hermitian manifolds, Ann. Math., II. Ser. 47, 85–121, Zbl.60,414

Chern, S.S. [1966] The geometry of G-structures, Bull. Am. Math. Soc. 72, 167–219, Zbl.136,178

Chevalley, C. [1946] Theory of Lie groups, Princeton Univ. Press, Princeton, NJ

Clifford, W.K. [1968] Mathematical papers, Macmillan, New York-London

Courant, R. [1950] Dirichlet's principle, conformal mappings and minimal surfaces, Interscience, New York, Zbl.40,346

Darboux, G. [1894–1896] Leçons sur la théorie générale des surfaces et les applications géométriques du calcul infinitesimal (4 vols.), Gauthier-Villars, Paris (2nd ed. 1925), Jbuch 25,1159

Dolbeault, P. [1956–57] Formes différentielles et cohomologie sur une variété analytique complexe, Ann. Math., II. Ser. 64, 83–130, Zbl.72,406; 65, 282–330, Zbl.89,380

Dubrovin, B.A., Novikov, S.P., Fomenko, A.T. [1979]. Sovremennaya geometriya, Nauka, Moscow. English transl.: Modern geometry – methods and applications, Part I, Springer-Verlag, New York-Berlin 1984, Zbl.433,53001

Efimov, N.V. [1948] Qualitative questions in the theory of deformations of surfaces, Usp. Mat. Nauk 3, 47–158 [Russian], Zbl.30,69

Efimov, N.V. [1964] Generation of singularities on surfaces of negative curvature, Mat. Sb., Nov. Ser. 64, 286–320, English transl.: Am. Math. Soc., Transl., II, Ser. 66, 154–190 (1968), Zbl.126,374

Efimov, N.V. [1966] Surfaces with a slowly changing negative curvature, Usp. Mat. Nauk, 21, No. 5, 3–58. English transl.: Russ. Math. Surv. 21, No. 5, 1–55, Zbl.171,199

Ehresmann, C. [1953] Introduction à la théorie des structures infinitesimales et des pseudo-groupes de Lie, Colloque de topologie, Strasbourg, 1952, 97–100, Zbl.52,398

Ehresmann, C. [1954] Structures locales, Ann. Mat. Pura Appl., IV. Ser. 36, 133–143, Zbl.55,420

Eisenhart, L.P. [1933] Continuous groups of transformations, Princeton Univ. Press, Princeton, NJ, Zbl.8,108

Eisenhart, L.P. [1949] Riemannian geometry, Princeton Univ. Press, Princeton, NJ, Zbl.41,294

Favard, J. [1957] Cours de géométrie différentielle locale, Gauthier-Villars, Paris, Zbl.77,150

Fenchel, W., Jessen, B. [1938] Mengenfunktionen und konvexe Körper, Danske Vid. Selsk., Math.-Fys. Medd. *16*, No. 3, 1–31, Zbl.18,424

Finikov, S.P. [1948] Cartan's method of exterior forms in differential geometry, Gostekhizdat, Moscow-Leningrad [Russian], Zbl.33,60

Fomenko, A.T. [1982] Variational methods in topology, Nauka, Moscow [Russian], Zbl.526.58012

Frankel, T.T. [1966] On the fundamental group of a compact minimal submanifold, Ann. Math., II. Ser. *83*, 68–73, Zbl.189,224

Frölicher, A., Nijenhuis, A. [1956–58] Theory of vector-valued differential forms. I, II, Indagationes Math. *18*, 338–359, Zbl.79,375; *20*, 414–429, Zbl.105,147

Frölicher, A., Nijenhuis, A. [1957] A theorem of stability of complex structures, Proc. Natl. Acad. Sci. USA *43*, 239–241, Zbl.78,142

Fubini, G., Čech, E. [1931] Introduction à la géométrie projective-différentielle des surfaces, Gauthier-Villars, Paris, Zbl.2,351

Galois, E. [1951] Oeuvres mathématiques d'Evariste Galois, Gauthier-Villars, Paris, Zbl.42,4

Goldschmidt, H. [1967] Integrability criteria for systems of non-linear partial differential equations, J. Differ. Geom. *1*, 269–307, Zbl.159,141

Goldschmidt, H. [1972–76] Sur la structure des équations de Lie. I, II, III, J. Differ. Geom. *6*, 357–373, Zbl.235.58011; *7*, 67–95, Zbl.273.58015; *11*, 167–223, Zbl.321.58021

Goldschmidt, H., Spencer, D.C. [1976–78] On the nonlinear cohomology of Lie equations. I, II, Acta Math., *136*, 103–239, Zbl.385.58015, Zbl.452.58025. III, IV, J. Differ. Geom. *13*, 409–526, Zbl.452.58026, Zbl.452.58027

Goursat, E. [1922] Leçons sur le problème de Pfaff, Hermann, Paris, Jbuch 48,538

Green, M. [1978] The moving frame, differential invariants and rigidity theorems for curves in homogeneous spaces, Duke Math. J. *45*, 735–779, Zbl.414.53039

Greene, R.E. [1970] Isometric imbeddings of Riemannian and pseudo-Riemannian manifolds, Mem. Am. Math. Soc. 97, 1–63, Zbl.203,240

Griffiths, P. [1974] On the Cartan method of Lie groups and moving frames as applied to existence and uniqueness questions in differnetial geometry, Duke Math. J. *41*, 775–814, Zbl.294.53034

Griffiths, P., Harris, J. [1978] Principles of algebraic geometry, Wiley-Interscience, New York-Chichester-Brisbane-Toronto, Zbl.408.14001

Gromoll, D., Klingenberg, W., Meyer, W. [1975]. Riemannsche Geometrie im Grossen, 2nd ed., Lect. Notes Math., *55*, Zbl.155,307

Gromov, M.L. [1970] Isometric embeddings and immersions, Dokl. Akad. Nauk SSSR, *192*, 1206–1209. English transl.: Sov. Math. Dokl. *11*, 794–797, Zbl.214,504

Gromov, M.L., Rokhlin, V.A. [1970] Embeddings and immersions in Riemannian geometry, Usp. Mat. Nauk, *25*, No. 5, 3–62. English transl.: Russ. Math. Surv., *25*, No. 5, 1–57, Zbl.202,210

Guillemin, V. [1966] A Jordan-Hölder decomposition for certain classes of infinite-dimensional Lie algebras, J. Differ. Geom. *2*, 313–345, Zbl.183,261

Guillemin, V., Sternberg, S. [1964] An algebraic model of transitive differential geometry, Bull. Am. Math. Soc. *70*, 16–47, Zbl.121,388

Guillemin, V., Sternberg, S. [1966] Deformation theory of pseudogroup structures, Mem. Am. Math. Soc. *64*, Zbl.169,530

Guillemin, V., Sternberg, S. [1967] The Lewy counterexample and the local problem for G-structures, J. Differ. Geom. *1*, 127–131, Zbl.159,234

Hartshorne, R. [1977] Algebraic geometry, Springer-Verlag, New York, Zbl.367.14001

Helgason, S. [1962] Differential geometry and symmetric spaces, Academic Press, New York, Zbl.111,181

Helgason, S. [1978] Differential geometry, Lie groups and symmetric spaces, Academic Press, New York-San Francisco-London, Zbl.451.53038

Herglotz, G. [1943] Über die Starrheit der Eiflächen, Abh. Math. Semin. Univ. Hamburg *15*, 127–129, Zbl.28,94

Hermann, R. [1964] Cartan connections and the equivalence problem for geometric structures, Contrib. Differ. Equations *3*, 199–248, Zbl.141,386

Hermann, R. [1965] Cartan's geometric theory of partial differential equations, Adv. Math. *1*, 265–317, Zbl.142,71

Hirzebruch, F. [1966] Topological methods in algebraic geometry, Springer-Verlag, New York, Zbl.138,420

Hodge, W.V.D. [1952] The theory and applications of harmonic integrals, Cambridge Univ. Press, Zbl.48,157

Husemoller, D.H. [1966] Fibre bundles, McGraw-Hill, New York, Zbl.144,448

Janet, M. [1926] Sur la possibilité de plonger un espace riemannien donné dans un espace euclidien, Ann. Soc. Math. Pol. *5*, 38–43, Jbuch 53,699

Kagan, V. G. [1933] Riemann's geometrical ideas and their modern development, Gostekhizdat, Moscow-Leningrad [Russian]

Kagan, V. G. [1947–48] Foundations of the theory of surfaces (2 vols.), Gostekhizdat, Moscow-Leningrad [Russian], Zbl.41,487

Kähler, E. [1934] Einführung in die Theorie der Systeme von Differential-gleichungen (Hamburg. Math. Einzelschr., No. 16), Teubner, Berlin, Zbl.11,161

Kárteszi, F. [1976] Introduction to finite geometries, North-Holland Publ. Co., Amsterdam-Oxford; Elsevier, New York, Zbl.325.50001

Killing, W. [1885] Die nicht-Euklidischen Raumformen in analytischer Behandlung, Teubner, Leipzig, Jbuch 17,508

Klein, F. [1872] Vergleichende Betrachtungen über neuere geometrische Forschungen (Erlangen Program), Erlangen

Klein, F. [1926a] Vorlesungen über höhere Geometrie, Springer-Verlag, Berlin, Jbuch 52,624

Klein, F. [1926b] Vorlesungen über die Entwicklung der Mathematik im 19. Jahrhundert, Band 1, Springer-Verlag, Berlin, Jbuch 52,22

Kobayashi, S. [1972]. Transformation groups in differential geometry, Springer-Verlag, Berlin-Heidelberg-New York, Zbl.246.53031

Kobayashi, S., Nagano, T. [1964–65–66] On filtered Lie algebras and geometric structures. I–V, J. Math. Mech., *13*, 875–907; *14*, 513–521, 679–706; *15*, 163–175, 315–328, Zbl.142,195; Zbl.163,281

Kobayashi, S., Nomizu, K. [1963, 1969] Foundations of differential geometry (2 vols.), Interscience, New York-London-Sydney, Zbl.119,375 and Zbl.175,485

Kodaira, K., Spencer, D.C. [1957] On the variation of almost-complex structure, in: Algebraic Geometry and Topology, A symposium in honor of S. Lefschetz, Princeton Math. Ser. 12, 139–150, Zbl.82,154

Kodaira, K., Spencer, D.C. [1958–60] On deformations of complex analytic structures. I, II, III, Ann. Math., II. Ser. *67*, 328–466; *71*, 43–76, Zbl.128,169

Kolmogorov, A.N., Yushkevich, A.I. (eds.) [1981] Mathematics of the 19th century. Geometry. Theory of analytic functions, Nauka, Moscow [Russian], Zbl.492.00001

Kuiper, N.H. [1955] On C^1-isometric imbeddings. II, Indagationes Math. *17*, 683–689, Zbl.67,396

Kuiper, N.H., Yano, K. [1955] On geometric objects and Lie groups of transformations, Indagationes Math. *17*, 411–420, Zbl.67,398

Kumpera, A. [1975] Invariants différentiels d'un pseudogroupe de Lie, J. Differ. Geom. *10*, 347–416, Zbl.346.58012

Kumpera, A., Spencer, D.C. [1972] Lie equations, Vol. 1: General theory. Ann. Math. Stud., No. 73, Princeton, Zbl.258.58015

Kuranishi, M. [1957] On E. Cartan's prolongation theorem of exterior differential systems, Am. J. Math. *79*, 1–47, Zbl.77,297

Kuranishi, M. [1959] On the local theory of continuous infinite pseudogroups. I, Nagoya Math. J. *15*, 225–260, Zbl.212,565

Kuranishi, M. [1961] On the local theory of continuous infinite pseudogroups. II, Nagoya Math. J. *19*, 55–91, Zbl.212,565

Kuranishi, M. [1962] Lectures on exterior differential systems, Tata Inst. of Fundamental Research, Bombay

Levi-Civita, T. [1917] Nozione di parallelismo in una varietà qualunque e consequente specificazione geometrica della curvatura Riemanniana, Rend. Circ. Mat. Palermo *42*, 173–205, Jbuch 46,1125

Levi-Civita, T. [1928] Der absolute Differentialkalkül, Springer-Verlag, Berlin, Jbuch 54,754

Levi-Civita, T., Ricci, G. [1901] Méthodes de calcul différentiel absolu et leurs applications, Math. Ann. *54*, 125–201

Lewy, H. [1938] On the existence of a closed surface realizing a given riemannian metric, Proc. Nat. Acad. Sci. USA *24*, 104–106, Zbl.18,88

Liber, A.E. [1938] On a class of Riemannian manifolds of constant negative curvature, Uch. Zap. Sarat. Gos. Univ. Ser. Fiz.-Mat. *14*, No. 2, 105–122

Lichnerowicz, A. [1955] Théorie globale des connexions et des groupes d'holonomie, Ed. Cremonese, Rome, Zbl.116,391

Lichnerowicz, A. [1958] Géométrie des groupes de transformations, Dunod, Paris, Zbl.96,160

Lie, S. [1922–37] Gesammelte Abhandlungen, (6 vols.), Teubner, Oslo, Jbuch 53,26; Jbuch 50,2; Jbuch 48,9; Jbuch 55,16. Zbl.9,318; Zbl.12,412; Zbl.17,180

Lie, S., Engel, F. [1888–93] Theorie der Transformationsgruppen, (3 vols), Teubner, Leipzig, Jbuch 21,356; Jbuch 23,364; Jbuch 25,623

Lobachevskij, N.I. [1956] Foundations of geometry, GITTL, Moscow [Russian], Zbl.71,358

Lychagin, V.V. [1975] Local classification of non-linear first order partial differential equations, Usp. Mat. Nauk *30*, No. 1, 101–171. English transl.: Russ. Math. Surv. *30*, No. 1, 105–175, Zbl.308.35018

Malgrange, B. [1972] Équations de Lie. I, II, J. Differ. Geom. *6*, 503–522, *7*, 117–141, Zbl.264.58009

Manin, Yu. I. [1977] A course in mathematical logic, Springer-Verlag, New York, Zbl.383.03002

Manin, Yu. I. [1984] Gauge fields and complex geometry, Nauka, Moscow. English transl.: Springer-Verlag, Berlin 1988, Zbl.576.53002

McKean, H.P., Singer, I.M. [1967] Curvature and the eigenvalues of the Laplacian, J. Differ. Geom. *1*, 43 69, Zbl.198,443

Medolaghi, P. [1897] Sulla theoria dei gruppi infiniti continui, Ann. Mat. Pura. Appl. *25*, 179–217, Jbuch 28,320

Milnor, J. [1963] Morse theory, Princeton Univ. Press, Princeton, NJ, Zbl.108,104

Milnor, J. [1968] A note on curvature and the fundamental group, J. Differ. Geom. *2*, 1–7, Zbl.162,254

Milnor, J., Stasheff, J. [1974] Characteristic classes, Ann. Math. Stud., No. 76, Princeton, Zbl.298.57008

Minkowski, H. [1903] Volumen und Oberfläche, Math. Ann. *57*, 447–495, Jbuch 34,649

Monge, G. [1850] Analyse appliqué à la géométrie, Paris

Morrey, C.B. [1966] Multiple integrals in the calculus of variations, Springer-Verlag, Berlin, Zbl.142,387

Morse, M. [1934] The calculus of variations in the large, Am. Math. Soc., New York, Zbl.11,28

Moser, J. [1961] A new technique for the construction of solutions of nonlinear differential equations, Proc. Natl. Acad. Sci. USA *47*, 1824–1831, Zbl.104,305

Nash, J.F. [1954] C^1-isometric imbeddings, Ann. Math., II. Ser. *60*, 383–396, Zbl.58,377

Nash, J.F. [1956] The imbedding problem for Riemannian manifolds, Ann. Math. II., Ser. *63*, 20–63, Zbl.70,386

Newlander, A., Nirenberg, L. [1957] Complex analytic coordinates in almost complex manifolds, Ann. Math., II. Ser. *65*, 391–404, Zbl.79,161

Nijenhuis, A., Richardson, R.W. [1966] Cohomology and deformation in graded Lie algebras, Bull. Am. Math. Soc. *72*, 1–29, Zbl.136,305

Norden, A.P. [1956] Theory of surfaces, GITTL, Moscow [Russian], Zbl.75,164

Ochiai, T. [1966] On the automorphism group of a G-structure, J. Math. Soc. Japan *18*, 189–193, Zbl.172,470

Ovsyannikov, L.V. [1978] Group analysis of differential equations, Nauka, Moscow [Russian], Zbl.484.58001

Palais, R. [1965] Seminar on the Atiyah-Singer index theorem, Princeton Univ. Press, Princeton, NJ, Zbl.137,170

Pogorelov, A.V. [1951] The unique determination of general convex surfaces, Izdat. Akad. Nauk UkrSSR, Kiev [Russian], Zbl.44,361

Pogorelov, A.V. [1961] Some results on geometry in the large, Izdat. Kharkov. Gos. Univ. Kharkov [Russian], Zbl.126,378

Pogorelov, A.V. [1969] Extrinsic geometry of convex surfaces, Nauka, Moscow. English transl.: Am.

Math. Soc., Providence, RI, 1973, Zbl.311.53067

Pogorelov, A.V. [1975] The Minkowski multidimensional problem, Nauka, Moscow. English transl.: Halstead Press, John Wiley & Sons, New York-Toronto-London, 1978, Zbl.387.53023

Pollak, A.S. [1974] The integrability problem for pseudo-group structures, J. Differ. Geom. *9*, 355–390, Zbl.281.53030

Pommaret, J.F. [1978] Systems of partial differential equations and Lie pseudogroups, Gordon and Breach, New York-London-Paris, Zbl.401.58006

Pontryagin, L.S. [1949] Some topological invariants of closed Riemannian manifolds, Izv. Akad. Nauk SSSR Ser. Mat. *13*, 125–162. English transl.: Am. Math. Soc. Transl., II. Ser. *7*, 279–331 (1962), Zbl.37,106

Pontryagin, L.S. [1973] Topological groups, 3rd ed., Nauka, Moscow. English transl. of 2nd ed.: Gordon and Breach, New York-London-Paris, 1966, Zbl.659.22001.

Pontryagin, L.S. [1976] Smooth manifolds and their applications in homotopy theory, 2nd ed., Nauka, Moscow. English transl. of 1st ed.: Am. Math. Soc. Transl., II. Ser. *11*, 1–114 (1959), Zbl.64,174

Postnikov, M.M. [1965] Variational theory of geodesics, Nauka, Moscow. English transl.: Saunders Comp., Philadelphia-London (1967), Zbl.134,395

Proceedings [1982] of the 1980 Beijing Symposium on Differential Geometry and Differential Equations, (3 vols.), Science Press, Gordon and Breach, Beijing, Zbl.509.00015

Rashevskij, P.K. [1947] Geometrical theory of partial differential equations, Gostekhizdat, Moscow-Leningrad [Russian], Zbl.36,64

Rashevskij, P.K. [1967] Riemannian geometry and tensor analysis, 3rd ed., Nauka, Moscow. German transl.: Berlin (1959), Zbl.52,388; Zbl.92,392

Reifenberg, E.R. [1960] Solution of the Plateau problem for m-dimensional surfaces of varying topological type, Acta Math. *104*, 1–92, Zbl.99,85

de Rham, G. [1955] Variétés différentiables. Formes, courants, formes harmoniques, Hermann, Paris, Zbl.65,324

Riemann, B. [1868] Über die Hypothesen, die Geometrie zugrunde liegen, Nachrichten Ges. Wiss. Göttingen *13*, 1–20, Jbuch 1, 22

Schläfli, L. [1871–73] Nota alla memorici del Sig. Beltrami sugli spazi di curvature constante, Ann. Mat. Pura Appl. (2) *5*, 170–193

Schoen R., Yau, S.-T. [1979] On the structure of manifolds with positive scalar curvature, Manuscr. Math. *28*, 159–183, Zbl.423.53032

Schoen, R., Yau, S.-T. [1982] Complete three-dimensional manifolds with positive Ricci curvature and scalar curvature, Ann. Math. Stud., No. 102, 209–228, Zbl.481.53036

Schouten, J.A., Struik, D.J. [1935–38] Einführung in die neueren Methoden der Differentialgeometrie, (2 vols.), Berlin, Zbl.11,174; Zbl.19,183

Séminaire "Sophus Lie" [1955] Théorie des algèbres de Lie. Topologie des groupes de Lie, Paris

Shafarevich, I.R. [1972] Basic algebraic geometry, Nauka, Moscow. English transl.: Springer-Verlag, New York-Heidelberg, 1974, Zbl.258.14001

Shirokov, P.A. [1966] Selected works on geometry, Izdat. Kazan. Univ., Kazan [Russian]

Shnider, S. [1970] The classification of real primitive infinite Lie algebras, J. Differ. Geom. *4*, 81–89, Zbl.244.17014

Simon, L. [1983] Survey lectures on minimal submanifolds, Ann. Math. Stud. 103, 3–52, Zbl.541.53045

Simons, J. [1968] Minimal varieties in Riemannian manifolds, Ann. Math., II. Ser. *88*, 62–105, Zbl.181,497

Singer, I.M., Sternberg, S. [1965] On the infinite groups of Lie and Cartan. I, J. Anal. Math. *15*, 1–114, Zbl.277.58008

Spencer, D.C. [1985] Selecta (3 vols.), World Scientific, Philadelphia-Singapore, Zbl.657.01016

Sternberg, S. [1964] Lectures on differential geometry, Prentice Hall, Englewood Cliffs, NJ (2nd ed. 1983, New York), Zbl.129,131

Thomas, T.Y. [1934] The differential invariants of generalized spaces, Cambridge Univ. Press, Cambridge, Zbl.9,85

Tomkins, C. [1939] Isometric imbeddings of flat manifolds in Euclidean space, Duke Math. J. *5*,

58–61, Zbl.20,397

Toponogov, V.A. [1959] Riemannian manifolds with curvature bounded below, Usp. Mat. Nauk *14*, No. 1, 87–130, English transl.: Am. Math. Soc. Transl., II, Ser. 37, 291–336 (1964), Zbl.114,375

Tresse, A. [1894] Sur les invariants différentiels des groupes continus de transformations, Acta Math. *18*, 1–88, Jbuch 25,641

Veblen, O. [1927] Invariants of quadratic differential forms, Cambridge Univ. Press, Cambridge

Veblen, O., Whitehead, J.H.C. [1932] The foundations of differential geometry, Cambridge Univ. Press, Cambridge, Zbl.5,218

Vessiot, E. [1904] Sur l'intégration des systèmes différentiels qui admettent des groupes continus de transformations, Acta Math. 28, 307–350, Jbuch 35,343

Vinogradov, A.M. [1984] The category of differential equations and its significance for physics, Differ. Geom. Appl., Proc. Conf. Morave/Czech. 1983, 289–301, Zbl.558.35060

Vinogradov, A.M. [1986] The geometry of differential equations, secondary differential calculus and quantum field theory, Zv. Vyssh. Uchebu. Zaved., Mat. 1986, No. 1, 13–21. English transl.: Sov. Math. 30, No. 1, 14–25, Zbl.616.58009

Vinogradov, A.M., Krasil'shchik, I.S., Lychagin, V.V. [1986] Introduction to the geometry of non-linear differential equations, Nauka Moscow [Russian], Zbl.592.35002

Vlehduts, S.G., Manin, Yu. I. [1984] Linear codes and modular curves, Itogi Nauki Tekh., Ser. Sovrem. Probl. Mat. *25*, 209–257. English transl.: J. Sov. Math. *30*, 2611–2643 (1985), Zbl.629.94013

Weitzenbock, R. [1923] Invariantentheorie, Noordhoof, Groningen, Jbuch 49,64

Weyl, H. [1915a] Das asymptotische Verteilungsgesetz der Eigenschwingungen eines beliebig gestalten elastischen Körpers, Rend. Circ. Mat. Palermo *39*, 1–50

Weyl, H. [1915b] Über die Bestimmung einer geschlossenen konvexen Fläche durch ihr Linienelement, Viertel. Natur. Gesell. Zürich *61*, 40–72

Weyl, H. [1923] Raum, Zeit, Materie, Berlin, Jbuch 49,616, 7ed. 1988, Zbl.642.53001

Weyl, H. [1929] Electron and gravitation, Z. Phys. *56*, 330–352, Jbuch 55,513

Weyl, H. [1939a] The classical groups. Their invariants and representations, Princeton Univ. Press, Princeton, NJ, Zbl.20,206

Weyl, H. [1939b] On the volume of tubes, Am. J. Math. 61, 461–472, Zbl.21,355

White, B. [1982] Research announcement, Notices Am. Math. Soc. *3*

Wu, H.H. [1982] The Bochner technique, Differential geometry and differential equations, Proc. 1980 Beijing Symp., vol. 2, 929–1071, Zbl.528.53042

Yang, C.N., Mills, R.L. [1954] Conservation of isotopic spin and isotopic gauge invariance, Phys. Rev. *96*, 191–195

Yano, K., Bochner, S. [1953] Curvature and Betti numbers, Princeton Univ. Press, Princeton, NJ, Zbl.51,394

Yau, S.-T. [1978] On the Ricci curvature of a compact Kähler manifold and the complex Monge-Ampère equation. I, Commun. Pure Appl. Math. *31*, 339–411, Zbl.362.53049

Author Index

Subject Index

Encyclopaedia of Mathematical Sciences
Editor-in-chief: R. V. Gamkrelidze

Algebra

Volume 11: **A. I. Kostrikin, I. R. Shafarevich** (Eds.)
Algebra I
Basic Notions of Algebra
1990. V, 258 pp. 45 figs. ISBN 3-540-17006-5

Volume 18: **A. I. Kostrikin, I. R. Shafarevich** (Eds.)
Algebra II
Noncommutative Rings. Identities
1991. VII, 234 pp. 10 figs. ISBN 3-540-18177-6

Volume 73: **A. J. Kostrikin, I. R. Shafarevich** (Eds.)
Algebra VIII
Representations of Finite-dimensional Algebras
1992. Approx. 220 pp. 98 figs.
ISBN 3-540-53732-5

Topology

Volume 12: **D. B. Fuks, S. P. Novikov** (Eds.)
Topology I
General Survey. Classical Manifolds
1992. Approx. 300 pp. 79 figs.
ISBN 3-540-17007-3

Volume 24: **S. P. Novikov, V. A. Rokhlin** (Eds.)
Topology II
Homotopies and Homologies
1993. Approx. 235 pp. ISBN 3-540-51996-3

Volume 17: **A. V. Arkhangel'skij, L. S. Pontryagin** (Eds.)
General Topology I
Basic Concepts and Constructions. Dimension Theory
1990. VII, 202 pp. 15 figs. ISBN 3-540-18178-4

Analysis

Volume 13: **R. V. Gamkrelidze** (Ed.)
Analysis I
Integral Representations and Asymptotic Methods
1989. VII, 238 pp. 3 figs. ISBN 3-540-17008-1

Volume 14: **R. V. Gamkrelidze** (Ed.)
Analysis II
Convex Analysis and Approximation Theory
1990. VII, 255 pp. 21 figs. ISBN 3-540-18179-2

Volume 26: **S. M. Nikol'sk ij** (Ed.)
Analysis III
Spaces of Differentiable Functions
1991. VII, 221 pp. 22 figs. ISBN 3-540-51866-5

Volume 27: **V. G. Maz'ya, S. M. Nikol'skij** (Eds.)
Analysis IV
Linear and Boundary Integral Equations
1991. VII, 233 pp. 4 figs. ISBN 3-540-51997-1

Volume 19: **N. K. Nikol'skij** (Ed.)
Functional Analysis I
Linear Functional Analysis
1992. Approx. 300 pp. ISBN 3-540-50584-9

Volume 20: **A. L. Onishchik** (Ed.)
Lie Groups and Lie Algebras I
Foundations of Lie Theory. Lie Transformation Groups
1992. Approx. 235 pp. ISBN 3-540-18697-2

Encyclopaedia of Mathematical Sciences
Editor-in-chief: R. V. Gamkrelidze

Geometry

Volume 28: **R. V. Gamkrelidze** (Ed.)
Geometry I
Basic Ideas and Concepts of Differential Geometry
1991. VII, 264 pp. 62 figs. ISBN 3-540-51999-8

Volume 29: **E. B. Vinberg** (Ed.)
Geometry II
1992. Approx. 260 pp. ISBN 3-540-52000-7

Volume 48: **Yu. D. Burago, V. A. Zalgaller** (Eds.)
Geometry III
Theory of Surfaces
1992. Approx. 270 pp. ISBN 3-540-53377-X

Volume 70: **Yu. G. Reshetnyak** (Ed.)
Geometry IV
1992. Approx. 270 pp.

Algebraic Geometry

Volume 23: **I. R. Shafarevich** (Ed.)
Algebraic Geometry I
Algebraic Curves. Algebraic Manifolds and Schemes
1992. Approx. 300 pp. ISBN 3-540-51995-5

Volume 35: **I. R. Shafarevich** (Ed.)
Algebraic Geometry II
Cohomological Methods in Algebra. Geometric Applications to Algebraic Surfaces
Approx. 270 pp.

Volume 36: **A. N. Parshin, I. R. Shafarevich** (Eds.)
Algebraic Geometry III
Approx. 270 pp.

Volume 55: **A. N. Parshin, I. R. Shafarevich** (Eds.)
Algebraic Geometry IV
Approx. 310 pp.

Several Complex Variables

Volume 7: **A. G. Vitushkin** (Ed.)
Several Complex Variables I
Introduction to Complex Analysis
1990. VII, 248 pp. ISBN 3-540-17004-9

Volume 8: **A. G. Vitushkin, G. M. Khenkin** (Eds.)
Several Complex Variables II
Function Theory in Classical Domains. Complex Potential Theory
1992. Approx. 260 pp. ISBN 3-540-18175-X

Volume 9: **G. M. Khenkin** (Ed.)
Several Complex Variables III
Geometric Function Theory
1989. VII, 261 pp. ISBN 3-540-17005-7

Volume 10: **S. G. Gindikin, G. M. Khenkin** (Eds.)
Several Complex Variables IV
Algebraic Aspects of Complex Analysis
1990. VII, 251 pp. ISBN 3-540-18174-1

Volume 69: **W. Barth, R. Narasimhan** (Eds.)
Several Complex Variables VI
Complex Manifolds
1990. IX, 310 pp. 4 figs. ISBN 3-540-52788-5